Handbook of Polymer Blends and Composites

Volume 4B

Editors:

C. Vasile
and
A.K. Kulshreshtha

Rapra Technology Limited

Shawbury, Shrewsbury, Shropshire, SY4 4NR, United Kingdom
Telephone: +44 (0)1939 250383 Fax: +44 (0)1939 251118
http://www.rapra.net

First Published in 2003 by

Rapra Technology Limited

Shawbury, Shrewsbury, Shropshire, SY4 4NR, UK

©2003, Rapra Technology Limited

A catalogue record for this book is available from the British Library.

ISBN for Volume 4: 1-85957-304-5
ISBN for Complete Set: 1-85957-309-6

Typeset by Rapra Technology Limited
Covers printed by The Printing House, Crewe, UK
Printed by Rapra Technology, Shrewsbury, UK

9 Polyacrylic-Based Polymer Blends

Aurica P. Chiriac

9.1 Introduction

Because new polymers have been synthesised and chemical modifications made to conventional polymers, the importance of polymer blends has increased tremendously to satisfy the growing needs of new materials with specific properties.

Polymer blends are typically formulated to achieve a property balance. Some properties, most notably toughness near the brittle - ductile transition, are highly dependent upon blend composition, and slight fluctuations can effect large property differences. The mixing process can also have a profound influence on blend performance.

The possibility of producing materials with special properties is dependent on the availability of preparation techniques able to control the phase structure of the final polymer blends.

One of the most numerous classes of monomers is the acrylics, namely esters of acrylic and methacrylic acid with general formula:

$$CH_2 \!=\! CH - C \overset{\displaystyle /\!\!/ O}{\underset{\displaystyle \diagdown OR}{}}$$

These monomers polymerise and copolymerise extremely readily, therefore by judicious selection of monomers and polymerisation conditions it is possible to develop a great variety of acrylic and methacrylic polymers. While esters of acrylic acid give soft and flexible polymers, methyl methacrylate (MMA) produces an extremely hard polymer, but the polymers in the alkyl methacrylate series gradually soften and become more flexible with increasing length of alkyl chain up to C_{12}. This is one of the characteristics taken into consideration when these monomers are copolymerised or polymers are produced from blends. To obtain specific, optimised physical properties of acrylic polymers, copolymers or often polymer blends, which have different and/or more suitable physical properties than the homopolymers alone, are the solution.

It is well-known that polyacrylics and polymethacrylics are commercially desirable materials for numerous end uses in various industries such as automotive, electronics, telecommunications, lighting, optics, business machines and decorative products for specific products such as lighting fixtures, automobile light lenses, dials, video discs, ophthalmic lenses and other articles where durability, wettability, excellent optical properties, weatherability properties and low cost are desired.

Blends with methacrylate resins in their composition have improved solvent craze resistance. Methacrylate resins are widely used in producing sheet, moulded parts and articles having known, desirable 'acrylic properties'.

Acrylic products are modified to give special properties best for particular end-uses. For example acrylic fibres are unique among synthetic fibres because they have an uneven surface, even when extruded from a round-hole spinneret. Among the characteristics of the acrylics are: superior resistance to sunlight degradation, resistance to moths, oil, and chemicals, flexibility, acrylic fibres can be dyed to bright shades with excellent fastness and they retain their shape, etc. Acrylic fibres are the generic name of man-made fibres derived from acrylic resins (minimum of 85% acrylonitrile (ACN)). The benefits of acrylic blends are: superior moisture management or wickability, quick drying time (75% faster than cotton), easy care, shape retention, excellent light fastness, takes colour easily, bright vibrant colours, odour and mildew resistant. Thus, a 0.9 denier acrylic micro fibre is MicroSupreme (Sterling Chemicals) available in staple and tow, natural and fluoropolycarbonate (PCF) for yarn spinning, sliver knits, industrial and technical applications.

The Japan Chemical Fibers Association mentions in their report on composition by fibre (**Figure 9.1**) the extent of acrylic fibres, their constant evolution and development [1].

The flexibility of many acrylic polymers eliminates the need for adding plasticisers. Since the polymers can be produced by any of the common manufacturing processes, they are available in bulk, solution, or emulsion grades, from which can be selected the type most convenient for an application.

Acrylic elastomers are among the synthetic rubbers, which not only exhibit oil resistance but also lack residual unsaturation in the polymer chain. Between these types of acrylic elastomers can be specified Nipol Nitrile Elastomers with their range of products such as Nipol Polyblends and Pre-Plasticised Nitrile Elastomers or Zetpol 1020.

There are well known industrial applications of solutions and/or dispersions of acrylic polymers:

- as coatings or impregnants,

- paints and finishes in coatings industry,

- as effective thickeners of polymer latex systems,

- as adhesives and binders,

- as modifiers and compatibilisers of other polymers,

- in leather finishing,

- in a number of textile-processing and textile-finishing operations,

- in paper industry as pigment saturants and fibre binders.

Practically there is no field that acrylic polymers do not have applications. Polyacrylics, including polymethacrylics, as constituent of a blend, can be divided in two categories,

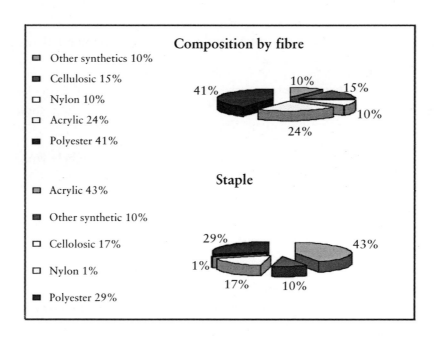

Figure 9.1 The extent and development of acrylic fibres according to the Japan Chemical Fibres Association

acrylics that necessitate the improvement of their characteristics, e.g., increase of adhesivity by blending with polyurethane or rubber, improvement of the ability of the resulting blend to absorb dyes, and so on, and acrylics which bring with their presence specific properties, e.g., compatibility, hardness, adhesivity, wetability, and so on.

Between acrylic homopolymers such as polymethyl methacrylate (PMMA), poly acrylic acid (PAA), polyacrylonitrile (PAN), binary or ternary copolymers such as acrylonitrile/ butadiene/styrene (ABS), styrene/acrylonitrile (SAN), MMA-glycidyl methacrylate-ethyl acrylate (MGE) are well known as products owing to their particular properties. At the same time, other acrylics are used as a result of the properties that offer.

Blending technologies offer many possibilities as according to the phase structures and morphologies developed, different systems may be realised being a cheaper alternative way to avoid the high cost synthesis of new polymers and also to optimise the properties of polymer components and in some cases to realise new materials with unexpected properties. Normally blending is used to combine the properties of two or more polymers. For a successful blend the following conditions should be fulfilled:

- The polymer mixture must be stable under the normal conditions for its use and no de-mixing should occur.

- The dispersed phase must have strong bonding to the surrounding polymer.

Most of the work in the past was necessarily empirical, and only in recent years has considerable effort has been devoted to the development of a basic knowledge in the field. In this area, research should be focused mostly on the following lines, covering fundamental as well as technological aspects of polymer blends:

- More appropriate theories of miscibility and modelling of phase behaviour.

- Development of new methods of characterisation of phase structure and interfaces.

- Molecular aspects (molecular mass, distribution of molecular mass, constitutional and conformational structure) - physical property relationships.

- Mechanisms and kinetics of phase separation with special care to processes of phase transitions characterised by an interplay of spinodal decomposition with other kinds of transitions (such as crystallisation, glass transition and microphase transition of block copolymers).

- New methods of blend preparation especially 'controlled reactive blending'.

- Innovative methodologies for polymer-polymer compatibilisation.

- Melt phase structure - rheological behaviour and flow induced morphology.

The final objective is to find new blend technologies aimed at realistic systems, acrylic polymer-based included, characterised by a tailored, stable and reproducible overall phase structure giving constant in time end use performances.

9.2 Methods of Obtaining Acrylic Polymer Blends

The most common technique adopted for preparing polymer blends are melting, mixing, casting from common solvents, freeze-drying, and mixing via reaction.

9.2.1 Casting Method and Specific Interactions in Binary and Ternary Blends Containing Acrylic Polymers

The blend miscibility is directly connected with the method of preparation. For example, the miscibility of the PMMA with polyvinyl acetate (PVAc), when obtained by casting, is known to depend on the solvent. Homogeneous solutions and transparent cast films were obtained when chloroform was used as a solvent and opalescent films were always obtained by casting from homogeneous solutions of *N,N*-dimethyl formamide (DMF) or tetrahydrofuran (THF).

The presence of interactions between atoms, or groups of atoms, of dissimilar polymers is essential to obtain a miscible polymer blend [2]. The existence or not of the interactions in PMMA/PVAc blends was tested. Inverse gas chromatography studies showed that the polymer-polymer interaction parameter, X, have small positive values, so the absence of specific interactions has been inferred [3]. Also Song and co-workers suggested, based on their Fourier transform infrared (FTIR) spectroscopy results, that conformation changes in PMMA/PVAc blends, in solution, from hydrogen interactions, are involved in miscible blends [4]. Dilute solution viscometry has been also utilised to evaluate the presence of interactions in polymer A – polymer B – solvent system.

As described by Sun [5], the α parameter, calculated using the equation: $\alpha = K_{BLEND} - K_1$, can be utilised to evaluate the presence of interactions. Thus, for example:

$$K_1 = \frac{K_{PMMA}[\eta]^2_{PMMA} W^2_{PMMA} + K_{PVAc}[\eta]^2_{PVAc} W^2_{PVAc} + 2\sqrt{K_{PMMA}K_{PVAc}}\, W_{PMMA}[\eta]_{PMMA} W_{PVAc}[\eta]_{PVAc}}{\left(W_{PMMA}[\eta]_{PMMA} + W_{PVAc}[\eta]_{PVAc}\right)^2}$$

where K_i is the Huggins constant and $[\eta]_i$ is the intrinsic viscosity of the *i*th pure polymeric component (generally K_{BLEND} is the Huggins constant of the blend and K_1 is calculated from viscometric parameters of solutions of each polymeric component).

If $\alpha \geq 0$, attractive forces occur between polymers, giving miscibility. If $\alpha < 0$, repulsive forces prevail and immiscibility is expected.

Differential scanning calorimetry (DSC) and viscosimetric experiments revealed that the miscibility of the PMMA/PVAc blends depends on the solvent and the solvent temperature [6]. The lower critical solution temperature (LCST) curve was determined by measurements of the intensity of transmitted light against temperature, at several compositions of PMMA/PVAc. Thus, Crispim and co-authors observed that blends were miscibile in chloroform at 30 °C and 50 °C, whereas in DMF at the same temperatures, the blends were found to be immiscible. Also in toluene the miscibility depends on the temperature: the blends are miscible at 30 °C and immiscible at 50 °C. Results from viscometry match very well those from DSC. The film of PMMA/PVAc (50/50) cast from chloroform presented only one thermogravimetric (TG) analysis (T_g). If cast from DMF, the blend presented two T_g values, close to those for pure polymers. The thermogram of the opalescent film, cast from DMF, was similar to the one obtained after a transparent film, cast from chloroform, was heated above the LCST curve.

Light scattering and phase-contrast optical microscopy has also been used to investigate the phase behaviour of blends of a liquid-crystal polymer (LCP) and PMMA as well as the phase behaviour of the blends of PMMA and PVAc [7]. The data show a similarity between the phase behaviour of blends of liquid – crystalline and of flexible amorphous polymers. The distinction consists of the absence of a linear regime of decomposition for LCP-PMMA blends.

The poor miscibility of PAN is rather surprising because it possesses a-hydrogens and cyano groups, both of which are capable of interacting with other functional groups [8]. Even so, two rare miscible PAN blend systems was reported [9]. The examples are PAN/cellulose and PAN/polyvinylpyrrolidone (PVP) blends [10]. The miscibility was considered to arise from dipole-dipole interactions between hydroxyl and cyano groups for the PAN/cellulose blends and from hydrogen-bonding interactions between α-hydrogens and carbonyl groups for the PAN/PVP blends. However, no spectroscopic evidence was provided to substantiate the suggestions.

Blends of PAN and poly(*p*-vinylphenol) (PVPh) [9] were prepared by solution casting from DMF. The miscibility of PAN with PVPh was shown by thermal and spectroscopic studies. A single T_g was found for all blend composition. FTIR studies showed that the hydroxyl band of PVPh and the cyano band of PAN shifted to lower frequencies upon blending, showing the existence of specific interactions between the two polymers. The involvement of cyano groups in specific interactions was further shown by the development of a high binding energy N1s peak (network interpenetrating network, first peak) in each blend from X-ray photoelectron spectroscopic studies. The miscibility behaviour of blends of PVPh has been extensively studied. The hydroxyl groups of PVPh are capable of undergoing intermolecular interactions with proton-accepting functional groups. As a result of such interactions, PVPh is miscible with a large variety of polymers including polyacrylates and polymethacrylates.

Solution blends of cellulose acetate (CA) with PAN and aminated PAN (with ethylene diamine (2DA), hexamethylene diamine (6DA) or dodecane diamine (12DA)) prepared in DMF were tested for miscibility as a function of the chain length of diamine [11]. The films were obtained by casting these blend solutions. It was observed that the morphology of PAN/CA blends does not vary significantly with composition, the dominant morphology being a particle-in-matrix structure. A significant reduction in domain size was proved with amination, and the effect became more pronounced with increasing chain length of diamine as well the improved miscibility. The FTIR measurements indicated that the interactions between CA and aminated PAN come from the hydrogen bond between ester groups of CA and the terminal amino groups of aminated PAN, which are facilitated by the longer aliphatic segments.

Some investigations were carried out on blends of collagen (C) with PAA (C/PAA), in which PAA was used as a model for studying the interactions between collagen and carboxylic functional groups in aqueous solution and in the solid state [12]. These studies showed it was possible to obtain a high degree of compatibility between these two polymers thanks to the strong ionic interactions that occur between them. Thus, in the C/PAA system, ionic interactions occur between the two oppositely charged macromolecules and lead to the formation of a polycomplex that precipitates as an insoluble aggregate because the charged groups responsible for solubility (carboxylic groups of PAA in the form, COO^- and aminic groups of collagen in the form of NH_3^+) are involved in the complex. The interactions between collagen and PAA show a complex of the two polymers formed in aqueous solution, within a narrow pH range, through electrostatic interactions involving the NH_3^+ groups of collagen and the COO^- groups of PAA.

9.2.2 Self-Propagating Frontal Polymerisation

An homogeneous solution of the monomer polymer solution is obtained and then the bulk polymerisation is performed and this is called self-propagating frontal polymerisation (propagating polymerisation fronts driven by the exothermicity of the polymerisation reaction and the transport of heat from the polymerised product to the monomer).

Chechilo and Enikolopyan first introduced frontal polymerisation in 1972 [13]. They studied travelling fronts in MMA polymerisation using reactors at high pressure. Pojman and colleagues investigated the movement of the polymerisation front in a variety of systems, including different monomers at ambient pressure, using both liquid and solid monomers and solution systems with high boiling point solvents [14]. The general mechanism of formation and propagation of the polymerisation front is based on the interaction between the production of heat by chemical reaction and its dispersion by thermal conduction.

Using a similar concept to the one described for polymer blending, frontal polymerisation has also been used to produce composite materials, where an inorganic powder is dispersed in the polymer matrix to provide special conductivity [15] or thermochromic [16, 17] properties. Better results were obtained using this method because the rapid polymerisation in the travelling front prevents phase separation.

Frontal polymerisation has also been investigated to produce simultaneous interpenetrating polymer networks (IPN): two independent and non-interfering monomers can be consequently crosslinked within a single travelling front, due to the high temperature reached in the front, which makes the rates of the two reactions nearly equal.

By using frontal polymerisation starting from a monomer/polymer solution, MMA/ polystyrene (PS), was obtained as a blend having 'freeze-in', the metastable phase where the polymer macromolecules are homogeneously dispersed. The advantage over the bulk polymerisation process is that because the polymerisation front is fast, the local transformation of monomer into polymer is sufficiently rapid to freeze the phase structure; this also occurs in the case of macromolecules with relatively high mobility. This technique, which is particularly simple and not energetically intensive, is expected to provide the means to produce blends that cannot be synthesised in other ways.

9.2.3 IPN Method

Acrylics and methacrylic polymers because of their available functional groups and because of the characteristic properties conferred are efficient partners in obtaining the blends by the IPN technique.

Fully IPN based on bisphenol A diglycidyl ether (DGEBA) and ethylene dimethacrylate (EDMA) and semi-IPN based on DGEBA and 2-hydroxyethyl methacrylate (HEMA) were prepared using isophronediamine and benzoyl peroxide (BPO) as curing agent [18]. It was shown that the methacrylic component acts as a plasticiser. Thus for EDMA/ DGEBA with a ratio of 18:82, the determined T_g was about 66 °C while at a ratio of 25:75 the T_g value was diminished at 54 °C.

The phase structure of the poly(hydroxy ether of bisphenol A)-based blends containing SAN, prepared through *in situ* polymerisation, i.e., the melt polymerisation between the DGEBA and bisphenol A in the presence of SAN, was characterised through DSC, dynamic mechanical analysis (DMA) and scanning electron microscopy (SEM) [19]. All the blends exhibited well-distributed phase-separated morphology. Mechanical tests effected by the authors showed that the blends containing 5-15 wt% SAN displayed a substantial improvement of tensile properties and Izod impact strength, which were in marked contrast to those of the materials prepared via conventional methods.

Acrylonitrile butadiene rubber (NBR)/phenolic resin (PH) blends are known to have excellent mechanical properties. Poly(alkyl methacrylates) are known to have a very high wetting behaviour compared to many other polymers. As the phenolics contain hard segments, their introduction into soft NBR may lead to formation of a constrained layer in the matrix. A constrained layer may cause high wetting (over a narrow range of frequency or temperature) and, at the same time, may retain high mechanical strength over a wide and useful temperature range. Thus, IPN synthesised from several NBR-PH blends using poly(alkyl methacrylates) as the third component are of higher strength compared to the corresponding NBR-PH blends. The strength of the PMMA-based IPN was found to be maximum. Thus, for NBR/PH/PMMA (51:3:46) the tensile strength was 13.42 MPa compared to 9.97 MPa for NBR/PH/PEMA (53:3:44) or 4.42 MPa for NBR/PH/(polybutylmethacrylate) PBuMA (46:3:51) [20].

The formation of a bi-continuous semi-IPN between atactic PS and a mixture of acrylic monomers (2-hydroxypropylmethacrylate/2-hydroxyethylacrylate/ethoxyethyl-methacrylate/methoxyethylmethacrylate/1, 3 butanedioldimethacrylate) by polymerisation and crosslinking into two steps is discussed by De Graaf and co-workers [21]. The systems achieved are mechanically stable and the stages of spinodal composition in polymer solutions have been shown.

Polymers with specific functionalities can be obtained by changing some groups on existing polymers or copolymers with suitable reactants. It has been determined that the polymers containing amidoxime groups have a tendency to form a complex with metal ions. Since there are no easily available monomers with pendant amidoxime groups, these types of polymers were synthesised by polymer-polymer conversion reactions where acrylonitrile type polymers were reacted with hydroxylamine hydrochloride [22].

IPN of PAN and PVP were prepared using amidoximated PVP networks [23]. The blend was synthesised by irradiation of dissolved PVP in acrylonitrile solution with a ^{60}Co-γ source. Gamma irradiation of acrylonitrile solutions of PVP yields easily IPN of PVP and PAN. The conversion of nitrile groups in the PAN backbone can be achieved at rather mild reaction conditions with an almost complete change into amidoxime groups.

Amidoximated PVP based IPN show good swelling behaviour in aqueous media due to the presence of PVP chains and the functional amidoxime groups are effective in the adsorption of various metal ions.

A wide variety of linear polymers, e.g., polymethacrylates, polyurethanes, and modified celluloses, have been used as interpenetrants in hydrogel semi-IPN. The semi-IPN technique can be successfully applied to a wide range of hydrophilic monomers. The properties of IPN hydrogels were found to be dependent on the composition and the type of IPN. The synthesis of sequential and simultaneous semi-interpenetrating hydrogel networks based on polyacrylamide (PAAM) and poly(itaconic acid) (PICA) is described in [24]. PAAM was used as a host polymer. The extraction results showed that the grafting level between the two components of PAAM/PICA in simultaneous semi-IPN is higher than that observed in sequential IPN. These results confirm that the decrease in swelling ratios with PICA contents is probably due to the entanglement effect. Both sequential and simultaneous IPN have quite different morphologies compared to the pure PAAM host hydrogel.

Semi-interpenetrating hydrogel networks based on PAAM and PICA were obtained to reinforce the PAAM gel and to produce materials that are stiffer than the pure hydrogel with similar water content [25]. Recently the acrylamide/itaconic acid hydrogels were prepared by γ irradiation to investigate the interaction of nicotine and its pharmaceutical derivatives. One of the main advantages of using IPN techniques in the preparation of hydrogels is to enable the design of materials with well-defined molecular structure. It has been proved that PAAM/PICA IPN hydrogels have considerable potential in the rapidly developing biomaterials field because of the ionisable groups of PICA that can interact with pharmaceuticals.

There are many articles regarding simultaneous interpenetrating networks (SIN) of polyurethane (PU), and an acrylic or styrenic polymer. Among these articles, the most visible are the PU/PMMA SIN, because this system gives good mechanical and wetting properties, while also serving as an excellent model SIN system [26]. For example, Kim and co-workers [27] studied PU/PMMA SIN and found that phase domains were finer in SIN than in the corresponding linear blends and that the two glass transitions (T_g) were shifted inward. Meyer and co-workers [28] found even larger inward shifts in the T_g of their SIN of the same system, e.g., 43 °C *versus* 15 °C, for about 50w/50w compositions. Significant broadening of the T_g were also noted along with improvements in some mechanical properties. Later, Hur and co-workers [29] obtained extremely broad, single T_g in their investigation. Akay and Rollins [30] found that varying the overall SIN composition could give from two shifted T_g to a single broad T_g. Lastly, SIN from the current work show two distinct T_g with no measurable inward shifting.

9.2.4 Functionalising Chains Method

New acrylic and allylic resins were prepared by functionalising perfluoro-polyethers chains, which assumed peculiar surface properties. Blends having different amounts of the acrylic and of the allylic systems were cured in air by UV irradiation [31]. Thus a resin having in its composition a perfluoropolyether chain (Fluorobase Z1030, Swan Tek) with isophorone diisocianate end groups has been functionalised with different hydroxy – unsaturated reagents, trimethylolpropan allylether and or hydroxypropyl-acrylate, respectively, to obtain the photocurable resins. The researchers have synthesised a macromer functionalised with acrylic group and the allylic one. The cure of the films was found to be dependent on the ratio between the two resins. It was found that the best blend contains 57% w/w of acrylic oligomer and 43% w/w of the allylic resin. The film obtained showed a high gel content (90%), was fully cured in the bulk and presented a good surface quality after 150 seconds of curing.

9.2.5 Aggregation Method

Production of crew-cut aggregates of block copolymers has received much attention [32-35]. The aggregates are formed through self-assembly of highly asymmetric amphiphilic block copolymers in which the insoluble core-forming blocks are much longer than the soluble blocks formed by corona treatment. Due to the fact that the insoluble blocks comprise a large fraction of the total chain, the aggregates are usually prepared by indirect methods. The procedures involve first dissolving the copolymers, such as PS-b-PAA, in a common solvent, such as DMF, and then adding water (water being a nonsolvent for polystyrene blocks), to a polymer solution to induce aggregation of the polystyrene segments. One of the noteworthy phenomena associated with the crew-cut aggregate systems is the existence of multiple morphologies, these include spheres, rods, vesicles, lamellae, large compound vesicles, a hexagonally packed hollow hoop structure, and several others [36, 37]. The most important morphogenic factors are the copolymer concentration in solution, the added water content, the nature of the common solvent, the type and concentration of added ions (salt, acid, or base), and the temperature. The formation of crew-cut aggregates is mainly controlled by minimisation of the interfacial energy which, in turn, is balanced by both the increase of the stretching of the core-forming blocks in the core and the repulsion among the chains. The morphogenic effects of the before mentioned parameters can, in general, be ascribed to their influence on the force balance during the formation of the aggregates. The effect of the water content in solution is related to the change of the interfacial energy. The effect of changing the common solvent on the morphology can be ascribed to the change in the dimensions of the core-forming blocks in the core due to the changes of both the solvent content in the core and the chain interactions.

The formation of crew-cut aggregates from blends of PS - b - PAA diblock and homopolystyrene in solution, by using the direct dissolution of aggregate preparation method has been studied [38]. The experimental results showed the aggregation behaviour of these polymer blends depends strongly on the relative volume fraction of the block copolymer and homopolymer. This difference in aggregation behaviour was the result of the difference in their solubilities that in turn, influences their miscibility and consequently the morphology of the resulting crew-cut aggregates.

The results of studies on the blends of phenoxy polymer and SAN prepared by bulk step-growth polymerisation between bisphenol A and the DGEBA in the presence of SAN involving *in situ* phenoxy formation, has been published [39]. Previously, the miscibility and the phase behaviour of binary blends of poly(hydroxy ether of bisphenol A) (phenoxy) and SAN has been reported by Vanneste and Groeninckx [40]. On the basis of their investigation of solution casting, the phenoxy/SANx blends (composition range = 5-34 wt %) obtained were recognised to be partially miscible. Through mechanical mixing, Okada and co-workers [41] recently prepared the multiphase phenoxy/SAN blends. These materials did not exhibit improvement in mechanical properties, especially in Izod impact properties.

It is well known that the physico-mechanical properties of polymer composites can be regulated by small additions of surfactants, the introduction of which alters the conditions of interaction at the phase boundary. However, their maximum effectiveness is seen when a high-strength surfactant monolayer is produced at the phase boundary, but many difficulties were encountered in the practical implementation of this approach. More promising is the introduction of reactive surfactants with functional groups capable of entering into chemical interaction with active groups of other components of the composite. Thus, to increase the adhesive strength of acrylate composites, monoallylurethane – an adduct of allyl alcohol and toluylene diisocyanate in a molar ratio of 1:1 – has been used as a surfactant of this kind. But it is difficult to increase the number of components of the composite, to introduce an active isocyanate group in the acrylic composites beforehand. The reactive surfactant was diallylurethane produced from a macrodiisocyanate based on a polyester and allyl alcohol in a molar ratio of 1:2 and it was used to modify an acrylate composite based on a solution of PbuMA in MMA [42]. Being introduced beforehand into the base of the composite, it does not shorten the storage time and exhibits reactivity only after the introduction of a polymerisation initiator, a paste of BPO, during composite preparation. A polymerisation accelerator, dimethylaniline, was a supplementary ingredient.

9.2.6 Ternary Blends

Ternary blends are gaining importance in the field of polymers. Paul and co-workers [43] reported the first example of ternary blends based on polyepichlorohydrin (PECH), PMMA, and poly(ethylene oxide)(PEO) where all three binary pairs are miscible. In 1990, two other examples of completely miscible ternary blends were reported, one containing PMMA, PECH/PMMA/PVAc [44, 45]. Recently, other examples of completely miscible ternary blends were further reported by Grannock and Paul [46]. Completely miscible ternary blends were composed of poly(vinylidene fluoride)(PVDF), PMMA and PVAc [47].

Kwei reported the first systematic study on ternary blends [48]. In his study, the addition of PVDF to the immiscible pair PMMA/poly(ethyl methacrylate) (PEMA) was studied and found to be miscible. In nearly all these blends, a third component either a homopolymer or copolymer is added to homogenise an immiscible pair. Miscibility is often achieved in cases where this third component is miscible with other polymers [49]. At the same time PVDF and PMMA blends are examples of blends with miscibility between a semi-crystalline polymer and an amorphous one. These blends remain homogeneous, and thus amorphous, even at low temperature provided that the PVDF content is smaller than approximately 50 wt.%. Otherwise PVDF crystallises from the melt with formation of two or three phases. Some discrepancy in the experimental values of T_g for the PMMA/PVDF blends has been noted, and more likely reflects the effect of the technique used for the sample preparation on the intimate phase morphology.

Isotactic, atactic, and syndiotactic PMMA (designed as i, a, and s-PMMA) with approximately the same molecular weight were blended with SAN (containing 25 wt% of acrylonitrile) in THF to cast into films [50]. T_g of the polymers were measured. aPMMA and sPMMA were found to be miscible with SAN because all the prepared films were transparent and had a single composition dependent T_g. An investigation of the miscibility of ternary blends composed of iPMMA, aPMMA, and SAN was carried out. iPMMA was mixed with aPMMA to form a binary blend in the weight ratios of 1/3, 1/1 or 3/1. The ternary mixture of iPMMA, aPMMA and PSAN in different ratios was also obtained. Certain compositions of the ternary blends of iPMMA, aPMMA and polystyrene-acrylonitrile (PSAN) showed some opacity indicating possible immiscibility. aPMMA was found to act as a 'cosolvent' between PSAN and iPMMA. The T_g of the ternary mixtures were determined calorimetrically. The approximate phase diagram of the ternary blends was established based on calorimetry data and a single T_g was used as the criterion for determining miscibility. The results indicated that the formation of miscible ternary blends is favoured for a high content of PSAN and/or aPMMA in the blends. The only blend with one T_g after quenching was found to be iPMMA/aPMMA/PSAN (12.5:37.6:49.9).

Three other polymer blend systems have been made with isotactic, atactic, and syndiotactic PMMA that were mixed with poly(styrene-*co*-*p*-hydroxystyrene) (PHS) containing 15 mol% of hydroxystyrene in 2-butanone [51]. PMMA has been preferred to show the effect of tacticity on miscibility. DSC and FTIR spectroscopy were used to study the miscibility of these blends. The three polymer blends were found to be miscible. All the prepared films were transparent and there was a single T_g for each composition of the blends. T_g elevation (above the additivity rule) was observed in all the three blends mainly because of hydrogen bonding. It has also been shown that iPMMA and aPMMA have a higher degree of hydrogen bonding with PHS than sPMMA.

9.2.7 Reactive Blending Using Acrylic Monomers

An effective way to gain control of the morphology and to strengthen the interfacial zone is to form block or graft copolymers *in situ* during blend preparation via interfacial reaction of added functionalised polymeric components. *In situ* compatibilised polymer blends based on copolymers containing glycidyl methacrylate (GMA) monomer have attracted much attention because of their potentially broad applications. Chung and co-workers patented a polymer based on polycarbonate (PC), polyethylene terephthalate (PET), ABS, and styrene-acrylonitrile (SAN)-glycidyl methacrylate copolymer [52]. The product presented excellent low-temperature impact properties. Akkapeddi and co-workers [53] reported that polyethylene (PE)-*g*-GMA acted as a good compatibiliser of blends of PC with PET and various polyolefins. Yin and co-workers [12] reported a series of reactive compatibilisations of many blending systems based on GMA-containing copolymers. PS-*co*-GMA was used as a compatibiliser for ABS/Nylon 1010 (PA1010) blends [54-56]. The rheological, thermal, and morphological properties of ABS/Nylon 1010 blends with or without compatibiliser were investigated. ABS and PA1010 are immiscible and incompatible with poor interfacial adhesion and large domains. This incompatible blend has been converted into a compatible one by incorporating a PS-*co*-GMA copolymer, which itself does not function as a compatibiliser but will become a compatibiliser after reacting with PA1010 end groups during melt blending. The addition of PS-*co*-GMA does not change the rheological behaviour of the blending system too much but the viscosity is increasing compared to the noncompatibilised system. The crystallinity of the compatibilised system is a little lower than that of the noncompatibilised system due to the reactions between glycidyl groups with PA1010 end groups. The morphological studies show that PS-*co*-GMA copolymer behaves as an effective interfacial agent by improving the adhesion between ABS and polyamide (PA) and decreasing the dimensions of dispersed phase.

The terpolymers, MMA/GMA/ethyl acrylate (EA) (MGE–10, reactive compatibiliser) have been used to compatibilise polybutylene terephthalate (PBT)/ABS blends [53]. All blend components were mixed together in powdered form before extrusion. The effects of specimen and notch geometry, ABS content and reactive compatibilisation has been examined for understanding their influence on the fructure properties of PBT/ABS blends.

9.2.8 Non-Conventional Methods for Obtaining Blends

Magnetic effects can be of technological interest because they offer a new way to perform radical processes. This is because through very weak perturbations of the magnetic field, one can control chemical kinetics and thus the course and the rate of reactions that normally require much higher chemical energies.

Beside the modifications brought to the evolution of the polymerisation process, the magnetic field influences the properties of the resulting polymers obtained through unconventional methods. The changes in the properties of the polymers synthesised in the magnetic field are attributed to the catalytic effect of the field on the molecules that can be re-shaped through growing of distance interactions and modification of angles between bonds.

In grafting studies, the emphasis is generally laid on improving the textile, chemical, hydrophobic, hydrophilic or thermoplastic properties without significantly altering the others. Numerous methods have been developed for the grafting of acrylic monomers onto the lignocellulosic materials, to make the cellulose-acrylic polymer mixture compatible, with the aim of obtaining composites with good thermal stability and UV radiation resistance. The difficulties that appear during the grafting reactions are due to the monomer sensitivity, which leads to homopolymerisation and irreproducibility.

An increased efficiency of the grafting process from the viewpoint of minimising the concurrent homopolymer formation can be achieved through the development of the reactions in a continuous magnetic field. Thus a series of grafting processes in a continuous external electromagnetic field have been realised to improve the course of the syntheses and also to ameliorate the characteristics of the products obtained in the field [57-63].

Cellulose powder obtained from oak furfural lignocellulose by a nitration reaction was grafted with acrylamide using a radical initiation [58]. The grafting reactions were carried out and compared both in presence and in absence of a continuous external magnetic field with the intensity extended between 0.15-0.35 T. The effect of monomer and initiator concentration, time and temperature of reaction and magnetic field intensity onto the grafting process were examined. The most important increase of conversion has been registered for the grafting reaction in a magnetic field, namely about 25% compared to 15% for the growth process in the absence of the field, see **Table 9.1**. That corresponds

Table 9.1 Conversions and nitrogen content determined for the graft products obtained under different reaction conditions

Grafting Process	Parameter			Conversion, %			Nitrogen Content*, %		
	I	II	III	I	II	III	I	II	III
Monomer/Cellulose, (g/g) ([K$_2$S$_2$O$_8$] = 0.154 mol/L, T = 60 °C, B = 0.25 T, cellulose = 10 g/l)									
Classic	1:1	2:1	3:1	18.2	23.2	25.6	3.2 (3.6)	4.4 (4.6)	4.85 (5.05)
Magnetic field	1:1	2:1	3:1	21.2	24.1	30.1	4.15 (4.2)	4.5 (4.75)	5.75 (5.9)
Initiator/Cellulose, (g/g) ([AAm] = 1.4 mol/l, T = 60 °C, B = 0.25 T, cellulose = 10 g/l)									
Classic	18/100	3/10	42/100	15.6	22.0	23.2	3.0 (3.07)	4.2 (4.35)	4.5 (4.57)
Magnetic field	18/100	3/10	42/100	16.5	23.1	24.1	3.0 (3.18)	4.5 (4.55)	4.65 (4.75)
Reaction temperature, (°C) [AAm] = 1.4 mol/l, [K$_2$S$_2$O$_8$] = 0.154 mol/L, B = 0.25 T, cellulose = 10 g/l)									
Classic	50	60	70	22.7	23.2	33.1	4.4 (4.48)	4.5 (4.57)	6.45 (6.53)
Magnetic field	50	60	70	23.2	24.1	37.9	4.4 (4.57)	4.6 (4.75)	7.45 (7.47)

*: Calculated values in parentheses
Parameter I: Monomer/cellulose: 1:1; initiator/cellulose: 18:100
Parameter II: Monomer/cellulose: 1:2; initiator/cellulose: 3:10
Parameter III: Monomer/cellulose: 1:3; initiator/cellulose: 42:100 B: magnetic field intensity (0.25 T)

to the variant having 30:100 weight ratio initiator/cellulose, compared to 18:100 weight ratio of initiator/cellulose.

A magnetokinetic study of the graft polymerisation of acrylamide onto a ligno-cellulosic support from softwood bark, using different methods of obtaining free-radicals, was also carried out [59]. Application of the continuous external magnetic field in the grafting heterogeneous processes of the lignocellulosic material with acrylamide resulted in an increased efficiency of the grafting reactions as can be observed in **Figure 9.2**. The modifications observed in the course of grafting reactions are attributed to the dual character of the magnetic field effects exerted on the dynamics of radicals spins and also on the dynamics of molecular movement. Growth of the macroradical's lifetime, owing to the singlet – triplet transitions through the magnetic field effects, is materialised in the increase of the graft reaction rate and also of the graft product yield.

The cellulose microfibrils respond to the action of an external continuous magnetic field because of their structure. At the same time, the field influence upon the radicals can be conceived as a growth of the macroradical's lifetime.

The grafting reaction of MMA onto polyvinylalcohol (PVA) in a continuous magnetic field was obtained using UV-radiation and benzophenone as catalyst. The graft ratio as well as stereoregularity of grafts in both the cases increased with magnetic field reaching the maximum stereoregularity at 85% graft ratio at the field of 12 kG (1.2 T)was exhibited.

The emulsion polymerisation under the magnetic field corresponds to the graft copolymerisation of MMA onto PVA in the presence of the magnetic field and this explains the phenomenon revealed. The cage in this case is formed by PVA macromolecules and it contains the monomer and initiator radicals. The excited benzophenone triplet, obtained under UV-irradiation, abstracts a hydrogen atom from the tertiary carbon of PVA and produces a triplet radical pair consisting of ketyl and PVA. Radical pairs in the triplet state due to the magnetic field presence and caused by the Zeeman effect, determine the higher graft ratio of the copolymer. At the same time, the grafted copolymers obtained under magnetic field showed improved moisture and heat resistance properties. This was attributed to the increased graft ratio and greater stereoregularity of the copolymers obtained under the field.

9.3 Characterisation of Blends with Polyacrylics in Composition

Binary blends of high nitrile polymers such as ACN/methyl acrylate/butadiene terpolymer (Barex 210, Geox Corporation) and poly(ethylene-alt-maleic anhydride) have been reported to be miscible based on optical transparency and DSC data [64-67]. T_g of this

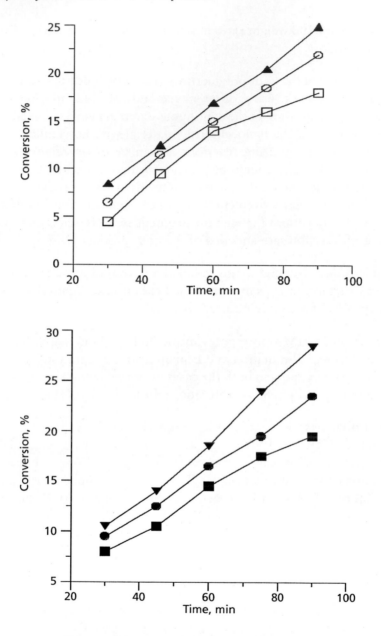

Figure 9.2 Conversion of the graft reaction of acrylamide onto lignocellulosic material initiated by: (□, ■) H_2O_2, (O, ●) $K_2S_2O_8$, (▲, ▼) $K_2S_2O_8/(FeSO_4x7H_2O)$ in (■, ●, ▼) or out (□, O, ▲)the magnetic field presence [58]

Reproduced with permission from A.P. Chiriac, I. Neamtu and C.I. Simionescu, Die Angewandte Makromolekulare Chemie, 1999, 273, 75. Copyright 1999, John Wiley & Sons blend system vary monotonically with composition, following Gordon - Taylor and Kwei equations [48]. Solid-state cross-polarisation nuclear magnetic resonance (NMR) demonstrated the monophase behaviour of these blends. The systems also exhibit single component rotating frame spin-lattice relaxation times intermediate in value as compared to the pure components. The formation of such a homogeneous monophase blend was attributed to a favourable heat of mixing arising from some specific interactions between the two component polymers. The FTIR study conducted on ACN/methyl acrylate/ butadiene terpolymer/poly(ethylene-alt-maleic anhydride) blends showed a shift in the nitrile absorption peak of Barex 210 and in carbonyl absorption peak of PEMA. This indicates the involvement of these groups in nonbonding interactions.

9.3.1 Acrylic/PVC – Blends

The deformation and relaxation behaviour of compatible blends of chain molecules can be used as method to elucidate aspects of the fundamental and applied science of blends. The deformation of polymer melts gives rise to an elongation of the coiled molecules which in turn leads to an orientation of their segments. Optical birefringence measurements, X-ray investigations have been used among other methods to characterise the induced orientational order. The chain deformation, on the other hand, gives rises to a restoring force, which is predominantly of entropic nature, so the deformation tends to relax as the force is remove. The binary blend of PMMA/polyvinyl chloride (PVC) has been characterised from this point of view on the basis of the friction coefficient [68]. The friction coefficient does not only depend on the respective polymer chains in blends, but also on the coupling to their surroundings, the strength of which is controlled by the interaction parameter. Consequently the central point in this interpretation is that a negative interaction parameter gives rise to an enhanced friction coefficient as compared to the pure polymer systems. The friction coefficient, in turn, sets the timescale of the orientational relaxation process.

The high PMMA friction coefficient is attributed to the bulky methyl groups [69] which are strongly engaged in specific interactions with PVC. Due to the strong interactions one expects an influence of blending on the orientational behaviour of PMMA. The PMMA/PVC system was also chosen for testing the theoretical concept on the segmental orientational process in polymer blends owing to small difference in the T_g. That small difference makes the ambiguity between the individual T_g and the T_g of the blend relatively unimportant, but it cannot mask the large difference between the friction coefficients of the blend component's. The results obtained for the orientation of the PMMA in the blend are consistent with the concept of an enhanced friction coefficient caused by a negative interaction parameter and thus by a stronger coupling to the surrounding.

Thermally stimulated depolarisation current (TSDC) measurements have been used to study the relaxation properties of dielectric materials in the solid state, especially because it permits working in ranges of very low equivalent frequencies as well as resolving multicomponent or distributed peaks by special techniques such as fractional polarisation or partial cleaning. The TSDC technique was used to characterise PVC and PEMA blends [70]. It was concluded that PEMA/PVC polyblends are not compatible enough because the TSDC spectrum of polyblend includes more than one peak, i.e., different modes of relaxations have been observed. This broadness has been explained on the basis of the existence of a distribution in activation energies and relaxation times as revealed by applying the thermal stimulation (TS) technique. When the TS technique has been applied to the complex spectrum of 50/50 wt% polyblend, a number of elementary peaks and their relaxation parameters that characterise different processes existing in the polyblend has been obtained. A linearity between the pre-exponential factor τ_0 the activation energy E_a was observed indicating the operation of the compensation phenomena.

It is well established that, under degradation conditions, considerable interactions may occur between the components in the polymer blend and their degradation products. The type of interaction will depend on the degree of miscibility of the components, as well as on their ratio in the blend. On blending PVC with ABS, polymer modifier, the processability and impact strength of PVC are improved. Optimal improvement of the blend's processing characteristics is achieved by adding 15% to 20% ABS. However, by modifying the polymer material, the thermal stability of the blend is also changed.

The thermal and thermo-oxidative degradation of PVC/ABS blends of different compositions was investigated by means of isothermal thermogravimetric analysis at temperature of 210-240 °C in an airflow atmosphere [71-73]. The thermo-oxidative degradation in the range of 5%-30% conversions was studied through Flynn equation, the method of stationary point, and kinetic equation using the Prout-Tompkins model. The apparent activation energy E and pre-exponential factor Z were calculated for all compositions of PVC/ABS blends. The ratios E/ln Z are constant for pure and modified PVC and point to the unique mechanism of the degradation process. Upon increasing the ratio of ABS in the PVC/ABS blend up to 50%, only the rate of the process is changed; the mechanism remains unchanged. The influence of different ratios of ABS on the stability of PVC in oxidising atmosphere it was also investigated through the isothermal thermogravimetric (TG) analysis. The calculated apparent activation energies of thermo-oxidative degradation of PVC/ABS blends are 126-129 kJ mol^{-1} and pre-exponential factors are 1.60-3.56 x 10^{12} min^{-1}.

The study of the properties of the binary polymer mixture containing PMMA and PVC or PVAc has been the subject of numerous reports. It was shown that a limited compatibility that depends on the method of obtaining the blend and also on the characteristics of the constituents. The behaviour of these systems by TG and IR methods was investigated too [74]. The PVC/PMMA system was characterised by pseudocompatibility due to the interaction between functional groups, having two maximum compatibility degrees, at PVC/PMMA (10:90) and PVC/PMMA (80:20). For 1:1 composition the system was found to be incompatible. The PVAc/PMMA system was also described as being chain pseudocompatible for the composition range 40-90 wt % PVAc, with a maximum degree of compatibility for PVAc/PMMA (80:20), at which interpenetration of the two component chains takes place.

Techniques such as dynamic mechanical, impact resistance, and SEM have also been used to study the compatibility of ethylene-propylene terpolymer (EPDM) – PVC and MMA grafted EPDM rubber (MMA-g-EPDM) – PVC (graft contents of 4, 13, 21, and 32%) blends [75]. The authors found that all the regions of viscoelasticity showed two T_g for blends, and the T_g increased with increasing graft content, indicating the incompatibility of these blends. The tan δ curve showed three dispersion regions for all blends arising from the α, β, and γ transitions of the molecules. The sharp α transition peak shifted to higher temperatures with increasing concentration of the graft copolymer in the blends. EPDM showed less improvement while six-fold increase in impact strength was noticed with the grafted EPDM. The authors concluded that blends of EPDM – PVC and MMA-g-EPDM – PVC were incompatible. Also, it was found the maximum of impact strength of PVC was at 20% EPDM and 13% grafting. The impact strength of PVC was increased six-fold when blended with MMA-g-EPDM rubber with 13% grafting as compared to EPDM rubber, which improves the *impact* strength by two-fold.

9.3.2 PC/SAN – Copolymers

Blends of PC with SAN, are used as a model for commercially important blends of PC with ABS materials. The effects of PC molecular weight, ACN content of the SAN, concentration of the dispersed SAN phase, and processing procedure on the blend morphology was examined [76]. The effect of reactive compatibilisation using a SAN – amine on the morphology was investigated in relation to the other blend variables. Compatibilisation by block or graft copolymers improves morphological stability primarily by introducing a steric hindrance to coalescence. Unlike the case for polyamides or polyesters, schemes for formation of block and graft copolymer by *in situ* reactions during melt processing of PC blends have not been available until recently.

A novel chemical approach for formation of SAN-g-PC copolymers at the PC/SAN interface was mentioned [77]. The SAN-amine polymer was synthesised by reacting a styrene/ACN/maleic anhydride terpolymer (67:32:1) with 1-(2-aminoethyl) piperazine

(AEP) in a reactive processing scheme. The graft copolymer formed was capable of reducing the SAN dispersed phase particle size and stabilising the blend morphology. Dispersed phase particle size was found to increase significantly with SAN concentration in the uncompatibilised blend due to coalescence during mixing, whereas the particle size was relatively independent of blend composition for compatibilised blends which were SAN-amine terminated.

Miscibility diagrams of copolymer – copolymer can be described via the Flory-Huggins theory combined with the binary interaction model for copolymers and have been used to delineate the shape of the miscibility regions for copolymer – copolymer blends in terms of the molecular weights of the components and the magnitude of the pairwise interaction energies. Information about interaction energies obtained from a variety of sources was used to interpret the miscibility behaviour of blends of tetramethyl polycarbonate – hexafluoropolycarbonate (TMPC-HFPC) copolymers with polystyrene-*co*-methyl methacrylate (SMMA) copolymers and the observed regions of miscibility between SMMA copolymers with TMPC-HFPC copolymers [78]. The blends were found to be miscible in two, narrow, disjointed regions, contrary to predictions based on prior estimates of the six binary interaction energies involved.

9.3.3 PS and Styrene Copolymer/Acrylics Systems

Influence of the temperature on the compatibility of the ABS/PS/cyclohexanone has been studied by light scattering and viscosity measurement [79]. It was established that the compatibility of components was changed in the temperature range where the conformational transitions taken place. At the same time the systems presented different degrees of compatibility depending on the mixture composition. At higher temperatures, the compatibility of the components was changed.

PS and PMMA form an important and unique pair of polymers in that they are chemically different; however, they display compatibility over different composition ranges at room temperature. Thus the T_g of a series of polymer blends containing various ratios of PMMA and PS indicated a single T_g over the compositions studied. The effect of gamma radiation on T_g behaviour of polymer blends composed of a radiation-sensitive polymer (PMMA) and a radiation-resistant polymer (PS) has been studied [80]. The authors found that upon exposure of the PS/PMMA blends a further depression in T_g was observed, particularly with polymer blends rich in PMMA ratios. The drop in T_g was attributed to the degradation of the PMMA component. The relative rise in T_g of blends was attributed to grafting between the blend components. This behaviour was attributed to the restrictions of the molecular motion of the polymer chains, hence, increasing the T_g. The calculation of the activation energy and order of reaction for PS/PMMA blends exposed to gamma radiation indicated that PS offers protection against oxidative degradation to these blends.

Time-resolved anisotropy measurements (TRAMS) have been used to study macromolecular mobilities both in solution and bulk phases. The measurements are powerful means of measuring macromolecular dynamics. The use of phosphorescent labels allows the relaxation behaviour of polymers in the bulk state to be studied and 'opens the door' to investigations of individual macromolecular species within polymer blends, colloidal lattices, IPN, etc. Time-resolved energy transfer has been used to study polymer compatibility and interdiffusion in blends of PS-PMMA [81]. The measurements have confirmed the 'totally immiscible' nature of PS-PMMA blends and have demonstrated the tendency of terminal segments of polymer chains to congregate in the interphase regions of phase-separated blends.

SEM has been used to investigate the composite droplet morphology development in a polymer blends, particular a high density polyethylene (HDPE)/PS/PMMA ternary blend model, which allows the PS and the PMMA dispersed phase to be dissolved away [82]. The authors found that the composite droplet morphology occurs within the first two minutes of mixing and remains stable in size and shape thereafter. They demonstrated also quantitatively that the presence of sub-inclusions generates a swelling effect, i.e., an increase of PS droplet size.

Opportunities in the field of optoelectronic technology raise interest in the design of miscible blends of transparent polymers like PS, PMMA, and various PC. Tetramethyl bisphenol-A polycarbonate (TMPC) was found to be miscible with PS whereas hexafluorobisphenol-A polycarbonate (HFPC) is miscible with PMMA. These properties enable copolycarbonates of TMPC and HFPC to be blended with styrene-methyl methacrylate copolymers to form a range of transparent materials.

Blends that are not miscible at high molecular weights may become compatible by lowering the molecular weight of one or both components below critical values. Knowledge of these critical molecular weights can be valuable for evaluating miscibility of the components [83]. When miscible blends are obtained, the interaction energy can be quantified if the mixture shows phase separation upon heating (LCST) or cooling upper critical solution temperature (UCST). Low molecular weight polysulfone (PSU) was synthesised for this purpose and the low molecular weight oligomers were tested. Modified PSU were blended with oligomers of PS, poly(α-methyl styrene) (PαMS) and PMMA as well as with styrenic copolymers to determine their interaction energies quantitatively. The study is based on the Flory-Huggins theory and the Sanchez-Lacombe equation of state theory. Modifications of PSU consisted into methyl substitution on the phenyl rings and fluoro substitution at the isopropylidene unit, and combinations of these. It was observed that the modifications of PSU that favour miscibility with PS and Pα-MS are generally unfavourable for miscibility with PMMA and *vice versa*. Increasing the polarity of the phenyl rings of PSU while maintaining a positive outer surface charge on the isopropylidene connector group is favourable for miscibility with PS. Substitutions at the isopropylidene connector group of PSU decrease the polarity of the phenyl rings and facilitate miscibility with PMMA. Blends of tetramethyl polysulfone (TMPSU) and SAN display miscibility windows for which size depends on the molecular weight of TMPSU. TMPSU and poly(α-methyl styrene-co-acrylonitrile) (α-MSAN) copolymer blends showed similar miscibility boundaries to those for TMPSU/SAN. The minimum in the interaction energy with ACN content coincides with the solubility parameter determinations by Shaw. End groups have been found to significantly affect the interaction energies and were taken into account. For molecular weights up to 12,000, the effect of the end group (which is determined by multiple polar end groups per chain, that can interact strongly and significantly affect the interaction energies) such as the hydroxyl groups on the PSU oligomer, is significant. For one nonpolar end group per chain, such as the butyl group on the styrene oligomer, the effect appears to be insignificant at molecular weights above 3,000. As a conclusion the change in interaction energy density is considered to depend on the type, size, and concentration of the end group on each chain rather than of the change in molecular weight itself.

The effects of miscibility and blend ratio on uniaxial elongational viscosity of polymer blends were studied by preparing miscible and immiscible samples of the same composition by using PMMA and SAN [84]. The study established a method to make miscible and immiscible samples of the same composition using different combinations of PMMA and SAN, for elongational viscosity. Miscible polymer blend samples, for the elongational viscosity measurement, were prepared by using three steps: solvent blends, cast films, and hot press. Immiscible blend samples were prepared by maintaining the prepared miscible samples at 200 °C, which was higher than the cloud points on the basis of a LCST phase diagram. The miscibility was determined by the observation of whether the fibre was transparent or opaque. It was found that strain-hardening property of miscible blends in the elongational viscosity was only slightly influenced by the blend ratio, and it was the case with immiscible blends.

The phase behaviour of blends involving several polymethacrylates (PMA) and a series of poly(styrene-co-maleic anhydride) (SMA) copolymers has been investigated by DSC and density measurements by Feng and co-workers [85]. The results of Paul and co-workers [74] indicated miscible blends of PMMA with SMA containing MA in the range of 6%-8% to 33%-47% by weight. It was shown that the cloud point behaviour of SMA and PMA blends is similar to that of SAN and PMA blends. The cloud point of SMA/PMA blends decrease in the order PMMA > PEMA > polybutyl methacrylate (PBMA). Blends of PMMA with SMA [50 wt.% maleic anhydride (MA)] are miscible on a molecular level. Feng concludes that the MA content in SMA needed for forming miscible blends with PMMA is higher than that given by Paul and co-workers. The difference was attributed to the molecular weight of SMA used in that study being an order lower than that used by Paul and co-workers. It has been indicated that strong intermolecular interactions between phenyl groups in SMA and carbonyl groups in PMMA. This interaction instead of the intramolecular repulsion force within the SMA copolymer that makes SMA/PMMA blends miscible. The strength of the intermolecular interactions depends on the composition of the blends.

Blends of a homopolymer and a copolymer or of two copolymers can be miscible in the absence of specific interactions, provided there is sufficient repulsion between the comonomers. The miscibility/immiscibility behaviour of the blends in this category includes: SAN/poly(ethyl or methyl methacrylate), SAN/poly(styrene-*co*-maleic anhydride) [86]. To obtain a miscible blend, ACN units in poly (styrene – *stat* – ACN) have been partially converted to the 2,4-diamino-1,3,5-triazine (DAT). It produced a range of terpolymers of poly(styrene – *stat* – acrylonitrile – *stat* – vinyl 2,4-diamino-1,3,5-triazine) that can be used to prepare blends with copolymers of methyl acrylate and maleimide. The triazine rings and the maleimide units formed complementary pairs of donor-acceptor-donor and acceptor-donor-acceptor, triple hydrogen bonding sites. It was found that miscible, one-phase, polymer blends could be formed if the DAT content of the modified SAN terpolymers was $\geq$ 10 mol% and the melt index (MI) of the methyl acrylate and maleimide copolymers was > 22 mol %, or alternatively if maleimide content was 9-14 mol% and the corresponding DAT content in the modified SAN was $\geq$ 21 mol%. Once formed these secondary interactions were stable up to the onset of degradation of the blends and there was no evidence of a lower critical temperature for phase separation.

Exothermic interactions such as hydrogen bonding, ionic and charge transfer, etc., and 'copolymer effect' are commonly used to induce miscibility in immiscible blends. The efficacy of these methods in promoting miscibility in poly(benzyl methacrylate) (PBzMA) – PS immiscible blends has been studied by suitably modifying the structure of the component polymers [87]. It has been shown that these results can be extended to the blends of poly(acrylates) or poly(methacrylates) with PS. It has been found that the hydrogen-bonding approach is most advantageous among these approaches as it involves the need for minimum interacting sites. To promote miscibility through ionic interactions, sulfonic acid groups in PS chains and pyridine basic groups in PBzMA chains were introduced. Cherrak and co-authors found that miscibility in PbzMA/PS blends have been achieved through a copolymer effect when a minimum of 14 mol % of ACN units were incorporated in PS. Through ionic interaction required only 4 mol % of the interacting sites in both the components for miscibility, both their structures ought to being modified. Besides, at higher concentrations of the interacting sites, insoluble crosslinked gels were obtained. Miscibility through hydrogen bonding mechanism required only 5 mol % of the phenolic units to be incorporated in PS. Their results on PbzMA/PS blends has been extended for other poly(acrylate)/PS or poly(methacrylate)/PS blends as these systems broadly behave in the same fashion. Thus, PEMA has been shown to be miscible with PS when the vinyl pyridine units and the sulfonic acid groups of 5 mol% concentrations were incorporated in the respective polymers. Similarly, it has been shown that the presence of 1 mol% of phenolic groups in PS was sufficient to cause miscibility with PMMA and its ethyl analogue, whereas for poly(butyl methacrylate) 3 mol% of phenolic groups was needed. PMMA required nearly 17.5 mol% of ACN units to be incorporated in PS chains to induce miscibility through the copolymer effect mechanism. As a conclusion, polyacrylates are not miscible with SAN copolymers, and the comparison of the results edifies that hydrogen-bonding approach is the most convenient and potential applicable way to induce miscibility.

One of the most sensitive techniques to prove the compatibility is to detect the separate, smaller domain [88]. The assessment of the level of homogeneity depends on the technique used in the investigation. Thus, it is fundamental to use suitable and coupled techniques in studies of miscibilities of polymer blends. Significant importance has been put on the behaviour of polymer blends under the effect of a flow field. It is well known that the flow can change the thermodynamic state of the system and perturb the phase diagram of polymer mixtures. Changes in the phase behaviour of a polymer blend that occur as a consequence of the stress associated with melt flow are likely to influence significantly the ultimate morphology of the processed and solidified materials. In this context, the effect of simple shear flow on the phase behaviour of a PMMA/SAN blend was studied [89]. It was observed that there was a shear-induced miscibility for all of the measured samples, and the shear effect was found to be composition and molecular weight dependent. The morphology of the blend under different shear rate values has also been studied to understand the previous results for the PMMA/SAN blend and to obtain deeper insights into the shear rate – morphology relationship. The magnitude of the elevation of the cloud points under shear was found to be composition dependent. Thus, the cloud points of the polymer blends of PMMA/SAN indicated a high sensitivity to shift under shear rate, higher than the sensitivity of both polymer solutions and low molecular weight mixtures by about two and five orders of magnitude, respectively.

The miscibility of poly(styrene-*co*-hydroxystyrene) containing 5 mol% hydroxystyrene monomer units (MPS-5) with iPMMA or with sPMMA was studied by Jong [90] with ^{13}C solid state NMR complemented with cloud point measurements and DSC. sPMMA is more miscible with MPS-5 than iPMMA. T_g of the sPMMA/MPS-5 blends are lower than the predictions of the Fox equation or the weight average. This contrasted with the results found for the blends of aPMMA with MPS-5. In the latter blends, T_g values are higher by 1-4 °C than the weight average predictions [91]. Jong found that two compositions (50% and 70% iPMMA) of the iPMMA/MPS-5 blends show three transitions, indicative of multiple phases present in the blend.

9.3.4 PSU/Acrylic Blends

Blends of bisphenol-A-polysulfone (bisA-PSU) and PMMA were found to be either miscible or partially miscible depending on the molecular weights used. Blends of bisA-PSU/PMMA (12:1:21) and PMMA (2.4) were optically clear when heated to 300 °C and had T_g which approximated well with the Fox equation. Blends of TMPSU and PMMA were found to be partially miscible but showed phase homogenisation on heating. Blends of hexamethyl biphenol polysulfone (HMBIPSU) and PMMA also showed partial miscibility. Blends of hexafluoro bisphenol A polysulfone (HFPSU) and PMMA oligomers were found to be miscible, but they phase separated when heated. At the same time blends of tetramethyl hexafluoro polysulfone (TMHFPSU) and PMMA were found to be partially miscible and showed UCST behaviour. Blends of bisA-PSU with α-MSAN were found to be immiscible. At equilibrium, all PSU and α-MSAN copolymer blends showed two T_g similar to the pure component values and were visually cloudy. Blends of TMPSU and poly (styrene–*co*–acrylonitrile) were found to display a miscibility window that is dependent on the molecular weights of the components. However, high molecular weight blends of TMPSU and SAN were found to be totally immiscible. Oligomeric blends of TMPSU and α-MSAN presented a miscibility window for which ACN limits depend on the molecular weight of the components.

It has been found that oligomers of bisA-PSU can be used to manipulate the phase behaviour of polymer components and it is a useful method to obtain interaction energies from components that are extremely incompatible, i.e., TMPSU – ACN. Thus, blends that have been considered immiscible in the past may actually be miscible at reduced molecular weights that are still high enough to retain desired properties: impact strength, yield stress, etc.

9.3.5 Acrylates/Other

The technique used to demonstrate the miscibility of poly(vinyl methylketone) (PVMK) with poly(2-hydroxyethyl methacrylate) (PHEMA) and PECH was DSC for glass transition behaviour [92]. Homogeneous films from blends were obtained. The existence of a single, composition-dependent T_g reveals that the blend forms a homogeneous single phase. An FTIR study revealed that the hydrogen-bonding interaction occurs between these two components, which is slightly stronger than that in pure PHEMA. PVMK is miscible with PECH and the interaction between PVMK and PECH is weak. Miscibility of PVMK with PECH was ascribed to the hydrogen-bonding interactions between the components.

With the aid of Flory-Huggins theory, which incorporated temperature-dependent interaction parameters, the phase behaviour of a binary blends of poly(cyclohexyl methacrylate-*co*-methyl methacrylate) (P(CHMA-*co*-MMA)) with poly-α-methylstyrene (PaMS) and poly-*p*-methylstyrene (PpMS) was determined experimentally and compared with that of the P(CHMA-*co*-MMA)/PS system [93]. The enthalpic and entropic parts of the segmental interaction parameters for each binary monomer pair involved were determined from a fit of the LCST behaviour of these systems. The results showed that the strength of the interaction with the P(CHMA-*co*-MMA) decreased in the order PaMS, PS, PpMS. These differences were also illustrated in the T_g-composition curves of the above systems as well as in analogous calorimetry results and scanning electron microscopy studies.

9.3.6 Specific Interactions in Blends Containing Acrylics

Generally the miscibility of polymer blends is ascribed to specific interactions between the constituents of the blend, such as poly(ethylene oxide)/poly(styrene-*co*-acrylic acid), poly(ethylene oxide)/poly(ethyl methacrylate-*co*-methacrylic acid), poly(ethylene-*co*-methacrylic acid)/poly(2-vinylpyridine) or poly(styrene-*co*-2-vinylpyridine), poly[styrene-*co*-2-(*N*, *N*-dimethylamino)ethyl methacrylate]/poly (isobutyl methacrylate-co-acrylic acid), poly(styrene-*co*-4-vinylphenol)/poly(ethyl or methyl methacrylate), styrene-*co*-*p*-(hexafluoro-2-hydroxy-2-propyl)styrene/poly(ethyl or methyl methacrylate).

Copolymerisation of interacting species such as 4-vinylpyridine and methacrylic acid with PS or poly(butyl methacrylate) led to extensive miscibility [94]. The miscibility of these blends depends on the nature, the number, and the strength of the interacting species.

Investigations of poly (styrene-*co*-4-vinylphenyl-dimethylsilane) and poly(*n*-butyl methacrylate) blends by DSC and FTIR spectroscopy showed that miscible blends were formed only for the copolymers containing 9-34 mol% of the latter polymer [95].

Owing to differences in glass transition behaviour, different authors concluded different levels of interactions and mixing between the two phases [96]. The differences were shown through the corresponding mechanical property studies as well as morphology studies. Noticeable differences have been obtained in the morphology and the wetting behaviour of PU/PMMA after changing the relative reaction kinetics of the two polymers

SIN obtained. Thus, depending on the concentration of the catalyst for the PU, for the same overall SIN composition, different morphologies were obtained, ranging from very fine phase dispersion to more or less individualised domains. The results were confirmed by dynamic mechanical spectroscopy (DMS) showing either a single broad transition or two well-defined peaks in tan δ. DSC was used to demonstrate that samples realised by IPN - technique changed slowly from a single phase to a two-phase morphology over a period of eight months. The initially miscible state was termed a metastable state, which reflects the high degree of entanglement of the linear PS chains with the poly(carbonate-urethane) (PCU) network. That was attributed to high degree of entanglement of the linear PS chains with the PCU network. The net effect was a limited molecular mobility and greatly retarded phase separation in the otherwise immiscible PCU/PS system. By contrast, the PCU/PMMA SIN were found to be stable, one-phased materials for up to 2.5 years.

The time stability of binary and ternary copolymers, as well as of mixture of binary copolymers of ACN with vinyl acetate and/or α-MS has been studied to obtain information regarding the compatibility of such polymers [97]. The binary copolymers and incompatible mixture are stable with time, while ternary copolymers and pseudocompatible mixtures evolve towards an equilibrium limiting state of phase separation, much faster in solutions than in solid state.

9.3.7 PMMA/PEO Blends

Macroscopic properties of blends being strongly influenced by their microscopic structure, an important task of blend characterisation is to study phase separation and microstructure changes induced by preparation. Blends of PEO with PMMA rank among the most extensively studied systems. They consist of a crystallisable and an amorphous component. Much effort has been devoted to the study of the compatibility of the PEO/PMMA systems. The investigation of melting point depression, T_g changes, measurement of spherulitic growth and crystallisation rates showed that the components are compatible in the melt and in the amorphous phase. The determination of binary interaction parameter χ gave small and negative values, which indicate a weakly interacting miscible system. A single T_g was reported in DSC measurements for blends of PEO with aPMMA and sPMMA, while two T_g were detected in a blend with iPMMA. The results indicate that the compatibility decreases in order sPMMA > aPMMA > iPMMA.

Cimmino and co-workers [98] have found by small-angle X-ray scattering (SAXS) and DSC that the system PEO/iPMMA with a relatively low molecular weight of PEO (20,000) is incompatible. Other papers show incompatibility of PEO with aPMMA. No changes of crystallinity, melting temperature (T_m) and T_g were mentioned in mixtures of PEO

($M_n \sim 2,000$ and $125,000$) with aPMMA ($M_n \sim 500,000$). The increase in T_g depression was observed for blends with increasing molecular weight of low-molecular-weight fractions of PEO ($<7,000$) and with increasing molecular weight of PMMA. Using low-molecular-weight PEO ($8,500$) with aPMMA ($46,800$), segregation has been found at certain compositions by inverse gas chromatography studies. A strong influence of molecular weight on compatibility for higher molecular weights ($>10^5$) of aPMMA and PEO in blends was observed. Recent investigations report on a local demixing in the disordered phase of solid PEO/aPMMA blends [99].

One of the important issues is the location of amorphous diluent in the supramolecular structure of solid mixtures. As shown by SAXS, the amorphous component can reside between crystalline lamellae (sPMMA, aPMMA), or is excluded from interlamellar amorphous regions. It was found that segregation of the amorphous diluent (PMMA) strongly depends on T_g and molecular weight. The high T_g diluent resides exclusively in the interlamellar regions, whereas the low T_g diluent is excluded at least partially from these regions. SAXS and wide-angle X-ray scattering (WAXS) measurements also showed that with increasing weight average molecular weight (M_w) of PMMA, the content of PMMA in the interlamellar regions decreased [100]. The studies led to the idea that there exists an interface at the surface of PEO lamellae from which the PMMA is excluded. The changes of compatibility strongly influence the structural behaviour of the system. It may be concluded that despite all research efforts, there are conflicting reports regarding the level of mixing for PEO/PMMA, in particular concerning molecular weights of components and the tacticity of PMMA. It has been shown that chains of low-molecular-weight fractions ($< 10^5$) of PEO form crystalline lamellae. This has been supported by quantified change in lamellar thickness under varying crystallisation conditions. The most stable crystals are formed at low supercoolings without any fold. Therefore extended chain crystals are formed.

9.3.8 PBT/ABS Blends

The fracture properties of blends of PBT with ABS materials, compatibilised by a MMA–GMA–EA terpolymer, MGE, have been characterised by Izod impact and single-edge notch, three-point bend (SEN3PB) type tests [100, 101]. Thus, moderate amounts of GMA functionality in the compatibiliser ($>5\%$) and small amounts of compatibiliser in the blend ($<5\%$) significantly improve low temperature impact properties and ABS dispersion. Higher amounts of GMA in the blend increase the room temperature impact strength with little effect on the ductile-brittle transition temperature and increase blend viscosity. Hale and co-workers remark that crosslinking reactions occur in the PBT/ABS/MGE system in addition to the desired graft copolymer formation. They consider that residual acid components in certain emulsion-made ABS materials catalyse ring-opening polymerisation of the epoxide groups in MGE or other possible mechanisms. The formation of a gel fraction leads to a reduction in the room temperature impact strength. Studies revealed that the toughness and morphology of PBT/ABS blends materials are very sensitive to the PBT molecular weight or melt viscosity. High melt viscosity of PBT was shown to lead to improve ABS (and SAN) dispersion that at its turn improves low temperature fracture toughness of PBT/ABS blends. Compatibilisation by MGE improves ABS and SAN dispersion and reduces the ductile-brittle transition temperature. Addition of compatibiliser broadens the domain of acceptable processing temperatures. The minimum quantity of ABS required to toughen PBT was lower, the higher the PBT melt viscosity was. The tensile properties of PBT/ABS/MGE blends were shown to be relatively insensitive to PBT melt viscosity or reactive compatibilisation.

Dlubek and co-workers studied the role of the free-volume holes in polymer blends and their changes as a function of composition using positronium lifetime spectroscopy [72]. The results were discussed in light of data from dynamic mechanical analysis, DSC, WAXS, polarised-light microscopy and mechanical testing performed by others co-authors on specimens from polyamide 6 (PA6) and ABS, knowing the blends as a combination of partially crystalline and amorphous components and are of practical interest because of the expected improvement in their mechanical properties.

The fracture of thin (3.18 mm) and thick (6.35 mm) specimens of PC/ABS blends with both standard and sharp notches was examined by standard Izod and SEN3PB instrumented Dynatup tests [102]. Significant morphology coarsening was seen in some PC/ABS (70/30) blends for thick (6.35 mm) injection moulded parts. Blends compatibilised with 1% of an SAN-amine polymer exhibited well-dispersed, stable morphologies.

Studies have shown the benefits of analysis of fracture data using a two-parameter model, which analyses the fracture energy per unit area as a function of the ligament size [103, 104]. Two mathematical approaches for analysing fracture behaviour as a function of ligament size were introduced. Each is based on the idea first introduced by Broberg that the region at the tip of a crack consists of an end region where actual fracture occurs and an outer region where energy is plastically absorbed during crack propagation [105].

9.3.9 Blends with PBzMA

DSC and FTIR was used to study the miscibility of poly(vinyl methyl ether) PVME and PBzMA [106]. It was shown that solvents and casting temperatures strongly affect the kinetics of the formation of blends and lead to different phase structures of the blends. The thermal transition behaviour of the PVME/PBzMA blend suggests a marginal miscibility. Various scales of molecular aggregation might exist in the blends. FTIR results show that intermolecular interactions are weak and not specific, which is in agreement with the T_g behaviour studies of the blends. The study showed that PVME is immiscible with any other polymethacrylate, such as PMMA, PEMA, poly(propyl methacrylate), poly(phenyl methacrylate), poly(cyclohexyl methacrylate), etc. The thermal transition behaviour of the PVME/PBzMA blends indicated that miscibility exists according to the conventional criterion of glass transition.

Miscibility of PBzMA and (poly) ε-caprolactone (PCL) was studied practically [107], with acrylic polymers that possess an aromatic pendant group as polyphenyl methacrylate (PPhMA) [108, 109]. The authors found that intermolecular interactions between the pairs were likely nonspecific, but sufficient to lead to a thermodynamic miscibility between the components. FTIR results do not indicate any specific interaction between C=O of PPhMA and C=O of PCL, or through other pairs of functional groups. The miscibility was attributed to the polarity and non-specific intermolecular interactions.

9.3.10 Acrylic/PO Blends

The effects of chain flexibility, concentration and molecular mass of the polymer, type of phase separation, and stress and shear rate on phase transitions in polymer solutions induced by the application of mechanical fields on phase diagram were studied on PE-PVA and PE-PMMA blends using the methods of cloud point, viscometry, polarisation microscopy, and DSC [110, 111]. It was confirmed that the application of shear stress determines weakens of the interaction between the components, resulted by an increase of the temperature of crystalline phase separation and for PE-PVA system, the shapes of the boundary curves change.

The compatibilisers which can be block or graft copolymers may improve the properties of blend materials such as polyolefins and polar polymers that exhibit poor mechanical properties because of poor interfacial adhesion. Random copolymers with a strong specific interaction with each individual polymer can be used as efficient compatibilisers. Ethylene-methacrylic acid random copolymer (EMA) contains both ethylene segments and methacrylic acid segments, and so may be used as a compatibilising agent. The role of EMA in compatibilisation was investigated in binary blends of EMA and polyolefins, and EMA and polar polymers [112]. Crystallisation of the linear low density polyethylene (LLDPE) and EMA random copolymer blends was studied by DSC, WAXS and excimer fluorescence. Then Zhao and co-workers found that ratio of the excimer emission intensity to the emission intensity of the isolated monomer decreases upon addition of EMA copolymer, indicating that PE segments of EMA copolymer interpenetrate mainly into the amorphous phase of LLDPE but also in crystalline phase. As a result was suggested the use of EMA copolymer as a compatibilising agent in recycling of polymer waste.

Polypropylene (PP) is limited in several applications due to its low surface energy, lack of reactive sites, difficulty to dye, extremely poor hygroscopicity, poor barrier to oxygen and many organic solvents, low melting and sticking temperatures, low impact strength, sensitivity to oxidation, and poor compatibility with polar polymers. Additional properties can be obtained by chemical modification, through grafting unsaturated polar groups onto the backbone of non-polar resins. Among polar monomers that have been reported in the literature for grafting have been acrylic (AA) and methacrylic acid (MAA) and their esters [113]. AA and its esters were also grafted on the surface of other polyolefins to improve the wettability and adhesion. The PP/PP-*g*-AA blends have been used as an effective moisture barrier external layer. The effect of PAA has been as an oxygen resistant central layer imparting at the same time an adequate interfacial adhesion. Contact angle measurement is an adequate method for characterising surface polarity of polymer blends. The method was used coupled with FTIR to characterise smooth surfaces of PP/PP-*g*-AA blends, so confirming that AA was grafted onto molecular chains of PP. Surface polarity of PP/PP-*g*-AA blends was augmented by an increase of PP-*g*-AA content, which promotes a better wetting of the ink on the surface due to incorporation of carbonyl polar groups.

Sun and co-workers [114] has shown that PP/PBT blends can be compatibilised '*in situ*' by reactive extrusion. This is achieved by functionalising PP with three different monomers AA, MA, and GMA. Therefore, the chemical functions grafted onto PP upon reaction with the end groups of PBT (OH and COOH) ensure the compatibilisation of the blends. A comparative study showed that GMA is the more reactive monomer towards the end groups of PBT. The addition of GMA to PP/PBT (70/30 by weight) increases the elongation at break from 29% to 110% and the impact strength (at 0 °C) from 2 J to 19 J.

Stutz and co-workers [115] examined the reactive compatibilisation of PU/PP with PE-co-AA at various levels of acid. They reported compatibilisation with the ethylenic copolymer although no reaction of the carbonyl groups with the –NCO functionality was detected. In an extension of this work, Wallhainke and co-workers [116] examined the morphology evolution of PU/PP blends compatibilised with ethylenic copolymers containing varying contents of AA and butyl acrylate. No reaction between PU and the compatibiliser was detected. The property improvement in the ternary blends was attributed to dispersed-phase diminution and morphology stabilisation in the presence of compatibiliser. To overcome the incompatibility, PE was functionalised with groups able to interact with the terminal groups of polyamide. In this context the effect of methacrylic acid derivatives grafted on PE chains on the PA6/PE blends' structure has been studied [117]. Ethyl-, isobutyl- and hydroxyethyl-methacrylate were grafted into PE, so the compatibilising effect of ester and alcohol group was investigated. The functional groups grafted in the PE chains induced 'compatibilisation' phenomena in the blends made with the non-functionalised PE.

9.3.11 PMMA/Others

The amorphous Nylon (Am-Nylon)/PMMA/poly (*p*-vinyl phenol)(PVPh) miscibility was investigated by DSC, thermo-photometry, FTIR, and SEM [118]. Binary Am-Nylon/PMMA blends revealed phase separation by DSC analysis. Binary PVPh/Am-Nylon and PVPh/PMMA blends exhibited miscibility by DSC and thermo-photometry analysis. From quantitative FTIR analysis and using stoichiometric equations, the equilibrium constant (K_c = 3.2), describing hydrogen bonding between the OH group of PVPh and amide group of Am-Nylon, was determined. FTIR showed that this miscibility was attributed to intermolecular hydrogen bonding interactions. The miscibility of Am-Nylon/PMMA blend was improved by adding 20 wt% PVPh to the binary immiscible polymer. The miscibility was attributed to intermolecular hydrogen bonding interactions. Under the new conditions, the size of the dispersed PMMA phase was smaller than in the absence of PVPh. It was demonstrated that PVPh can acts as a compatibiliser for immiscible Am-Nylon/PMMA blends.

Compatible blends can be obtained from mixing diblock copolymers of THF and MMA (PTHF-β-PMMA) with a homopolymer of THF (PTHF) [119, 120]. The authors presented the PTHF-β-PMMA-PTHF blend as a very good system for the basic studies of microphase separation, morphology, and crystallisation, including co-crystallisation, in different confined regions. It was found that these blends could have an alternating PTHF and PMMA lamellar morphology, a PTHF matrix microphase with dispersed PMMA cylindrical or spherical microdomains, depending on the blend composition. For the blends with PMMA spheres, it was found that the long period of PTHF microphase increased with increasing PMMA weight fraction. This was attributed to the presence of PMMA microdomains that could slow down the PTHF crystallisation process, resulting in a larger long period. The morphology study of the blends showed that the crystalline morphology is strongly dependent on blend composition, copolymer composition, and PTHF block length, as well as the crystallisation temperature. In the blends where PMMA spherical or cylindrical microphases were formed, the crystalline morphology changed with the change in the PTHF microdomain size and PMMA interdomain distance. A study of the solution-crystallised morphology of the blends at different temperatures showed that the morphology was also dependent on the isothermal crystallisation temperature. This behaviour suggested that the PMMA microdomains could have different effects on the morphology formation when the blend was crystallised at different rates.

Poly-3-hydroxybutyrate (PHB)/PMMA blends are of interest as a result of the acceptance of PMMA in biomedical applications and from the reported biocompatibility of natural PHB. The miscibility of PMMA with bacterial PHB was investigated. The blends showed a solubility limit of PHB in PMMA around 20 wt % PHB. Contents of the natural polymer exceeding 20% were present in PHB/PMMA blends as a pure partially crystalline

PHB phase [121]. The thermal and dielectric properties of blends of PMMA with atactic PHB (aPHB) synthesised via anionic polymerisation have been reported. Unlike the natural polyester, which very easily crystallises due to its complete (100%) stereoregularity, the synthetic PHB, being atactic, is unable to develop a crystalline phase. In a aPHB/PMMA blends, were both components are amorphous, the phase behaviour is expected to be much simpler than when one of the components can crystallise and may phase separate even in miscible blends above the T_g. The blends of PMMA with aPHB have been found to be miscible while those using isotactic natural PHB showed limited miscibility with PMMA. The difference of miscibility behaviour observed has been attributed to the large molecular weight difference of the polyesters blended with PMMA, for example a high molecular weight natural PHB (M_n = 350,000), and a low molecular weight aPHB (M_n = 3,000). The authors concluded that blend miscibility increases when the molecular weight of the component polymers decreases.

Because immiscibility usually leads to low degrees of phase dispersion and poor mechanical properties, the control of the morphology of immiscible polymer blends has been realised by using polymeric emulsifiers, in the form of block and graft copolymers. Modification of blends by adding properly selected polymeric emulsifiers, often referred to as compatibilisers, generally provides better dispersion and increased interfacial adhesion; consequently the mechanical properties are improved. Block or graft copolymers having in composition acrylic and/or methacrylic monomers have been used as compatibilising agents to improve the interfacial characteristics between the different polymer phases. However, graft copolymers are less widely used as compatibilising agents than block copolymers because the molecular structures of these graft copolymers, which are commonly prepared by chemical grafting reactions, are difficult to characterise and the experimental techniques for controlling these grafting reactions are limited as well. As a consequence the possibility of preparing copolymers with uniform grafts using the macromonomer technique was intensively studied and handled.

The inherent brittle of the phenolic resins due to their higher crosslink density can be moderate by incorporating the elastomeric and/or thermoplastic polymers into phenolics. Matsumoto and co-workers [122] reported the modification of phenolic resin by a *p*-hydroxyphenylmaleimide/acrylic ester copolymer to improve both flexural strength and heat resistance. In the same context, the effect of the crosslinking density of phenolic resins and the molecular weight of PMMA on the miscibility and structure development of a phenolic derivative/PMMA mixture during cure was investigated [119]. Also the mixture of phenolic derivatives with three poly(methyl methacrylate-*co*-ethyl acrylate) (PMMA-*co*-EA) copolymers evidenced the effect of various MMA/EA compositions of PMMA-*co*-EA and the fraction of multifunctional phenol in total phenols on the structure development in phenolic/PMMA-co-EA mixtures [123, 124]. DSC measurements revealed that the higher the fraction of multifunctional phenols in the total phenols, the faster the conversion at gelation in the phenolic-rich phase. Optical microscopy showed that a continuous two-phase structure appearing after a certain time lag and coarsened to a spherical domain structure consisting of phenolic resin particles dispersed in the PMMA-*co*-EA matrix.

Mini emulsion copolymerisation was used to incorporate the PMMA macromonomer into polybutyl acrylate chains. Based on the difference in the hydrophilicity between PMMA macromonomer and *normal* butyl acrylate monomer (*n*-BA), the *n*-BA/PMMA macromonomer copolymers resided at the interface of the polybutyl acrylate particles. These copolymers have been utilised as compatibilising agents to improve the coverage of the core polybutyl acrylate particles by a PMMA shell in composite latex particles, when the polybutyl acrylate particles are used as seed in a second-stage emulsion polymerisation of MMA monomer [124]. These graft copolymers with PMMA and polybutyl acrylate side chains were used as compatibilisers in silicone-acrylate rubber blends with large phase separation.

PMMA has been considered as a potential compatibiliser for PC/PVDF polyblends, since PMMA is known to be miscible with PVDF and compatible with PC. PMMA might behave as a common 'solvent' for PC and PVDF in the melt. PC has been toughened by core-shell rubber particles with a PMMA shell. It was suggested that the compatibility of this PMMA shell with PC provided an interfacial adhesion strong enough for the rubber particles to cavitate rather than to be debonded. PC envelops PMMA domains in a ternary PC/PMMA/PBT blend. Thus, the interfacial tension between PC and PVDF as well as PMMA and a series of PVDF/PMMA blends have been measured by the imbedded fibre retraction method. The high interfacial tension between PC and PVDF rapidly decreased by premixing approximately 35 wt.% PMMA with PVDF. In parallel, the PC/PVDF interfacial adhesion increased as rapidly as the interfacial tension decreased in relation to PMMA content in PVDF/PMMA blends.

Blends of bisphenol-A PC and PMMA have been widely studied in the literature. The interactions between PC and PMMA are very weak as deduced from the difficulty of preparing homogeneous mixture and the very low temperature at which these mixtures phase separate on heating [125]. The improvement of the PC/PMMA blends with great use for the manufacture of optical data storage media mention the enhancement of the compatibility with acrylics or by changing the specific interactions between PC and PMMA [64]. Some researchers intended to enhance specific interactions between the two polymers by either changing the molecular structure of PC or by copolymerising the MMA units with other monomer [65-67]. The improvement of compatibility between PC and PMMA has been realised with imidised PMMA (polyglutarimide) when the imide concentration was increased. Thus, it was shown that a slight increase of compatibility between PC and acrylic polymers when the concentration of imide in acrylic increases. This was attributed to the specific interactions between the highly polar imide groups and PC or by internal repulsions in the copolymer between MMA and the glutarimide units as has been depicted in acrylic/SAN blends [126].

Optical Polymer Research, Inc., provides fluorescent fibre in a rainbow of colours and diameters, as 'Round Acrylic Core Optical Fibre'. The fibre is made with a core of PS surrounded by a clear acrylic cladding. Special fluorescent dyes are added to the core where they absorb UV light and emit visible light. Light that is within the fibre's capture angle is guided down the fibre to the end for viewing.

9.3.12 Ion-Containing Polymer Blends

Blends of ion-containing polymers have been extensively studied with the purpose to increase miscibility as result of ion-dipole, ion pair-ion pair or acid-base interpolymer interactions. The 'ionomerisation' of PMMA by potassium carboxylate groups and its effect on miscibility with PVDF, it was studied [127]. Miscibility was increased when PMMA was modified by small amounts of ionic groups (approximately 1-2 mol%).

One of the most commonly used ionomer to compatibilise certain incompatible blends is poly(ethylene-methacrylic acid-isobutyl acrylate) terpolymer. This ionomer has a small number of ionic groups (less than 15% mol where the acid groups are partially neutralised by metal ions) along nonionic backbone chains [128]. Thus it has been mentioned the addition of the ionomer into an immiscible blend to provide a lowering of the interfacial energy and an improvement of the interfacial adhesion between the two phases. The final morphological effect was a reduction in the particle size of the dispersed phase in the blend.

Because blends of PA1010 and butyl rubber (IIR) are considered to be immiscible due to their different polarities it has been chosen the poly(ethylene-methacrylic acid-isobutyl acrylate) terpolymer ionomer as compatibiliser for this binary system owing to the miscibility and interaction of the ionomer segments with the two blends components. The authors considered the fact that the ionomer hydrocarbon backbone being a common feature with IIR, molecular intermixing occurs, to some extent, in the amorphous molten state between the IIR and ionomer. The reactive groups of PA1010 have affinity to link with (COO$^-$) groups present in the side chains of the ionomer. Thus the ionomer act as a bridge holding the PA1010 and IIR components of the blend and give a compatibilised blend. The presence of poly(ethylene-methacrylic acid-isobutyl acrylate) terpolymer as compatibiliser improves the interfacial adhesion between the two phases and the dispersion of IIR domains in the PA1010 matrix. SEM observations shown a great decrease in the IIR domains size on addition ionomer. A ductile type of fracture was observed on the

401

fracture surface of the PA1010/IIR blend compatibilised by this terpolymer. The system presents the improvement in the notched impact strength of the blend after the addition of the ionomer.

The literature in the field mentions the use of acrylic ionomer poly(ethylene-methacrylic acid-isobutyl acrylate) terpolymer as a compatibiliser for incompatible PU/HDPE blends [129]. The authors found that Zn^{2+} ionomer added at moderate concentrations (10-15 wt.%) is an effective compatibiliser for melt-mixed and quenched PU/HDPE blends. At the same time, the ionomer strongly 'solvates' the PU component. It was considered that compatibilisation is caused by the strong adhesion of the PU/ionomer component due to specific forces and the 'mechanical' compatibility of the ionomer – polyolefin components.

A random copolymer of MMA and AA (6 mol. %) has been synthesised and neutralised to different extents by Zn cations. The effect of this PMMA modification on miscibility with PVDF has also been studied [109]. The interaction parameter has been estimating from melting point depression data, whereas T_g has been measured by dynamic mechanical analysis. Modification of PMMA by 5.7 mol.% AA (random copolymer) has a deleterious effect on miscibility with PVDF. But neutralisation of half the AA comonomer units by Zn cations changes completely the situation as χ_{12} decreases significantly becoming even smaller than χ_{12} for the original PVDF/PMMA pair. However, the complete neutralisation is very unfavourable to miscibility, which was explained by the tendency of the Zn carboxylated groups to self-associate into multiples and/or cluters (which is the typical behaviour of ionomers) rather than to contribute to specific interactions with PVDF.

Blends present interesting theoretical as well as technological and practical interest. Acrylic blends with their multiple applications in many fields such as, fibres, adhesives and binders, paints and finishes, synthetic rubbers. As a result the interest into developing these kinds of products is entirely justified.

References

1. Japan Chemical Fibers Association, www.fcc.co.jp/JCFA.

2. E. Corradini, A.F. Rubira and E.C. Muniz, *European Polymer Journal*, 1997, **33**, 1651.

3. E. Yilmaz, O. Yilmaz and H. Caner, *European Polymer Journal*, 1996, **32**, 927.

4. M. Song and F. Long, *European Polymer Journal*, 1991, **27**, 983.

5. Z. Sun, W. Wang and Z. Feng, *European Polymer Journal*, 1992, **28**, 1259.

6. E.G. Crispim, A.F. Rubira and E.C. Muniz, *Polymer*, 1999, **40**, 5129.

7. A.E. Nesterov, Y.S. Lipatov and V.V. Horichko, *Polymer International*, 1999, **48**, 117.

8. L. Zhang and A. Eisenburg, *Polymers for Advanced Technologies*, 1998, **9**, 677.

9. C. Honda, K. Sakaki and T. Nose, *Polymer*, 1994, **35**, 5309.

10. J. Ding and G. Liu, *Macromolecules*, 1997, **30**, 655.

11. Y.S. Oh and B.K. Kim, *Journal of Macromolecular Science B*, 1997, **36**, 5, 667.

12. Z.H. Yin, X.M. Zhang and J.H. Yin, *Journal of Applied Polymer Science*, 1997, **63**, 1857.

13. N.M. Chechilo and N.S. Enikolopyan, *Doklady Physical Chemistry*, 1975, **221**, 392.

14. J.A. Pojman, G. Curtis and V.M. Ilyashenko, *Journal of the American Chemical Society*, 1996, **118**, 3783.

15. J. Szalay, I.P. Nagy, L. Barkai and M. Zsuga, *Die Angewandte Makromolekulare Chemie*, 1996, **236**, 97.

16. I.P. Nagy, L. Sike and J.A. Pojman, *Advanced Materials*, 1995, **7**, 1038.

17. A. Tredici, R. Pecchini, A. Sliepcevich and M. Morbidelli, *Journal of Applied Polymer Science*, 1998, **70**, 13, 2695.

18. L.H. Sperling and V. Mishra, Proceedings of the 25th Anniversary Symposium of the Polymer Institute, University of Detroit-Mercy, 1994, Technomic Press, Lancaster, PA, USA, 1994.

19. E.V. Rusinova, S.A. Vshivkov and I.V. Zarudko, *Polymer Science Series A*, 1999, **41**, 676.

20. L.H. Sperling in *Interpenetrating Polymer Networks and Related Materials*, Ed., L.H. Sperling, Plenum Press, New York, NY, USA, 1981.

21. L.A. De Graaf, J. Beyer and M. Moeller, *Journal of Polymer Science: Polymer Physics Edition*, 1995, **33**, 1073.

22. H.S. Shin, S.Y. Kim and Y.M. Lee, *Journal of Applied Polymer Science,* 1997, **65**, 685.

23. B. Ramaraj and G. Radhakrishnan, *Journal of Applied Polymer Science,* 1994, **52**, 837.

24. Y.L. Guan, L. Shao and K.D. Yao, *Journal of Applied Polymer Science,* 1996, **61**, 393.

25. S. Turri, M. Scicchitano, G. Simeone and C. Tonelli, *Progress in Organic Coatings,* 1997, **32**, 305.

26. M. Krupers, P.J. Slangen and M. Moeller, *Macromolecules,* 1998, **31**, 2552.

27. S.C. Kim, D. Klempner, K.C. Frisch and H.L. Frisch, *Macromolecules,* 1976, **9**, 2, 263.

28. A. Morin, H. Djomo and G.C. Meyer, *Polymer Engineering and Science,* 1983, **23**, 7, 394.

29. T. Hur, J.A. Manson, R.W. Hertzberg and L.H. Sperling, *Journal of Applied Polymer Science,* 1990, **39**, 1933.

30. M. Akay and S.N. Rollins, *Polymer,* 1993, **34**, 9, 1865.

31. M. Krupers, P.J. Slangen and M. Moeller, *Macromolecules,* 1998, **31**, 2552.

32. L. Zhang and A. Eisenberg, *Macromolecules* 1996, **29**, 8805.

33. C. Honda, K. Sakaki and T. Nose, *Polymer,* 1994, **35**, 5309.

34. K. Iyama and T. Nose, *Polymer,* 1998, **39**, 651.

35. J. Ding and G. Liu, *Macromolecules,* 1997, **30**, 655.

36. L. Desbaumes and A. Eisenberg, *Langmuir,* 1999, **15**, 36.

37. Y. Yu, L. Zhang and A. Eisenberg, *Macromolecules,* 1998, **31**, 1144.

38. S. Kim and J.H. An, *Journal of Applied Polymer Science,* 1995, **58**, 3, 491.

39. S. Zheng, J. Huang, Z. Zhong, G. He and Q. Guo. *Journal of Polymer Science: Polymer Chemistry Edition,* 1999, **37**, 525.

40. M. Vanneste and G. Groeninckx, *Polymer,* 1994, **35**, 162.

41. M. Okada, K. Fujimoto and T. Nose, *Macromolecules*, 1995, **28**, 1795.

42. E.I. Fedorchenko, V.I. Pavlov, S.S. Ishchenko and E.V. Lebedev, *International Polymer Science and Technology*, 1999, **26**, T/77.

43. D.R. Paul, J.W. Barlow, C.A. Cruz, R.N. Mohn, T.R. Nassar and D.C. Wahrmund, *Coatings and Plastics Preprints*, 1977, **37**, 1, 130.

44. Q. Guo, *European Polymer Journal*, 1990, **26**, 1329.

45. Q. Guo, *European Polymer Journal*, 1990, **26**, 1333.

46. G.R. Brannock and D.R. Paul, *Macromolecules*, 1990, **23**, 5240.

47. Q. Guo, *European Polymer Journal*, 1996, **12**, 1409.

48. T.K. Kwei, *Journal of Polymer Science: Polymer Letters Edition*, 1984, **22**, 307.

49. N. Moussaif and R. Jérôme, *Polymer*, 1999, **40**, 6831.

50. N. Barbani, L. Lazzeri and C. Cristallini, *Journal of Applied Polymer Science*, 1999, **72**, 971.

51. W.P. Hsu and C.H. Yeh, *Journal of Applied Polymer Science*, 1999, **73**, 431.

52. J.Y.J. Chung, R.P. Carter and D. Neuray, inventors; Mobay Chemical Corporation, assignee; US Patent 4,554,315, 1985.

53. M.K. Akkapeddi, C.D. Mason and B. Van Burskirk, *Polymer Preprints*, 1993, **34**, 848.

54. M. Okada, K. Fujimoto and T. Nose, *Macromolecules*, 1995, **28**, 1795.

55. E.I. Fedorchenko, V.I. Pavlov, S.S. Ishchenko and E.V. Lebedev, *International Polymer Science and Technology*, 1999, **26**, T/77.

56. S. Bicakci and M. Cakmak, *Polymer*, 1998, **39**, 4001.

57. A.P. Chiriac and C.I. Simionescu, *Progress in Polymer Science*, 2000, **25**, 2, 219.

58. A.P. Chiriac, I. Neamtu and C.I. Simionescu, *Die Angewandte Makromolekulare Chemie*, 1999, **273**, 75.

59. A.P. Chiriac, I. Neamtu and C.I. Simionescu, *Synthetic Polymer Journal*, 1998, **5**, 43.

60. A.P. Chiriac, C.I. Simionescu, I. Neamtu and G. Cazacu, *Cellular Chemistry and Technology*, 2000, **34**, 3.

61. A.P. Chiriac, C.I. Simionescu and I. Neamtu, *Revue des Composites et des Materiaux Avances*, 1999, **9**, 1, 133.

62. A.P. Chiriac, C.I. Simionescu, I. Neamtu and M. Popa, *Cellular Chemistry and Technology*, 1998, **32**, 5-6, 425.

63. A.P. Chiriac, I. Neamtu, G. Cazacu, C.I. Simionescu and G. Rozmarin, *Die Angewandte Makromolekulare Chemie*, 1997, **246**, 1.

64. C.K. Kim and D.R. Paul, *Macromolecules*, 1992, **25**, 3097.

65. G.C. Eastmond and L.W. Harvey, *British Polymer Journal*, 1985, **17**, 275.

66. C.C. Hung, W.G. Carson and S.P. Bohan, *Journal of Polymer Science: Polymer Physics Edition*, 1994, **32**, 141.

67. M.E. Fowler and D.R. Paul, *Journal of Applied Polymer Science*, 1989, **37**, 513.

68. J. Zhao, B. Jasse and L. Monnerie, *Polymer*, 1989, **30**, 1643.

69. J.M. Willis, V. Caldas and B.D. Favis, *Journal of Materials Science*, 1991, **26**, 4742.

70. M.T. Ahmed and T. Fahmy, *Polymer Testing*, 1999, **18**, 589.

71. Y.C. Jean, *Microchemical Journal*, 1990, **42**, 72.

72. G. Dlubek, M.A. Alam, M. Stolp and H.J. Radusch, *Journal of Polymer Science: Polymer Physics Edition*, 1999, **37**, 1749.

73. M. Stolp, H.J. Radusch and A.K. Bledzki, *Kunststoffe*, 1993, **83**, 7.

74. G.R. Brannock, J.W. Barlow and D.R. Paul, *Journal of Polymer Science: Polymer Physics Edition*, 1991, **29**, 413.

75. D. Singh, V.P. Malhotra and J.L. Vats, *Journal of Applied Polymer Science*, 1999, **71**, 1959.

76. I. Fortelny, J. Kovar and M. Stephan, *Journal of Elastomers and Plastics*, 1996, **28**, 106.

77. G. Wildes, S.H. Keskkula and D.R. Paul, *Journal of Polymer Science: Polymer Physics Edition*, 1999, **37**, 71.

78. K. Takakuwa, S. Gupta and D.R. Paul, *Journal of Polymer Science: Polymer Physics Edition*, 1994, **32**, 1719.

79. M. Sabiliovschi, S. Ioan and C. Vasile, *Polymer Bulletin*, 1981, **5**, 53.

80. Y.C. Jean, *Microchemical Journal*, 1990, **42**, 72.

81. I. Soutar, L. Swanson, J.M.G. Cowie, I.C. Barker, N.J. Flint and M.J. Conroy, *Macromolecular Symposia*, 1999, **141**, 69.

82. S. Sanchez-Valdes, I.Yanez-Flores, L.F. Ramos de Valle, O.S. Rodrigues-Fernandez, F. Orona-Villarreal and M. Lopez-Qunitanilla, *Polymer Engineering and Science*, 1998, **38**, 127.

83. T.A. Callaghan and D.R. Paul, *Journal of Polymer Science: Polymer Physics Edition*, 1994, **32**, 1847.

84. T. Takahashi, J. Takimoto and K. Koyama, *Journal of Applied Polymer Science*, 1999, **73**, 757.

85. H. Feng, L. Shen and Z. Feng, *European Polymer Journal*, 1995, **31**, 243.

86. J.M.G. Cowie and D. Cocton, *Polymer*, 1999, **40**, 227.

87. D.E. Cherrak and S. Djadoun, *Journal of Applied Polymer Science, Applied Polymer Symposium*, 1992, **51**, 209.

88. S.A. Madbouly, M. Ohmomo, T. Ougizawa and T. Inoue, *Polymer*, 1999, **40**, 1465.

89. S.A. Madbouly, T. Chiba, T. Ougizawa and T. Inoue, *Journal of Macromolecular Science B*, 1999, **38**, 79.

90. L. Jong, E.M. Pearce, T.K. Kwei and L.C. Dickinson, *Macromolecules*, 1990, **23**, 5071.

91. C.P. Papadoupolou and N.K. Kalfoglou, *Polymer*, 1999, **40**, 905.

92. Q. Guo, J. Huang and X. Li, *European Polymer Journal*, 1996, **32**, 321.

93. A. Mugica, M. Barral, J.A. Pomposo and M. Cortazar, *Acta Polymerica*, 1999, **50**, 304.

94. L.C. Cesteros, E. Meaurio and I. Katime, *Macromolecules*, 1993, **26**, 2323.

95. C.A. Barron, S.K. Kumar, J.P. Runt and J. Fitzgerald, *Polymer Materials Science and Engineering*, 1991, **65**, 333.

96. V. Mishra, F.E. Du Prez, E. Gosen, E.J. Goethals and L.H. Sperling, *Journal of Applied Polymer Science*, 1995, **58**, 331.

97. C. Vasile, H.A. Schneider, A. Airinei and A.P. Chiriac, *European Polymer Journal*, 1976, **12**, 749.

98. S. Cimmino, E. Martuscelli and C. Silvestre, *Makromolekulare Chemie, Macromolecular Symposia*, 1988, **16**, 147.

99. S. Cimmino, E. Di Pace, E. Martuscelli, C. Silvestre, *Makromolekulare Chemie, Rapid Communications*, 1988, **9**, 261.

100. W. Hale, J.H. Lee, H. Keskkula and D.R. Paul, *Polymer*, 1999, **40**, 3621.

101. W. Hale, H. Keskkula and D.R. Paul, *Polymer*, 1999, **40**, 3353.

102. G. Wildes, H. Keskkula and D.R. Paul, *Polymer*, 1999, **40**, 7089.

103. Y. Kayano, H. Keskkula and D.R. Paul, *Polymer*, 1997, **38**, 1885.

104. Y. Kayano, H. Keskkula and D.R. Paul, *Polymer*, 1998, **39**, 821.

105. K.B. Broberg, *International Journal of Fracture Mechanics*, 1968, **4**, 11.

106. T.K. Mandal and E.M. Woo, *Macromolecular Chemistry and Physics*, 1999, **200**, 1143.

107. D. Debier, J. Devaux and R. LeGras, *Journal of Polymer Science: Polymer Chemistry Edition*, 1995, **33**, 407.

108. H. Li and Z. Li, *Journal of Applied Polymer Science*, 1998, **67**, 61.

109. H. Li, Y. Wang and Z. Li, *Polymer Preprints*, 1999, **40**, 173.

110. S.A. Vshivkov, S.G. Kulichikhin and E.V. Rusinova, *Uspekhi Khimii*, 1998, **67**, 261.

111. E.V. Rusinova, S.A. Vshivkov and I.V. Zarudko, *Polymer Science Series A*, 1999, **41**, 676.

112. H. Zhao, Z. Lei, Z. Wang and B. Huang, *European Polymer Journal*, 1999, **35**, 355.

113. S. Flores-Gallardo, S. Sanchez-Valdes, L.F. Ramos de Valle, Proceedings of Antec '99, New York, NY, USA, Volume 2, 2286.

114. Y-J. Sun, G-H. Hu, M. Lambla and H.K Kotlar, *Polymer*, 1996, **37**, 4119.

115. H. Stutz, P. Pötschke and U. Mierau, *Macromolecular Symposia*, 1996, **112**, 151.

116. K. Wallhainke, P. Pötschke and H. Stutz, *Journal of Applied Polymer Science*, 1997, **65**, 2217.

117. M. Xanthos, *Polymer Engineering and Science*, 1988, **28**, 1392.

118. S. Zheng, J. Huang, Z. Zhong, G. He and Q. Guo, *Journal of Polymer Science: Polymer Chemistry Edition*, 1999, **37**, 525.

119. L-Z. Liu, W. Xu, H. Li, F. Su and E. Zhou, *Macromolecules*, 1997, **30**, 1363.

120. L-Z. Liu and B. Chu, *Journal of Polymer Science: Polymer Physics Edition*, 1999, **37**, 779.

121. G. Ceccorulli and M. Scandola, *Journal of Macromolecular Science*, 1999, **36**, 327.

122. A. Matsumoto, K. Hasegawa, A. Fukuda and K. Otsuki, *Journal of Applied Polymer Science*, 1992, **44**, 205.

123. B.S. Kim, G-I. Nakamura and T. Inoue, *Journal of Applied Polymer Science*, 1998, **11**, 1829.

124. P. Rajatapiti, V.L. Dimonie and M.S. El-Aasser, *Journal of Applied Polymer Science*, 1996, **6**, 891.

125. C.K. Kim and D.R. Paul, *Polymer*, 1992, **33**, 4929.

126. N. Moussaif and R. Jerome, *Polymer*, 1999, **40**, 3919.

127. L.M. Leung and K.K.C. Lo, *Polymer International*, 1995, **36**, 339.

128. E.M. Woo and T.K. Mandal, *Macromolecular Rapid Communications*, 1999, **20**, 46.

129. C.P. Papadoupolou and N.K. Kalfoglou, *Polymer*, 1989, **26**, 7015.

10 Rubber Toughened Epoxies/Thermosets

B. Kothandaraman and Anand K. Kulshreshtha

10.1 Introduction

The first synthetic plastics like cellulose trinitrate and polyvinyl chloride (PVC) were rigid. PVC could not be processed unless it was flexibilised/toughened first. PVC and polystyrene could become more useful because of the development of methods to toughen/flexibilise them. The development of rubber modified thermosets and thermoplastics is an important contribution to commercial polymer industry. A small amount of discrete rubbery particles in a glassy plastic matrix can greatly improve the crack resistance and impact strength of the normally brittle plastic. This must be accomplished with a minimal loss in stiffness and thermal properties. Proper methods to toughen plastics and composites can lead to enhanced usage of such materials in aerospace applications where weight savings are of paramount importance.

Epoxy resin is the most widely used matrix material for aircraft structural composites. It has many good properties like stiffness, low shrinkage, good adhesion to glass/carbon fibre, etc. Unfortunately, the very factor contributing to its high stiffness and heat resistance leads to its main drawback (lack of toughness), i.e., its highly crosslinked structure. Aircraft structures are often subjected to impact loads during flight which can cause severe damage to the structure by delamination. Toughening can help in developing damage tolerant structural components for aircraft.

Fracture toughness of various materials is given in **Table 10.1**.

Table 10.1 shows that rubber toughened epoxy resins can match engineering thermoplastics in fracture toughness values. Composites made out of such matrix systems are expected to be tougher and hence more useful.

10.1.1 Various Approaches to Toughening Epoxy Resins

The need for toughening epoxy resins was felt in the mid 1960s. Initially, solid rubbers like nitrile and polysulfide rubbers were found to flexibilise epoxy resins. Such resin

411

Table 10.1 Fracture toughness of various materials (data from [1]).	
Material	Fracture toughness range (kJ/m^2)
Inorganic glass	Very low
Epoxy/polyester resins	10^{-1}
Polysulfones/rubber toughened epoxy	10
Metals	> 10

systems were found to be suitable for adhesives as these rubber modified resins could sustain larger deformations and hence gave better peel joints than the unmodified epoxy-based adhesive. However, as they caused unacceptable levels of deterioration of stiffness and thermal properties for composites, better tougheners were needed.

A few scientists started working on toughening epoxy resins by using reactive liquid rubbers synthesised by BFGoodrich Co., USA in 1965:

- Carboxyl terminated butadiene acrylonitrile copolymers (CTBN),

- Amine terminated butadiene acrylonitrile copolymers (ATBN),

- Hydroxyl terminated butadiene acrylonitrile copolymers (HTBN), etc.

These rubbers had molecular weights (MW) in the range of 1000-2000. They could be pre-reacted with the resin - they dissolve in the resin initially and precipitate during cure to form discrete particles of diameters about 1-5 µm. Often bimodal particle distributions are encountered where two types of particles are formed. Large particles (above 5 µm diameter cause crazing while small particles (1 µm or less in diameter) cause shear banding. Shear banding is the main toughening mechanism. This is amplified by particle cavitation, particle deformation and tearing. All these lead to strain energy release which causes toughening. Rigid particles like treated fillers and thermoplastics contribute to toughening by crack deflection or crack pinning mechanisms.

Most of the studies were confined to CTBN as it provided the most ideal morphology needed for effective toughening. CTBN has two active hydrogens in its end groups so that the resin underwent chain extension and was precipitated as spherical rubbery particles in a glassy epoxy matrix. ATBN has four active hydrogens hence it acted as an additional curative and provided a different morphology.

Until today, CTBN has been the most popular toughening additive for epoxy resins. However, any other rubber which can precipitate as small particles during cure can also toughen epoxies like polyacrylates, silicones (with reactive end groups) and latex based core-shell particles.

Rigid particles like alumina filler, alumina short fibre, etc., alone or in combination with rubbers can also toughen epoxy resins effectively [2]. Rigid particle and thermoplastic modification can toughen epoxies without decreasing the T_g, but the effectiveness on G_{1c} (fracture energy, see Section 10.1.2) is less pronounced compared with rubbers [2].

Reducing the crosslink density of the resin by chain extension with bis phenol-A, for example, can also improve the toughness of epoxy resins. Such systems are used widely in the adhesives field. Bertram and Burton [3] have argued that if such a chain extender has more rigid backbones for example, tetra bromo bisphenol-A (instead of bisphenol A), this can make the resin more ductile without decreasing the T_g of the resin. It has been found to be so.

Presence of microvoids, and more homogenous networks, i.e., uniform crosslink densities, can also lead to development of toughened epoxy resins [3].

For highly crosslinked resins, which were considered to be impossible to toughen, core-shell particles, uniform crosslink distribution, thermoplastic modification and chain extension are the approaches which are expected to lead to tougher resins and hence, tougher composites [3].

10.1.2 Measurement of Toughness

Before going into the subject, mention may be made about how fracture toughness is assessed. One definition is 'area under the stress-strain curve'. More commonly, fracture toughness is explained in terms of K_{1c}, (stress intensity factor) and G_{1c} (critical strain energy release rate). These are measured by making a notch in the specimen and subjecting it to tensile load. From the stress-strain data obtained, these parameters are calculated using standard formulae [4].

10.2 Modification of Epoxy Resins by Rubbers

10.2.1 CTBN

CTBN may be incorporated into an epoxy resin by a procedure described by BFGoodrich product data sheet:

For incorporating the rubber, an adduct between the epoxy and the CTBN needs to be formed. For example, for the diglycidyl ether of bisphenol A (DGEBA) resin, 62.5 parts of the resin may be mixed with 37.5 parts of CTBN in the presence of 0.25 parts of triphenyl phosphine catalyst and the mixture may be heated for about half an hour at 120 °C to form the adduct. The heating time may change with temperature. To incorporate the necessary amount of rubber, this adduct may be suitably diluted with the resin.

CTBN was found to be an excellent toughening additive for epoxy resin by around 1965. Epoxy resin mostly a DGEBA – CTBN – hardener system has been studied comprehensively over the years and a lot of data have been generated. The most important thing to remember is: the efficacy of the toughening additive depends on how effective the phase separation between the rubber and the resin, i.e., morphology of the rubber dispersion in the resin is. This in turn depends on the type of epoxy resin, curative, cure temperature, rubber composition, etc., and further, these factors are inter-related. So reproducibility of the morphology is a problem and hence, consistency in the mechanical properties of such system is also difficult to achieve [5].

The degree of phase separation may also be supplemented/opposed by another factor, i.e., adhesive bonding between the resin matrix and the rubber particles [6]. The effect of various parameters on the toughening effect will be considered next:

- **Crosslink density of the resin**

If the crosslink density of the resin (Mc) is less, more ductility should result. Thus, more plastic energy dissipation occurs during fracture. This gets magnified by the presence of rubber particles, because this causes the formation of a large number of energy dissipating shear zones. Thus, for an epoxy-CTBN-diamino diphenyl sulfone (DDS) hardener system, Kinloch [7] observed the following:

For two epoxy resins, of epoxy equivalents 172 g/mol, and 475 g/mol, G_{1c} values (in kJ/m^2) were almost equal. However the improvement in G_{1c} caused by 15% (by wt.) CTBN on the less densely crosslinked resin, i.e., with epoxy equivalent of 475, was up to 10 times approximately, compared with a very small increase for the resin with an epoxy equivalent of 172 g/mol. The improvement on a still less crosslinked resin (epoxy equivalent of 1600 g/mol), was found to be up to 50 times that of the unmodified resin.

- **Glass transition temperature (T_g) of the resin**

The T_g of the resin can vary with the crosslink density and the cure schedules. Thus, Kinloch [7] showed that for an epoxy resin-CTBN-piperidine system, for 16 hours of cure at 120 °C, crosslink density was observed to be about 610 g/mol for which T_g was

100 °C and G_{1c} was 2.23 kJ/ m², for 6 hours of cure at 160 °C, the crosslink density was found to be about 7 times the first case, T_g decreased by about 10 °C while the G_{1c} increased by more than 2.5 times that of the earlier case.

- **Effect of cure temperature**

According to Levita [8], epoxy resins, when cured at higher temperatures are tougher than those cured at lower temperatures, in spite of higher crosslink densities. This is attributed to free volume effects at higher temperature though loss of curatives by oxidation and volatalisation can also be cited as reasons. In epoxy - CTBN - piperidine systems, toughening is observed only with cure temperature, above 100 °C.

Levita suggested, that for formation of fine sized elastomeric particles, phase separation should occur before gelation, when the viscosity is sufficiently high to prevent particle coalescence, but still low enough to favour nucleation and growth. At low cure temperatures, phase separation will be prevented.

Montarnal and co-workers [9] studied these in more detail, with DGEBA, CTBN cured by 1,8-diamino-*p*-menthane (MNDA), a hindered amine hardener. The effect of cure schedules for this system, is shown in **Table 10.2**.

Kim and co-workers [10] showed that if phase separation rate is faster than cure rate, bimodal particle distribution will result, otherwise, unimodal. Thus, at higher cure temperature, viscosities will be lower - this will also favour unimodal particle distribution.

Table 10.2 Effect of cure schedules on properties of MNDA cured epoxy-CTBN (15%) systems. Adapted from Tables IV and VI of [9]				
Cure Schedule	Post cure	Tensile Strength (MPa)	Charpy Impact Strength (kJ/m²)	Change in T_g (°C) due to CTBN*
1 day, 50 °C	14 h at 190 °C	2.12	23	–10
7 h, 75 °C	14 h at 190 °C	2.09	32	–15
2 h at 100 °C	14 h at 190 °C	2.07	46	–15
5 °C until 190 °C was reached	14 h at 190 °C	2.15	33	–17
*T_g of the base resin was 152 °C				

Levita [8] showed that CTBN particles can incorporate the resin as rigid filler and some resin can dissolve molecularly in the rubber and form copolymers and increase strength. The presence of bisphenol-A (a chain extender) can magnify the toughening caused by CTBN, by causing bimodal particle distribution. This may be because, bisphenol-A can increase the rate of the crosslinking which can make some rubber remain uncrosslinked before cure leading to complex morphologies.

- **Nature of the rubber**

As far as the nature of the rubber is concerned, its volume fraction, particle size and its distribution, interfacial adhesion and its T_g can all affect the extent of toughening.

Small particles (about 5 μm diameter) can promote shear banding while large particles cause crazing around large particles. Thus a combination of both, gives better results which explains why attempts are made to get bimodal particle distributions [2].

Regarding the interfacial bonding between the rubber and the resin, this has also been studied in detail and a few results are discussed next.

Huang and co-workers [6] reported that, compared with a difunctional liquid rubber, which gave good interfacial strength, a nonfunctional rubber, i.e., one without reactive end groups, gave a much poorer toughening effect, as this gave poor interfacial bonding between the resin and the rubber particle. To compare the effect, a difunctional rubber gave a 70% improvement in G_{1c} against a nonfunctional rubber which provided only 20% improvement in G_{1c}.

This may suggest that with good interfacial strength, some contribution to fracture energy by crack bridging (by rubber particles) is possible. If there is poor interfacial strength (as is the case with a nonfunctional rubber) this contribution will be nil because rubber particle debonding will occur and once debonded, it cannot perform crack bridging.

Further, when we compare the interfacial bonding between two nonfunctional liquid rubber grades, the one with higher acrylonitrile (ACN) content gives better interfacial strength. The observed higher value of fracture energy with higher ACN content (nonfunctional) may be due to enhanced matrix ductility caused by greater compatibility between the rubber and the resin.

The rubber particles form copolymers with the epoxy resin but still retain their low T_g (–55 °C). This means, their Poisson's ratio may still approach 0.5, leading to a higher bulk modulus. So they may not encourage large scale plastic dilatation in the matrix unless the rubber particles undergo debonding or cavitation.

- **Effect of ACN content in the CTBN**

This has also been studied by many researchers. The ACN content in CTBN can vary.

For example in CTBN 1300 X 8 grade supplied by BFGoodrich Co., the ACN content is 8% and in 1300 X 13, it is 13%.

The change in ACN content affects the compatibility of the CTBN with the epoxy resin, as with ACN content, polarity changes. Lee and co-workers [11] obtained the following results for a DGEBA resin (MY 750 supplied by Ciba Geigy, cured by piperidine), modified by CTBN:

The unmodified resin (MY 750) at 20% of CTBN (by weight), reduced the T_g by about 2 °C, when the ACN content was 8%. The rubber containing 13% ACN reduced the T_g by 7 °C at any level of modification up to 30% CTBN by weight.

CTBN of 8% ACN content in this resin showed dispersed particles of 0.6-4 μm diameter. Higher amounts of CTBN (about 30%) caused phase inversion, the molecular weight of CTBN increased by chain extension (by resin) leading to larger particles which ultimately made the rubber, the continuous phase. In these studies, CTBN with 13% ACN content did not show dispersed particles, i.e., no phase separation. The higher ACN content made the rubber more polar and hence miscible with the resin.

10.2.2 Toughening by other Acrylonitrile – Butadiene Copolymers

ATBN and HTBN are also being produced as possible modifiers for epoxy resins. The studies with ATBN have been mainly confined to adhesive applications. The amount of ATBN that can be incorporated into epoxy resin can be much higher than that of CTBN, because the amine group in ATBN can work as an additional hardener unlike CTBN which can only undergo chain extension.

HTBN can be incorporated into epoxy resins with difficulty. It needs a coupling chemical like a diisocyanate. Sankaran and co-workers [12, 13] have toughened epoxy resins with HTBN using toluene diisocyanate as a coupler and reported an optimum performance with HTBN at 6% by weight. In this case, toughening was accompanied by a slight increase in modulus (and without a loss in T_g). This was explained as being due to an increase in crosslink density caused by the addition of the diisocyanate.

10.2.3 Use of a Few Unconventional Rubbers to Toughen Epoxies

10.2.3.1 Polymethyl Methacrylate (PMMA) Grafted Natural Rubber (Heveaplus MG)

Heveaplus MG is a well-known modified form of natural rubber (NR). This is essentially a non-polar backbone on which a polar PMMA has been grafted. This polymer has certain good attributes of NR, which could be of value, while toughening the epoxy resin, having high tear strength and high tensile strength and a lower T_g than CTBN. Heveaplus MG could be masticated in a two roll mill and dissolved in a solvent methyl ethyl ketone (MEK) and mixed with epoxy resin followed by solvent removal (in vacuum). This system was studied by Rezaifard and co-workers [14].

DGEBA cured by piperidine was the resin studied. Before mixing the resin, the unreacted NR and MMA were removed by extraction in some cases with the appropriate solvents. The cure temperature was 120 °C for 16 hours.

Heveaplus MG, both at low (34-39 x 10^4) and high (8.5-9 x 10^5) molecular weights were shown to toughen epoxy to a good extent. At levels of 5 phr, the low molecular weight Heveaplus MG, increased K_{1c} by about 2 times and the high molecular weight copolymer (at 5 phr of the additive) to a slightly lesser extent (by about 1.8 times). The corresponding increase in G_{1c} are about 5 times and 2.5 times, respectively). The results for CTBN (10 phr) were: 2-fold increase of K_{1c} and 4-fold increase of G_{1c} while Heveaplus MG (both high and low molecular weight) gave a 3.5-fold increase of K_{1c} and an 8-fold increase of G_{1c}. The results were also better for Heveaplus MG compared with CTBN at low temperature tests done on toughening. This could be due to a lower T_g of NR. Fracture surface scanning electron microscopy (SEM) showed rubber particle stretching and tearing. This could be a major contributor to toughening, as it is well known that NR exhibits strain induced cystallisation, prominently.

10.2.3.2 Solid CTBN

Gouri and co-workers [15] studied modification of epoxidised novolac resin by solid CTBN, by heating in the presence of triphenylphosphine in solution. Such systems were cured by DDS and dicyandiamide systems. Such systems reduced T_g by 19 and 10 °C, respectively. Such systems were used in adhesive formulations. The rubber was found to be dispersed at 100-200 µm particle sizes. They showed improved peel strengths.

10.2.3.3 Silicones

Silicones are immiscible with epoxy resins but low molecular weight silicone polymers with reactive end groups can be made to react with epoxy groups and form block copolymers. These can toughen epoxy resins. Yorkgitis and co-workers [16] studied poly dimethyl siloxane (PDMS) at low molecular weights as modifiers for epoxies. They were compatibilised with epoxy resins by using small amounts of PDMS in which some or all methyl groups were replaced by fluoroalkyl or phenyl substituents. The silicones were incorporated by reacting with the epoxy resin for 1 hour at 65 °C.

Some of these combinations were found to be as competitive as CTBN or ATBN in toughening DGEBA resin. The fall in modulus and T_g were significantly less than those found with CTBN/ATBN modified systems. These are to be expected because, silicones are incompatible with epoxy resins and hence, toughening will not be at the cost of modulus.

Other advantages expected of silicone modification are: much better low temperature performance because T_g of PDMS is –120 °C compared with around –50 °C for CTBN); better moisture resistance, weather resistance, high temperature performance and lower wear rates. The wear resistance was found to be better with silicone modification [16].

Kothandaraman [17] used amine terminated PDMS (molecular weight 10,000, supplied by Resil Chemicals Ltd, India) in DGEBA resin (LY 556), cured by methylene dianiline (MDA). The maximum level of silicone that could be added was about 15 phr; beyond this lack of solubility of the hardener in the blend was suspected. Incorporation of amine terminated resins could be done by warming DGEBA with the PDMS at 50 °C for about 30 minutes. The tensile and flexural strength of the modified resins were slightly higher than those of the corresponding CTBN modified resins. The results are shown in the **Table 10.3**.

In the corresponding glass cloth composite, the modification of the resin by silicone did not decrease the tensile and flexural strength, while in the CTBN modified composite, the strength was found to decrease. The optimum performance was observed at the 10 phr level of amine terminated PDMS. The unnotched Izod impact test of the composite gave similar results on those of CTBN modified composite. Further the silicone modified composites showed lesser moisture ingress under accelerated testing conditions compared with unmodified composites.

This unusual behaviour, i.e., toughening of a composite by a rubber without loss of strength, could probably be explained as due to the silicone acting as an additional coupling agent between the resin and the fibre (the results are shown in Section 10.6).

Table 10.3 Mechanical properties of CTBN and amine terminated PDMS modified epoxy resin (LY 556 cured by MDA) [17]

Modifier (in phr)	Flexural strength (MPa)	Flexural Modulus (GPa)
Nil	148.70	2.65
CTBN (5)	116.42	2.69
CTBN (10)	94.5	2.23
CTBN (15)	66.23	1.63
CTBN (20)	55.84	1.61
Amine terminated PDMS (5)	136.2	2.29
Amine terminated PDMS (10)	113.1	2.42
Amine terminated PDMS (15)	70.69	1.51

Takahashi and co-workers [18] studied the effect of PDMS (MW of 2000) on epoxy-based moulding compounds used for electronics industry (for Printed Circuit Boards (PCB), etc). Here too, flexural strength was found to increase by addition of PDMS to the moulding compound (they contain silica filler). The increase in strength (but not in modulus) was explained by an improved epoxy filler interaction due to the silicone (which could act as an additional coupling agent). Further the idea of improved epoxy filler interaction was reinforced by the observation of a reduction in coefficient of thermal expansion (CTE) by the added silicone, in the silica filled resin. Normally, a rubber increases the CTE values.

A similar study due by Lin and co-workers [19] in a trifunctional epoxy-based moulding compound for electronic application, modified by PDMS with reactive end groups also showed similar results of increased toughness with reduced CTE and maintenance of flexural strength by the added silicone. Another study [20] showed that PDMS with an amine end group improved toughness with a slight increase in flexural strength. The results were better with the rubber having a higher MW (about 5000).

Shimizu and co-workers [21] studied the effect of silicone gel on casting compounds containing epoxy resin and silica flour as filler. Silicone gel at 15 phr caused a 7-fold increase of K_{1c} compared to that of the neat resin and a 2-fold increase compared to that of the untoughened compound. For the filled compounds, the silicone modified compound showed a small decrease in modulus by about 20%.

Ochi and co-workers [22] synthesised block copolymers of carboxylated PDMS with aramid. They were used as compatibilisers for a blend of epoxy with room temperature vulcanisable (RTV) silicone (liquid silicone rubber). Silicone particles were seen dispersed in the toughened resin, but covered by these copolymers, providing particle matrix compatibility. This system improved toughness (G_{1c}) by 100%.

IPN containing dispersed silicone micro and nano phases were prepared by Buchholz and Mulhaupt [23]. *p*-Hydroxyl benzoate was introduced as a terminal group into PDMS. This was blended with epoxy resin and cured. At levels of 10%, K_{1c} improved to 2.5-fold compared to the unmodified resin, with a reduction in flexural modulus of 10%. The silicone was found to be dispersed in nanometer level particle sizes.

10.2.3.4 Acrylic Rubbers

Banthia and co-workers [24] synthesised telechelic acrylic ligomers for modifying epoxy resins. They polymerised ethyl hexyl acrylate monomer in the presence of azobis (4-cyano valeric acid). This system gave oligomers with COOH end groups.

This could be reacted with epoxy resin (DGEBA) in presence of 1% triphenyl phosphine and the resultant toughened epoxy, cured by bis (4-amine cyclohexyl) methane at 100 °C for this followed by post cure at 160 °C for 2.5 hours.

The MW of the rubber were varied and their effect on T_g were reported. MW in the range of 9500 to 15000, at 10% rubber content T_g did not vary much. At a MW of 11500, change in rubber contents for 4% to 10% did not vary the T_g. Improvement in impact strength was comparable to what could be obtained by ATBN toughening.

Kelly and Ptak [25] synthesised acrylic oligomers as latices with two different particle sizes and mixed these with epoxy resin in presence of a solvent and subsequently the solvent was evaporated. They also incorporated a reactive monomer (glycidyl methacrylate) in some cases so that they could react with the epoxy resin and provide good resin-rubber particle bonding. Large particles (reacted with the resin) showed lesser reduction in T_g compared with smaller ones. Particles reacted with resin showed lesser T_g reduction compared with unreacted particles. While fracture energies were the same for large particle modified epoxies (whether reacted or not) those by non reacted small particles was better than reacted ones. The optimum properties were seen at rubber content of 5%-10%.

10.2.3.5 Other Systems

Epoxy resins were also evaluated for toughening by other systems. For example, Frischinger and Dirlikov [26] studied epoxidised soybean oil as a toughener for DGEBA resin. At 20 phr levels, this modifier can increase K_{1c} by about 77%-80% for MDA cured resin and by about 50% for DDS cured resin. The effectiveness of this modifier was considered to be on par with that of CTBN. This modifier costs much less than CTBN and hence this provides scope for more studies. Similar results were shown by Ratna and co-workers [27]. They also showed that such systems can improve the performance of adhesive joints.

Bagheri and Pearson [28] showed that hollow latex particles, based on methacrylate-butadiene-styrene are as good as liquid rubbers for toughening epoxies.

Bagheri and co-workers [29] used surface treated scrap rubber particles at 2.5 phr along with CTBN at 7.5 phr improved K_{1c} of the neat resin by two times. Similar observations were also made by Boynton and Lee [30].

10.3 Toughening of High Performance Epoxy Resin

Toughening of a high performance epoxy resin like tetra glycidyl methylene dianiline (TGMDA) has been a greater problem. These resins have a high T_g (190 °C and upwards) due to high crosslink density. The normal CTBN toughening has not been successful in this case. Increase in G_{1c} has been modest and accompanied by a drastic fall in modulus. The effect of elastomeric additives on aluminium filled (50 phr) TGMDA (modified with diluents) tooling resin is shown in **Table 10.4**.

Pearce and co-workers [32] showed a way for improving the toughness of TGMDA resin. Instead of curing the resin, by piperidine, they used adducts of piperidine with isocyanate or an adduct of isocyanate with methyl piperidine as curatives. Those with CTBN (20% by weight) give much better improvement in G_{1c} (about 24 times compared to the unmodified resin). Piperidine seems to catalyse homopolymerisation which makes the resin more ductile. In this condition CTBN can magnify the effect.

10.4 Toughening by Preformed Core-Shell Particles

Preformed core-shell rubbery particles have also been studied as tougheners for epoxy resins. Sue and co-workers [33] studied such systems as possible tougheners for high performance epoxy resins. They are prepared by a two stage latex emulsion

Table 10.4 Variation of compressive strength and T_g with rubber additive content [31]		
Additive (in phr)	T_g (°C)	Compressive Strength (MPa)
Nil	165*	197.59
ATBN (30)	110	32.57
ATBN (50)	105	58.33
ATBN (70)	102	30.11
Amine terminated PDMS (20)	152	57.33
Amine terminated PDMS (30)	138	44.33

T_g values obtained from dilatometric traces by extrapolation. Heating rates were 5 °C/min
*This value is less than that given in the product literature, by about 25 °C - this may be due to the low heating rate

polymerisation. The core is an elastomeric polymer which can be grafted upon. This polymer is either crosslinkable or insoluble in the epoxy resin. The shell should have chemically attached polymer chains and non grafted chains. The shell polymer should have a MW in the range of 20,000-100,000. The shell should be compatible with the epoxy resin so that the epoxy toughener additive system remains as a stable colloid until curing is complete. For this a mixed monomer system is of help. Two types of rubbers have been studied, (a) styrene-butadiene latex based-graft rubber concentrated (GRC) and (b) dispersed acrylic rubbers (DAR). In the studies by Sue and co-workers [33], the shell system studied was styrene, methyl methacrylate (MMA), ACN and glycidyl methacrylate (GMA). The grafted shell thickness was raised by changing the core-shell ratio from 84:16 to 75:25 or 65:35. Polarity could be raised by varying the ACN content from 0% to 25% (of shell weight). To vary particle resin interfaced strength, GMA content was varied from 0% to 30% of shell weight.

10.4.1 Procedure for Blending

First, a latex of styrene butadiene rubber was prepared. This was placed under an inert atmosphere and in presence of a radical initiator, some quantities of the monomers were added while the latex was still warm (at about 70 °C). The remaining monomers were added over a period of about 1.5 hours with a slow increase in temperature to 80 °C and followed by heating for another 2 hours. To the resultant mixture was added MEK to form a dispersion. From this, the organic layer was separated and added to the epoxy resin and the solvent removed by distillation. This method produces a dispersion of core-shell particles in epoxy resin. This could be directly cured. The morphologies could be varied by varying the GMA and ACN contents. Minor addition of GMA led to good dispersion of the GRC particles, but, further addition led to multimodal distribution of GRC. Some of the results of these studies are shown in **Table 10.5**.

The toughening mechanism was cavitation of the particle followed by shear yielding of the matrix around the crack tip. This was irrespective of GMA content. GMA was used for bonding chemically, the particle with the resin. Without GMA, ACN content controls the dispersion level – this is important in toughening. At lower core-shell ratios, GMA, together with ACN, changes the dispersion from a random to a locally clustered dispersion (probably caused by reactions involving GMA groups). This increases K_{1c} by introducing a new toughening mechanism called the crack deflection mechanism.

Similarly, dispersed acrylic rubber particles have been synthesised and studied by Riew and co-workers [34]. In this case, inner cores (plastic) and middle elastomer layer and an outer shell (copolymers of MMA and ethyl acrylate (EA) with functional groups, e.g., GMA) were developed and used. These show mixed results as far as effectiveness in toughening is concerned. In some cases, they appear to toughen less effectively than CTBN but this, unlike CTBN does not reduct T_g, stiffness or weather resistance.

Sue and co-workers [35] found that DAR are less effective than core shell rubber (CSR) particles in toughening. DAR particles seem to remain unextended during loading and

Table 10.5 Variation of G_{1c} with variation in core-shell ratio and GMA contents (adapted from Tables I and II of [33]		
Core-shell ratio	GMA content (parts)	*Change in G_{1c} (J/m^2
84-16	4.8	+ 310
84-16	0.0	+ 400
75-25	3.75	+ 440
65-35	5.25	+ 460
* with reference to the base resin - G_{1c} of the base resin was 180 J/m^2 *Please note: the relative concentrations of styrene, MMA and CAN were not identified in these cases*		

hence not able to promote crack deflection and crack bifurcation. These mechanisms however are less effective than shear yielding for toughening. The studies show the following results for DER 332 epoxy resin – DDS, a highly crosslinked system. The resin, hardener and core shell particles were mixed at 130 °C, and cured at 180 °C for 2 hours and post cured at 220 °C for 2 hours.

An increase of G_{1c} to 3.5 times that of the unmodified resin could be seen with GMA and ACN contents in the shell being higher. The changes in T_g at all levels of CSR additions were negligible.

However, Becu-Longuet and co-workers [36] showed that CSR particles based on butyl acrylate showed better toughening than styrene-butadiene based CSR particles.

In studies by Jansen and co-workers [5], epoxy rubber particles were preformed by curing droplets of alipatic epoxy resin (digycidyl ether of polypropylene oxide DGE-PPO) in the aqueous layer and transferred into aromatic epoxy resin, and cured. For less crosslinked epoxies like DGEBA, these particles improved toughness to a lesser extent but for tightly cured systems (higher T_g resins: 190 °C) this system gave much better toughening.

The improvement in toughness is due to better ductility caused by compatibility of the rubber particle towards the resin. The swelling of the rubber particles by the resin offsets the lowering of the T_g (caused by the compatibility). Thus, compared to a T_g of 190 °C for the base resin, this system caused a reduction of only 5 °C in T_g against 14 °C caused by the usual CTBN toughening.

Sue and co-workers [37] studied three high T_g epoxy resin systems, i.e., DGEBA (chain extended with varying amounts of bisphenol A), DGTMBA-diglycidyl ether of tetra methyl bisphenol A with tetramethyl bisphenol A (TMBA), and diglycidyl ether of tetra bromo bisphenol A (DGTBBA) with TBBA, toughened with core shell particles. They found in the case of DGTBBA resin, more chain extender (TBBA) led to more rigid (less tough) resins. Actually in these systems, longer chain lengths between crosslinks led to lower T_g as usual but this did not translate to greater flexibility. This was because of the change in curing systems or changes in epoxy rigidity when changing from DGEBA to DGTBBA monomers.

10.5 Toughening of Epoxy Resin with Engineering Thermoplastics

While CTBN toughening has been effective in epoxy resins it suffers from two major drawbacks: (1) the improvement in toughness is at the cost of thermal properties and modulus and (2) CTBN toughening is found to be less effective in tightly crosslinked

425

resins, e.g., TGMDA resin. In 1983, thermoplastics were studied as possible tougheners for epoxy resins with some success [38].

The plastics studied for this purpose are: polyimides, polyether sulfones (PES), polyether ketones (PEK), polybutylene terephthalate (PBT).

The mechanisms which could operate in these systems are (i) crack pinning, (ii) particle bridge, (iii) crack path deflection, (iv) shear banding and (v) microcracking.

The efficiency of the toughening additive will depend on (i) size of the particles, (ii) strength of the particle, (iii) particle matrix adhesion and (iv) particle distribution.

PBT modification of epoxy resin was studied by Kim and Robertson [39]. PBT was dissolved in phenol, precipitated from methanol, ground to a powder and mixed, with the resin. To find the best method of blending, the PBT and epoxy were mixed at various temperatures, with and without mechanical mixing. The best improvement in fracture energy was observed by heating PBT particles with epoxy resin without curing at 10 °C/min to 193 °C and suddenly cooling to room temperature, and then the curative was added. Such a system showed a 4.5 fold improvement in fracture energy. The reason for the success seemed to be formation of an interpenetrating network.

The reduction in T_g was 5 °C and the loss of modulus was negligible.

PES was studied as a possible toughener by a number of authors, e.g., in 1983, Raghava and co-workers [2] reported a modest increase in fracture toughness due to PES.

Hendrick and co-workers [40] used two low molecular weight of hydroxyl terminated poly sulfones in DGEBA epoxy cured with DDS (MW 5000-10000) at 15 phr loading which showed an increase in K_{1c} of about 0.7 MPa.m$^{0.5}$. The polysulfone was found to be precipitated as particles of about 0.2 μm.

The toughening was thought to be as due to strengthening of the plastic epoxy interface. Polyimide was tried as toughener for epoxy resins by a few researchers. For example, Wu and co-workers [44] used a readily processible maleimide resin for toughening epoxy resin, which produced good results.

PEK at various MW showed variation in particle sizes - between 0.5 and > 100 nm. The improvements in K_{1c} values obtained were +0.2 MPa.m$^{0.5}$, molecular weight of about 7000, +0.6 MPa.m$^{0.5}$ at 15000 and +2 MPa.m$^{0.5}$ at 18000. The better toughness at higher MW was attributed to increased ductility of the thermoplastic at higher MW. In most of the successful cases, the thermoplastic underwent phase separation from the epoxy resin, during cure. It was also shown that phase separation is not sufficient, though essential, to improve toughness.

Some results are summarised in **Table 10.6**.

Above certain levels of addition of the thermoplastic, the blends of epoxies with engineering thermoplastics undergo phase inversion where the discontinuous phase is now the crosslinked epoxy resin. They may be thought of, as semi interpenetrating networks (IPN) between the thermoplastic and the epoxy resin. They too have good toughness, without decrease in T_g and high modulus (compared with the rubber modified systems). As far as processability is concerned these are better than thermoplastics as matrix materials for advanced composites. However, their resistance to creep, solvents, stress cracking, etc., are expected to be low and hence such systems may pose problems on these counts [3].

Table 10.6 Improvements in K_{1c} by use of thermoplastic additive. Adapted from Table 1 of [39]	
Resin system	Change in K_{1c} (MPa.m$^{0.5}$)
DGEBA-DDS-polysulfone	+ 0.7
TGMDA-anhydride-PES	+0.2
DGEBA-MDA-PBT	+1.1
DGEBA-DDS-PEK	+1.4
TGMDA-DDS-polyether imide	+1.0

10.6 Toughening of Polyester Resin and its Composites

The use of polyester resin in composites is more widespread than epoxy resins mainly because of their lower costs. Polyester resin composites in various forms are used in automotive, auto-electrical, civil engineering, and other areas. The major disadvantages of polyester resin and its composites are poorer chemical resistance (especially alkali), lesser mechanical strength and toughness. The toughness is so low that sometimes, products made by sheet moulding compound (SMC) technology, break even during demoulding or assembly in their systems. Unlike epoxy resins, a correct toughening additive (like CTBN) is yet to be specified though many papers and patents in this area are available.

The main difficulty in toughening polyester resin is that its very molecular architecture prevents development of the necessary morphology. Furthermore, the presence of a reactive solvent monomer makes the resin undergo greater shrinkage during cure and causes formation of microvoids which can counter-balance the advantages offered by a toughening additive (Suspene and co-workers in [41]).

However, many solid rubbers have been added as a solution in styrene, to the resin in SMC and they have shown improvements in impact strength and surface finish of the products. The improvement in impact strength is always at the expense of stiffness. The incorporation of such an additive is difficult as it may require the rubber to be milled in a mixing mill before mixing with the solvent. Grossman [42] argues that addition of solid rubbers will be of a greater advantage compared with liquid rubbers. Unlike glass-epoxy composite, in polyester composite, failure begins at the fibre-resin interface. So, if a rubber with lesser polarity but with polar end groups like amine or COOH is used, and if it can cure slower than the polyester resin, its high MW which was earlier thought of as a disadvantage will now become an advantage as it can contribute favourably to the work of adhesion between the resin and the fibre. Thus, Grossman observed that, out of a few solid rubbers like solid carboxylated nitrile rubber, solid amine terminated nitrile rubber, solid carboxylated ethylene-propylene terpolymer rubber (EPDM), etc., amine terminated nitrile rubber improved the toughness of SMC while slightly increasing the mechanical strength.

For a hand lay up glass fibre reinforced plastic (GFRP), presence of a solution of styrene-butadiene rubber up to 0.4% by weight in the resin was found to increase impact strength of the composite [43].

For liquid rubbers, HTBN was found to be a fairly good toughener. Tong and co-workers [44] showed its effectiveness can be increased drastically by prereacting it with the resin, i.e., free OH in the HTBN, can be reacted with the free COOH groups in the resin and form block copolymers with the resin.

Further, BFGoodrich Co., developed a butadiene ACN copolymer with vinyl terminal groups (VTBN) which can be physically blended but can react with polyester resin during peroxide cure and toughen the resin. However, Tong and co-workers [44] showed that chemically reacted HTBN was superior to VTBN in toughening the resin and its SMC.

Table 10.7 shows the advantages of forming block copolymers between polyester resin and HTBN.

Further, in the SMC composites, increase in the reacted HTBN content of up to 15% by weight, increased flexural strength by 2.5 times, tensile strength by 5 times and impact strength by 3 times, although modulus, hardness and heat distortion temperatures decreased with compared to the control resins.

Sometimes, changing the reactive monomer from styrene (partially or fully) to one which can give a rubbery polymer on polymerisation, for example, ethyl or butyl acrylate, can effectively flexibilise/toughen polyester resin [45, 46].

Table 10.8 shows the effect of EA when added to a general purpose polyester resin, in addition to the styrene already present in the resin [45].

Lin and co-workers [47] showed that a tougher polyester composite can be made by developing an interpenetrating network between epoxy and polyester resins. They mixed DGEBA resin (with *m*-xylene diamine as a curative) and a general purpose polyester resin (curative: benzoyl peroxide) and cured the resins together, after prereacting at 60 °C for some time and then cured at 120 °C for 6 hours. Epoxy-polyester ratios between 25:75 and 50:50 showed good balance in properties. This is attributed to interpenetrating network formation. Networks of both resins get interlocked during cure leading to parts of both resins getting partly cured. These parts act as plasticisers and cause more shear yielding, leading to toughening.

Table 10.7 Variation of a few properties of plain polyester resin by various additives (from [44])			
Property/additive	HTBN blended	HTBN reacted to form block copolymer	VTBN
Change in modulus due to the additive	-25%	-13%	-25%
Change in impact strength due to the additive	+10%	+80%	<+10%
Change in T_g (°C)	-1.5	-4.5	-7.5

Table 10.8 Effect of ethyl acrylate monomer (in addition to styrene) on flexural properties of general purpose polyester resin [45]				
Modifier (phr)	Flexural strength (MPa)	Flexural modulus (GPa)	Maximum deflection (mm)	Tensile elongation (%)
—	79.01	3.21	3.00	4.11
Ethyl acrylate (5)	74.96	2.88	8.00	16.7
Ethyl acrylate (10)	70.14	2.52	No breakage	21.80

Subramaniam and McGarry [48] found another novel system for toughening polyester resins: here too IPN were formed, but here it was a rubber toughened (ATBN) epoxy-polyester system. Polyester resin was precured with a catalyst and then the ATBN dissolved in styrene was added, followed by the epoxy resin and its hardener. The whole system was heated to complete the epoxy resin cure.

The rubber content was varied. The rubber content at 6.7% provided the best set of properties - tensile strength was maintained compared with the rubber-less composition, while strain-to-failure increased by about 50%, K_{1c} almost doubled, while G_{1c} increased to 4 times that of the rubber-less compositions. The variations of these properties with rubber content, was not smooth. This could indicate that at rubber contents such as 6.7% and 17.7% drastic changes in morphology could occur. Beyond 10% rubber content, toughening was observed to be accompanied by a drastic loss in stiffness.

A higher performance polyester resin called vinyl ester combines some of the advantages of epoxy and polyester resins. Siebert and co-workers [49] studied toughening of elastomer modified vinyl ester resin using a rubber called epoxy terminated ACN butadiene copolymer (ETBN). This is made by reacting CTBN with bisphenol-A in a 1:2 molar ratio.

ETBN was dissolved in styrene as a 50% solution and added to an elastomer modified vinyl ester resin at various levels. The T_g did not decrease much by the addition of the ETBN while at 15.6 phr ETBN content, G_{1c} increased to 4.5 times that of the unmodified rubber modified vinyl ester resin.

Toughening by simple rubber addition is always at the cost of T_g and mechanical properties.

10.7 Toughening of Composites

A lot of work on improving the toughness of fibre reinforced composites has been done. It is well known that the improvement in G_{1c} of the resin does not get transmitted fully into the composite. In composites containing low values of interlaminar fracture energies an improvement of fracture energies of the bulk resin can be translated to the composite but not in composites with high resin G_{1c} values. For the low toughness polymers, the deformation size becomes larger and interacts with fibres resulting in a less complete transfer of resin toughness to the composites. The fibres do not totally inhibit the crack tip deformation zone and deformation of the polymer around the fibre is an important dissipating mechanism that enhances interlaminar fracture energy. The fibres also contribute to the resistance to fracture by diverting the crack and bridging the opening behind the crack tip. This may be the reason for the higher values of interlaminar fracture energy being greater than the polymer bulk fracture, for the brittle resins.

The effect of CTBN modifier in filament wound GFRP (fibre content 72% by weight) is shown in **Table 10.9** (Epoxy resin: LY 556 cured by MDA, CTBN: Hycar CTBN 1300 X 8).

Some authors suggest provision of an unreinforced layer of plastic (epoxy or engineering thermoplastic) between two prepreg plies to improve the toughness of the composite. For example, Matsuda and co-workers [51], used an ionomer thermoplastic as an 'interleaf' material for this purpose, in carbon fibre reinforced plastic (CFRP) prepreg system. They found that the G_{1c} of the system increased by 4-5 times when the ionomer thickness increased up to 100 μm. The crack path was found to be in the interleaf layer. Above 100 μm thickness, the failure was seen in the interface. Permanent deformation of the ionomer was observed only near the crack surface. In general, interleaving improves toughness of composite but mostly at the expense of stiffness.

Work by Scott and Phillips [52] showed that it may not be possible to reproduce accurately, toughness measurements in CFRP due to a few possible misalignments in fibres leading to additional contributions from the resultant fibre pull out, debonding, etc.

Kim and co-workers [53] argued that delamination is the most prevalent life limiting damage mode which can cause unacceptable reduction in stiffness and other mechanical properties. Rubber modified resins give much better post impact compressive strength to the composite than unmodified resins. Impact by drop weight along thickness direction does lesser damage to a composite based on rubber modified resin through a little

\multicolumn{6}{c}{**Table 10.9 Effect of CTBN modification of the resin on properties of glass reinforced, filament wound composites [50]**}
CTBN (phr)	Tensile strength (MPa)	Modulus (GPa)	Elongation at break (%)	Toughness (KJ/m²)*	ILSS (MPa)
0	883	25.71	3.81	39.21	21.47
5	871	22.73	4.45	43.99	23.75
10	864	21.43	4.78	47.66	24.30
15	851	21.11	5.23	49.48	20.40
20	826	20.83	5.57	51.86	18.90
25	779	20	5.79	46.66	—
30	712	17.85	5.51	43.51	16.75

*Area under stress-strain curves
ILSS:Inter laminar shear stress-according to ASTM D2344/D2344M-00e1 [54]

delamination and intralaminar transverse shear cracks. For unmodified resin composites, fibre and resin fracture are extensive delamination in all plies. These studies further show that at small fibre volume fractions, the loss of strength caused by the rubbery additive was severe. This may be due to formation of small air bubbles in the resin-fibre interface (this could be aggravated by the high viscosity of the rubber additive). Application of pressure can reduce the bubbles but this will increase the fibre content and this will constrain the plastic deformation of the resin.

A few cases of work done on toughening of epoxy resin composites are discussed next.

McGarry [55] proposed that instead of mixing the rubber in the resin, the rubber could be coated onto the fibre and the composite made without the rubber in the resin. The results were encouraging for glass fibres but not as much for carbon fibres. The rubber used was ATBN. At 1.5% of ATBN (in resin, by weight) interlaminar shear strength increased by 40% and Izod impact strength increased by 30%. SEM studies showed some fibre pull out and that the volume of the distorted matrix around the fibre appeared to increase with coating thickness. Mostly cohesive failure was seen in the rubber layer.

In another study, Barcia and co-workers [56], grafted hydroxyl terminated polybutadiene (HTPB) on the carbon fibre surface and then prepared the composite with epoxy resin. The grafting was done by reacting the fibre with HTPB in the presence of a catalyst like tin octoate under heat. The HTPB was also reacted with the resin and the impact strengths were measured. Impact strength was found to improve drastically when the fibre was grafted with HTPB and also the resin contained 10% by wt of HTPB. This was due to improved polymer-fibre interaction.

This concept was highlighted in studies by Kothandaraman, see Section 10.2.2.3.

Table 10.10 shows the effect of amine terminated PDMS (incorporated into the epoxy resin) on hand lay-up composites with glass cloth (based on epoxy resin LY 556).

The MW of the silicone polymer was 10,000.

Table 10.11 shows the effect of high temperature with moisture on composites unmodified or modified (by 10 phr of amine terminated PDMS [17]).

Low and co-workders [57] studied the effect of elastomeric additive (CTBN) on short alumina fibre filled epoxy resins. The properties were found to be the best at 15% of CTBN and alumina fibre. The fracture energy values were also better for a wider temperature range for the rubber modified composite than the unmodified composite.

Table 10.10 Mechanical properties of amine terminated PDMS modification of epoxy resin (LY 556) in its bidirectional glass cloth composite [18]

Silicone content (phr)	Tensile strength (MPa)	Tensile modulus (GPa)	Elongation at break (%)	Fracture toughness (kJ/m²)*	Flexural strength (MPa)	Flexural modulus (GPa)
0	267.4	7.37	4.45	17.42	246.6	15.47
5	243.4	6.24	4.83	19.31	267.6	15.41
10	276.5	6.12	5.31	23.01	256.4	15.99
15	238.4	5.41	5.23	20.32	211.7	14.31

*Measured as area under stress-strain curves

Table 10.11 Effect of silicone on moisture resistance of glass-epoxy composites [17]

Time (h)	Flexural modulus retention (%) (unmodified)	Wt gain by moisture ingress (%) (unmodified)	Flexural modulus retention (%) (modified resin with 10 phr of PDMS)	Wt gain by moisture ingress (%) (modified resin, with 10 phr of PDMS)
0.5	91	1.017	120	0.621
1.0	85	1.432	85	0.92
1.5	73	——	83	1.25

Silicone provided by M/s Resil Chemicals Ltd, Bangalore, India

Jang and co-workers [58] studied a few methods to toughen CFRP by modifying the moulding techniques. They observed that polyamide and PBT as modifiers can improve toughness in such composites. Also, TGMDA-DGEBA mixtures were observed to be tougher than pure TGMDA as matrix materials for the composites.

Bascom and co-workers [59] observed that, thermoplastic modified thermoset composites can be made to perform as well as thermoplastic composites for damage tolerance. They were optimistic that use of thermoplastic composites will develop in time, as structures based on such systems would require costly equipment and hence toughened thermoset-based composites will still, be favoured for aerospace use for some time to come.

10.8 Summary

1. CTBN modification is the most studied and popular approach to toughen epoxy resin. Its main limitation is that, the necessary morphology may be achieved only by very careful control of the mixing and curing conditions. Hence it will be difficult to get highly reproducible results. CTBN toughening is always accompanied by a loss of thermal and mechanical properties.

2. Rubbers less compatible with epoxy resins, but with reactive end groups, like silicone oligomers can toughen epoxy resins but with less loss of stiffness and T_g.

3. For high performance epoxy resins like TGMDA, conventional liquid rubber toughening is less effective but other techniques could be tried.

4. Core shell particles are easy to mix with epoxy resins and can give consistent results. Toughening is not accompanied by a reduction in T_g nor is it dependent on processing/curing conditions. Synthesising them is complex. They are suitable for high performance epoxy resins.

5. Toughening can also be achieved by use of high performance thermoplastics like polysulfones, but their effectiveness is less than CTBN. They are suitable for high performance resins and here again, toughening is not accompanied by loss of stiffness and reduction in T_g.

6. For toughening high performance epoxy resins, approaches such as using a cure system which can provide uniform crosslink density, core-shell particles, use of chain extension with rigid side chains, etc., can be useful.

7. Other approaches like use of scrap rubber, solid rubbers, hollow spheres, combinations of rubbers with hard particles, hard particles have also been successfully demonstrated by many authors.

8. Polyester resin and its composites are more difficult to toughen for various reasons. Still some practical solutions are possible.

9. The improvement in toughness in the resin is not fully transmitted into the composite for various reasons. This can be solved in some cases by promoting resin-fibre

interactions. Other approaches like interleaving by fibre-free resins/flexible polymers can also help in development of tough composites.

10.Toughening of epoxy resins by liquid and solid rubbers can be applied successfully for improving the performance of adhesives.

Acknowledgements

The authors thank RAPRA Technology Ltd, for giving this assignment and for the literature survey. Thanks are also due to Professor A.J. Kinloch of Imperial College of Science, Technology and Medicine, London, UK for sending reprints of his papers, and Professor S. Bandyopadhyay of University of New South Wales, Australia, for giving information on papers published by him and his associates.

References

1. W.D. Bascom and D.L. Hunston in *Rubber Toughened Plastics,* Ed., C.K. Riew, Advances in Chemistry Series No.222, ACS, Washington, DC, USA, 1989, 135.

2. R.S. Raghava in *Reference Book for Composites Technology*, Ed., S.M. Lee, Technomic Publishing Company Inc., Lancaster, PA, USA, 1989, 77.

3. B.L. Burton and J.L. Bertram in *Polymer Toughening*, Ed., C.B. Arends, Marcel Dekker, New York, NY, USA, 1996, 339.

4. S.J. Shaw in *Rubber Toughened Engineering Plastics,* Ed., A.A. Collyer, Chapman and Hall, London, UK, 1994, 165.

5. B.J.P. Jansen, K.Y. Tamminga, H.E.H. Meijer and P.J. Lemstra, *Polymer*, 1999, **40**, 5601.

6. Y. Huang, A.J. Kinloch, R.J. Bertsch and A.R. Siebert in *Toughened Plastics-I Science and Engineering*, Eds., C.K. Riew and A.J. Kinloch, Advances in Chemistry Series No.233, ACS, Washington, DC, USA, 1993, 189.

7. A.J. Kinloch in *Rubber Toughened Plastics,* Ed., C.K. Riew, Advances in Chemistry Series No.222, ACS, Washington, DC, USA, 1989, 67.

8. G. Levita in *Rubber Toughened Plastics,* Ed., C.K. Riew, Advances in Chemistry Series No.222, ACS, Washington, DC, USA, 1989, 93.

9.   S. Montarnal, J.P. Pascault and H. Sauterau in *Rubber Toughened Plastics,* Ed., C.K. Riew, Advances in Chemistry Series No.222, ACS, Washington, DC, USA, 1989, 192.

10. S.C. Kim, M.B. Ko and W.H. Jo, *Polymer*, 1995, **36**, 2189.

11. W.K. Lee, K.A. Hodd and W.W. Wright in *Rubber Toughened Plastics,* Ed., C.K. Riew, Advances in Chemistry Series No.222, ACS, Washington, DC, USA, 1989, 263.

12. S. Sankaran and M. Chanda, *Journal of Applied Polymer Science*, 1990, **39**, 1459.

13. S. Sankaran and M. Chanda, *Journal of Applied Polymer Science*, 1990, **39**, 1635.

14. A.H. Rezaifard, K.A. Hodd and J.M. Barton in *Toughened Plastics-I Science and Engineering*, Eds., C.K. Riew and A.J. Kinloch, Advances in Chemistry Series No.233, ACS, Washington, DC, USA, 1993, 381.

15. C. Gouri, R. Ramaswamy and K.N. Ninan, *International Journal of Adhesion and Adhesives*, 2000, **20**, 4, 305.

16. E.M. Yorkgitis, C. Tran, N.S. Eiss, Jr., T.Y. Hu, G.C. Wilkes and J.E. McGrath in *Rubber-modified Thermoset Resins*, Eds., C.K. Riew and J.K. Gillham, ACS Advances in Chemistry Series No.208, ACS, Washington, DC, USA, 1984, 137.

17. B. Kothandaraman, P. Baskaran, V. Baskar and D. Jayaprakash, Proceedings of the Regional Conference on Polymeric Materials, University Sains Malaysia, Penang, Malaysia, 1998, 48.

18. T. Takahashi, N. Nakajima and N. Saito in *Rubber Toughened Plastics,* Ed., C.K. Riew, Advances in Chemistry Series No.222, ACS, Washington, DC, USA, 1989, 243.

19. L-L. Lin, T.H. Ho and C.S. Wang, *Polymer*, 1997, **38**, 8, 1997.

20. J.Y. Shieh, T.H. Ho and C.S. Wang, *Angewandte Makromolekulare Chemie*, 1997, **245**, 125.

21. T. Shimizu, M. Kamino, M. Miyagawa, N. Nishiwaki and S. Kida, Proceedings of Deformation, Yield and Fracture of Polymers Conference, Cambridge, UK, 1994, 76/1.

22. M. Ochi, K. Takemiya, O. Kiyohara and T. Nakamishi, *Polymer*, **41**, 195.

23. U. Buchkholz and R. Mulhaupt in *Proceedings of ACS Polymeric Materials Science and Engineering*, 1996, **74**, 339.

24. A.K. Banthia, P.N. Chaturvedi, V. Jha and V.N.S. Pendyala in *Rubber Toughened Plastics*, Ed., C.K. Riew, Advances in Chemistry Series No.222, ACS, Washington, DC, USA, 1989, 343.

25. F.N. Kelly and K.N. Ptak, *Polymers and Composites - Recent Trends*, Oxford and IBH Publishing Co., New Delhi, India, 1989, 219.

26. I. Frischinger and S. Dirlikov in *Toughened Plastics-I Science and Engineering*, Eds., C.K. Riew and A.J. Kinloch, Advances in Chemistry Series No.233, ACS, Washington, DC, USA, 1993, 451.

27. D. Ratna and A.K. Banthia, *Journal of Adhesion Science and Technology*, 2000, **14**, 1, 15.

28. R. Bagheri and R.A. Pearson, Proceedings of 9th International Conference on Deformation, Yield and Fracture of Polymers, Cambridge, UK, 1994, 106/1.

29. R. Bagheri, M.A. Williams and R.A. Pearson, *Polymer Engineering and Science*, 1997, **37**, 2, 245.

30. M.J. Boynton and A. Lee, *Journal of Applied Polymer Science*, 1997, **66**, 2, 271.

31. B. Kothandaraman, *Experimental Studies on Rubber Modified Thermosets and their Composites*, Anna University, 1998. [Ph.D Thesis]

32. P.J. Pearce, C.E.M. Morris and B.C. Ennis, *Polymer*, 1996, **37**, 1137.

33. H-J. Sue, E.I. Garcia-Meitin, D.M. Pickelman and P.C. Yang, in *Toughened Plastics-I Science and Engineering*, Eds., C.K. Riew and A.J. Kinloch, Advances in Chemistry Series No.233, ACS, Washington, DC, USA, 1993, 259.

34. C.K. Riew, A.R. Siebert, R.W. Smith, M. Fernando and A.J. Kinloch in *Toughened Plastics II, Novel Approaches in Science and Engineering*, Eds., C.K. Riew and A.J. Kinloch, ACS Advances in Chemistry Series No.252, ACS, Washington, DC, USA, 1996, 33.

35. H-J. Sue, E.I. Garcia-Meitin and D.M. Pickelman in *Polymer Toughening*, Ed., C.B. Arends, Marcel Dekker Inc., New York, NY, USA, 1996, 131.

36. L. Becu-Longuet, A. Bonnet, C. Pichot, H. Sautereau and A. Maazouz, *Journal of Applied Polymer Science*, 1999, **72**, 849.

37. H-J. Sue, P.M. Puckett and J.L. Bertram in *ACS Polymeric Materials Science and Engineering*, Fall Meeting, 1998, **79**, 216.

38. R.A. Pearson in *Toughened Plastics-I, Science and Engineering*, Eds., C.K. Riew and A.J. Kinloch, Advances in Chemistry Series No.233, ACS, Washington, DC, USA, 1993, 405.

39. J. Kim and R.E. Robertson in *Toughened Plastics-I, Science and Engineering*, Eds., C.K. Riew and A.J. Kinloch, Advances in Chemistry Series No.233, ACS, Washington, DC, USA, 1993, 427.

40. J.C. Hendrick, N.M. Patel and J.E. McGrath in *Toughened Plastics-I, Science and Engineering*, Eds., C.K. Riew and A.J. Kinloch, Advances in Chemistry Series No.233, ACS, Washington, DC, USA, 1993, 293.

41. L. Suspene, Y.S. Yang and J-P. Pascault in *Toughened Plastics-I, Science and Engineering*, Eds., C.K. Riew and A.J. Kinloch, Advances in Chemistry Series No.233, ACS, Washington, DC, USA, 1993, 163.

42. R.F. Grossman in *Rubber Toughened Plastics*, Ed., C.K. Riew, Advances in Chemistry Series No.222, ACS, Washington, DC, USA, 1989, 415.

43. P. Ghosh and N.R. Bose, *Journal of Applied Polymer Science*, 1995, **58**, 2177.

44. S.N. Tong and T.K. Wu in *Rubber Toughened Plastics*, Ed., C.K. Riew, Advances in Chemistry Series No.222, ACS, Washington, DC, USA, 1989, 375.

45. B. Kothandaraman, *Experimental Studies on Rubber Modified Thermosets and their Composites*, Anna University, 1998. [Ph.D Thesis]

46. C.F. Jasso, R. Sanjuan and E. Mendizabal in Proceedings of SPE Antec '94 Conference, San Francisco, CA, USA, 1994, Volume 1, 956.

47. M-S. Lin, C-C. Liu and C-T. Lee, *Journal of Applied Polymer Science*, 1999, **72**, 4, 585.

48. R. Subramaniam and F.J. McGarry, *Journal of Advanced Materials*, 1996, **27**, 2, 26.

49. A.R. Siebert, C.D. Guiley, A.J. Kinloch, M. Fernando and E.P.L. Heijnsbrock in *Toughened Plastics II*, Eds., C.K. Riew and A.J. Kinloch, ACS Advances in Chemistry Series No.252, ACS, Washington, DC, USA, 1996, 151.

50. B. Kothandaraman, *Experimental Studies on Rubber Modified Thermosets and their Composites*, Anna University, 1998. [Ph.D Thesis]

51. S. Matsuda, M. Hojo, S. Ochiai, A. Murakami, H. Akimoto and M. Ando, *Composites Part-A, Applied Science and Engineering*, 1999, **30A**, 11, 1311.

52. J.M. Scott and D.C. Phillips, *Journal of Materials Science*, 1975, **10**, 551.

53. J. Kim, C. Baillie, J. Poh and Y.W. Mai, *Composites Science and Technology*, 1992, **43**, 283.

54. ASTM D2344/D2344M-00e1, *Standard Test Method for Short-Beam Strength of Polymer Matrix Composite Materials and their Laminates*, 2000.

55. F.J. McGarry in *Rubber Toughened Plastics*, Ed., C.K. Riew, Advances in Chemistry Series No.222, ACS, Washington, DC, USA, 1989, 173.

56. F.L. Barcia, B.G. Soares, M. Gorelova and J.A. Cid, *Journal of Applied Polymer Science*, 1999, **74**, 6, 1424.

57. I.M. Low, Y-W. Mai and S. Bandyopadhayay, *Composites Science and Technology*, 1992, **43**, 3.

58. K. Jang, W-J. Cho and C-S. Ha, *Composites Science and Engineering*, 1999, **59**, 7, 995.

59. W.D. Bascom, S-Y. Gwein and D. Grande in *Toughened Plastics-I, Science and Engineering*, Eds., C.K. Riew and A.J. Kinloch, Advances in Chemistry Series No.233, ACS, Washington, DC, USA, 1993, 519.

11 Blends Containing Thermostable Heterocyclic Polymers

Maria Bruma and Ion Sava

11.1 General Background

High temperature polymer blends are being increasingly used in many applications, from the automotive to the aircraft industry. They have also attracted much interest from the academic world. In the recent years, temperatures in excess of 200 °C for periods of hundreds of hours and longer have become the stated requirements for some materials. Particularly severe in this regard are demands from aircraft industry for engine components where high service temperatures for long periods of time are often normal. One approach to meeting these material requirements is to synthesise new polymers. Another method is to tailor the material's properties is by blending two polymers with the aim of benefiting from the positive features of both materials while attempting to eliminate the negative features.

This chapter will describe blends of high temperature heterocyclic polymers with traditional ones, or blends in which both components are heterocyclic polymers, and it will present the advantages of producing such blends. The chapter will concentrate mostly on the progress made in this field in the last five years, since some reviews have been reported previously [1-4].

Usually, high temperature polymers are defined as materials which have a continuous service temperature of above 175 °C for tens of thousands of hours. Similarly, high temperature polymer blends are defined by the same use temperature. In this chapter polymer blends means physical mixtures of polymers, but not systems which react chemically in the mixed state, as have been described in other chapters. The physical state of mixing in polymer blends can be miscible, partially miscible or immiscible. It is not always possible to define the exact scientific criteria ij°olved with compatible blends. However, the term compatible is used to describe blends which possess a desirable physical property regardless of the actual state of mixing.

A miscible blend is a homogeneous, single-phase material, and in many respects it behaves as if it is a single polymer. The most widely accepted criterion for miscibility is the single glass transition temperature (T_g), provided that the difference between the two T_g of the respective partners is higher than 20 °C. It is generally assumed that a polymer blend which displays a single T_g is mixed at the molecular level. If the two components of the blend are miscible only within a certain composition range the blends are called partially miscible. An immiscible blend contains two or more distinct phases and it is heterogeneous in nature.

The reason for blending two polymers is to produce a material which has a property tailored to a certain performance level. Since the resultant physical properties are determined by the miscibility and phase behaviour it is very important to control the phase structure in blend. Most high temperature heterocyclic polymers possess restricted rotation in their backbone and therefore their T_g is quite high. Based on the rigid conformation of such polymers it is expected that the criteria for miscibility will be more severe than for random coil polymers. However, in the last few years several high temperature miscible polymers have been discovered. The reason for miscibility observed in such systems is the existence of functional groups in the polymer backbone, which can interact with the functional groups of the other polymer in the mixture. A very important aspect with miscible polymer blends is the temperature range of their miscibility. When such materials are to be processed in the melt state, there must be a 'temperature window' between the T_g and the phase separation temperature, which should be large enough for melt processing. If this processing window is too narrow, the blends will either phase separate or have a viscosity that is too high for melt processability. Another category of polymer blends contains immiscible systems. Two types of immiscible blends have been described. The first type is constituted by molecular composites, which are generally accepted as partially miscible blends but are not truly miscible at the molecular level. Molecular composites are based on the mixing of a rigid rod polymer and a random coil polymer. Most of the work on molecular composites has focused on delay of the phase separation with a desirable morphology before complete immiscibility occurs.

The second type of immiscible blend involves thermotropic liquid crystal polymers (LCP) blended with conventional thermoplastic polymers. The advantage is that the LCP behaves as a processing aid, lowering the overall melt viscosity. The groups of LCP molecules, called domains, slide past each other and the overall effect is lubrication of the polymer melt and a subsequent lowering of the melt viscosity of the blend which means that the blend can be processed at a lower temperature. The reduction of processing temperature leads to lower energy consumption and less degradation of polymers that are sensitive to high temperature. Another effect in blends of LCP with thermoplastic polymers is the utilisation of LCP as reinforcement agent for flexible polymers, which gives improved mechanical properties. However the use of LCP as reinforcement agent has a drawback, namely the poor adhesion to the matrix.

11.2 Polyimide Blends

Polyimides (PI) were the first of the thermally stable heterocyclic polymers introduced in the mid 1960s. The most common synthetic method for preparing PI consists of a two-step reaction between a tetracarboxylic dianhydride and a diamine yielding a soluble polyamic acid (**I**) which can be converted to polyimide (**II**) by thermal or chemical loss of water [5-7] (see **Figure 11.1**).

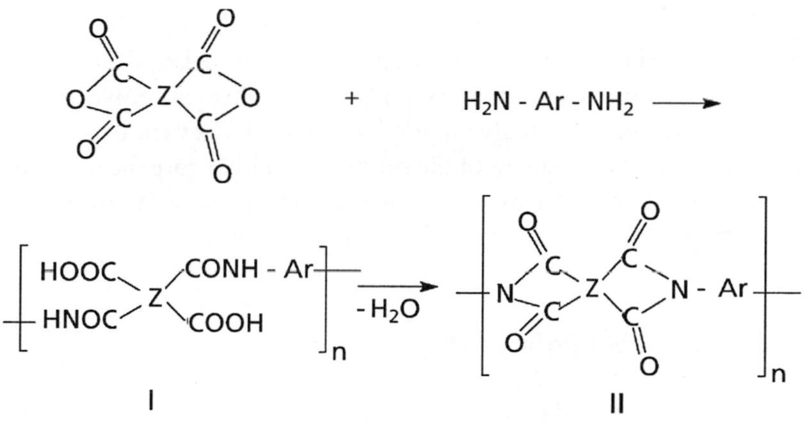

Figure 11.1 Synthesis of polyimides by the two-step method. Z: tetravalent aromatic radical

Wholly aromatic polyimides do not behave as thermoplastics even though they have a linear structure. The polymer backbone comprised a rigid, inherently stable, aromatic phenylene and the imide ring imparted excellent thermooxidative properties, but at the same time made these products extremely difficult to process because of their very high and sometimes lack of melting point. In the beginning, researchers' interest was concentrated mainly upon thermal stability with little regard to cost, ease of fabrication and end-use temperature. Lately, emphasis has shifted towards more practical, specific applications which made PI in these days, among the most common commercial high performance materials. Modified PI containing various flexibilising groups such as ether, ester, amide, sulfone, sulfide, hexafluoroisopropylidene and others, have been studied intensively to improve their processing properties such as solubility, flexibility, T_g, melt viscosity, cost and ease of fabrication although with a little sacrifice of their thermal stability [8-20]. Their outstanding properties include continuous service at temperatures

443

above 200 °C, wear resistance, low friction, good strength, toughness, thermal stability, high dielectric strength and radiation resistance, and low outgassing. Polyimides are available as precured films and fibres, curable enamels, adhesives, and resins for composite materials. In addition, due to their stable backbone structure, and resistance to acid hydrolysis and bacterial attack, PI are used as membranes for various processes.

Polyimides have been blended with conventional homopolymers and copolymers including polytetrafluoroethylene (PTFE), polyvinylidenefluoride (PVF), epoxy and phenolic resins, polyphenylenesulfide, polysulfone, polyesters, polycarbonates, polyamides, polyethers, polyether-ketones, polyaniline, polyurethanes and with elastomeric materials such as nitrile rubber; most of these blends will be discussed later. Also, PI have been blended with other heterocyclic polymers such as polybenzimidazoles, polyquinoxalines, and polyoxadiazoles. Blends have been also made by mixing PI of various structures. The PI are either a major or minor component of the mixture. In either case the inherent properties of the major blend component are amended by the addition of a minor component until the properties of the blend have attained the optimum desired.

11.2.1 Blends of Different Polyimides

Aromatic PI are increasingly used in microelectronic devices because of their good dielectric properties and high thermal stability. The requirements of polymer properties for more advanced microelectronic applications are quite severe, so that they can not be usually met by a single component polymer. Some of these requirements include high T_g, above 300 °C, and low thermal expansivities along with mechanical toughness and good adhesion properties. While PI with rigid rod-like conformation provide the required thermo-mechanical properties, these materials have inherently poor adhesion characteristics. One approach to obtain such a combination of desired properties is blending a rod-like polymer with a flexible one which could provide enhanced adhesion and increased toughness.

Blends of PI having various structures have been prepared mostly by mixing the solutions of the two polyamic acids in *N*-methylpyrolidone (NMP) or dimethylacetamide (DMAc), followed by casting into thin films, and drying and thermal imidisation at high temperature. The imidisation degree of PI-PI blends was evaluated with the aid of perylentetracarboxydiimide which shows unique fluorescence when molecularly dispersed in polyamic acid/imide mixtures. Their fluorescence is efficiently quenched in the polyamic acid films, due to amide groups functioning as a quencher, and increases abruptly when the imidisation process proceeds. On the other hand in mixtures of polyamic acid – PI based on biphenyltetracarboxylic dianhydride, the fluorescence is strong while in those based on pyromellitic dianhydride (PMDA) it is weak. Thus, perylentetracarboxydiimide provides a valuable indicator of the imidisation and miscibility on the molecular level [21-23].

Mixtures of rigid polyimides (**III**) based on 3,3′,4,4′-biphenyltetracarboxylic dianhydride and *p*-phenylene diamine and flexible polyimides (**IV**) based on 3,3′,4,4′-biphenyltetracarboxylic dianhydride and 4,4′-oxy-dianiline (ODA), which were made by blending solutions of the respective polyamic acid precursors. Investigation of such blends showed that the components are dispersed at molecular level, based on the observation of a single T_g and the transparency of the films (see **Figure 11.2**).

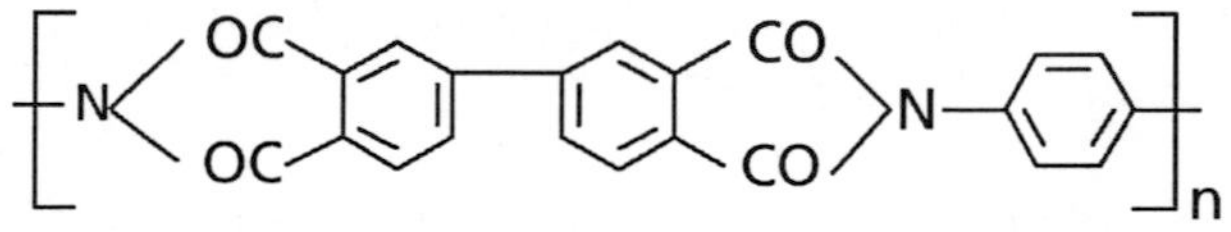

Figure 11.2 Chemical structures of a polyimide based on 3,3′,4,4′-biphenyl tetracarboxylic dianhydride and *p*-phenylene diamine (**III**) and of a polyimide based on the same dianhydride and ODA (**IV**)

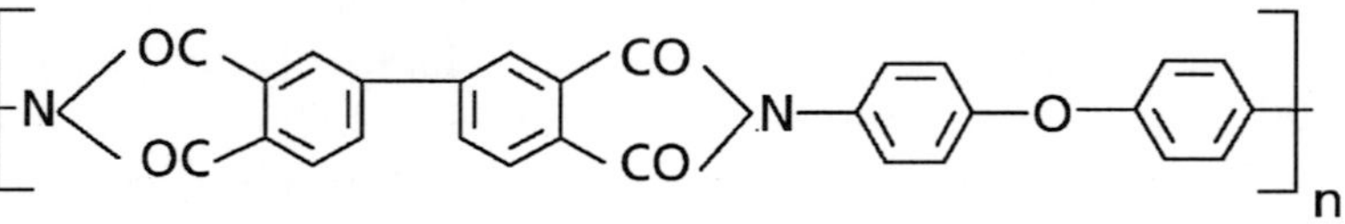

However, later studies showed that mixtures of polyamic acids can lead to the formation of segmented blockpolymers via a transamidation reaction rather than molecularly mixed blends of the two homopolyimides [24]. To prevent the transamidation reaction and obtain better miscibility of the blends, the ethyl or methyl esters of polyamic acids were used [25].

A blend was prepared from a polyamic acid based on 3,3′,4,4′-biphenyltetracarboxylic dianhydride and *p*-phenylene-diamine and another polyamic acid based on PMDA and ODA precursors to polyimides (**III**) and (**V**), respectively. After imidisation it provided blister-free films having high modulus and low thermal expansion coefficient for use in electronic applications [26] (see **Figure 11.3**).

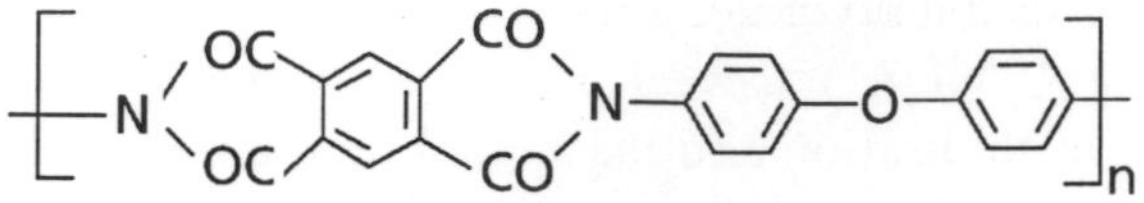

V

Figure 11.3 Chemical structure of a polyimide based on PMDA and ODA (V)

Other blends are based on polyimides prepared from ODA and various dianhydrides such as 3,3′,4,4′-diphenyloxytetracarboxylic dianhydride, 3,3′,4,4′-benzophenonetetracarboxylic dianhydride or 3,3′,4,4′-diphenyltetracarboxylic dianhydride, and are characterised by high hardness and high softening temperature as well as improved frictional behaviour [27].

A blend containing a polyimide (**VI**) based on PMDA and 4,4′-bis(3-aminophenoxy) biphenyl and another polyimide (**VII**) based on biphenyltetracarboxylic dianhydride and 3,3′-bis(*p*-aminophenoxy) benzene, prepared by melt mixing followed by injection moulding, produced a test piece showing dimensional change of only 0.55% after thermal treatment at 300 °C for 2 hours and flexural modulus of 6500 MPa at 250 °C [28-29] (see **Figure 11.4**). Such blends are useful for electric and electronic parts, office equipment and automobile parts and industrial machine components.

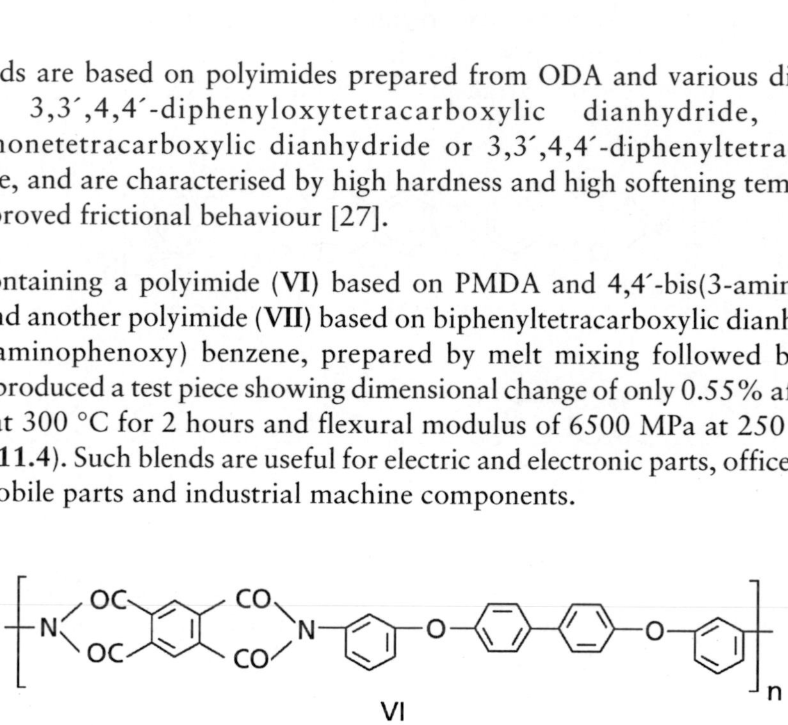

VI

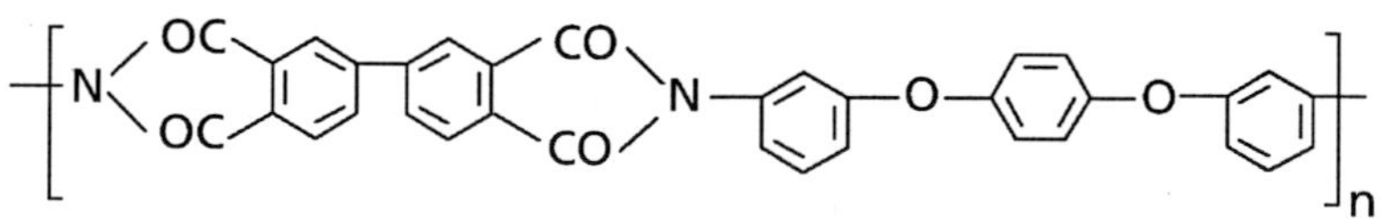

VII

Figure 11.4 Chemical structures of a polyimide based on PMDA and 4,4′-bis (3-aminophenoxy) biphenyl (**VI**) and of a polyimide based on biphenyltetracarboxylic dianhydride and 3,3′-bis(*p*-aminophenoxy) benzene (**VII**)

The use of silane coupling agents with blends made from diphenylsulfone tetracarboxylic dianhydride based polyimides provided excellent adhesives for PI films [30]. Blends prepared by melt mixing followed by injection moulding of such PI showed no change after treatment with 35% H_2SO_4 for ten days and distortion of 7.2% at 250 °C for 5 hours [31].

Because of their high inherent thermal stability and excellent electrical properties, fluoro-containing aromatic PI have excellent potential for use in aerospace composites, aircraft wire and cable, adhesives, and flexible wire boards [32, 33]. The PI containing hexafluoroisopropylidene (6F) groups gave very interesting blends, fully miscible, when they were based on the same 6F-containing dianhydride and the diamine was 4,4′-6F-dianiline or 3,3′-6F-dianiline, (**VIII**) and (**IX**), respectively, (see **Figure 11.5**). Their miscibility arises from the fact that their dianhydride parts have identical bond angle, rigidity and length and their diamine segments also have 6F groups [34, 35].

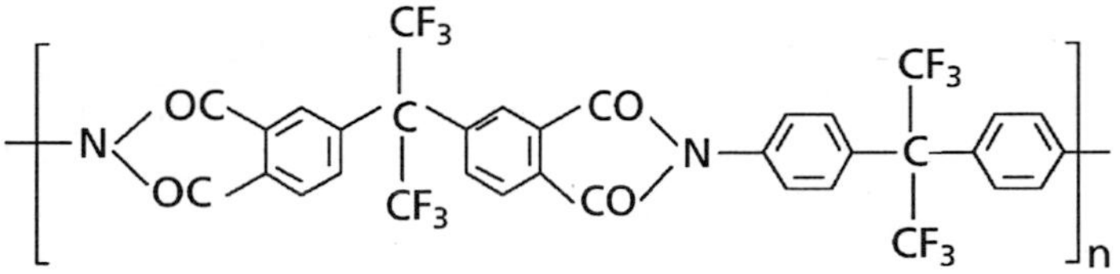

VIII

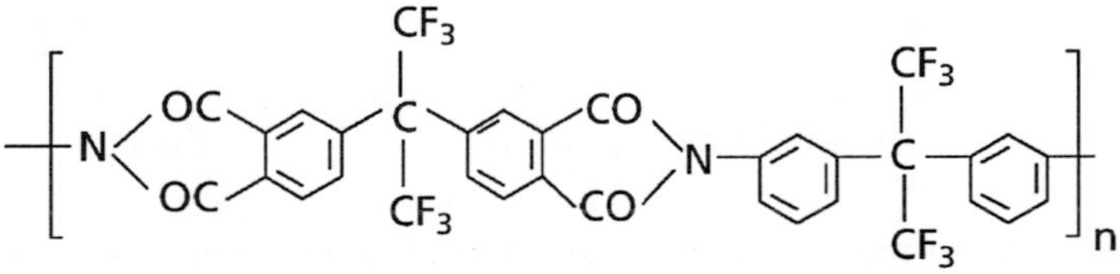

IX

Figure 11.5 Chemical structure of a polyimide based on hexafluoroisopropylidene diphthalic anhydride and 4,4′-hexafluoroisopropylidene dianiline (**VIII**) and of a polyimide based on the same dianhydride and 3,3′-hexafluoroisopropylidene dianiline (**IX**)

A rod-like PI, prepared by the reaction of PMDA with *m*-phenylene diamine, was blended with a flexible polyimide based on 6F dianhydride and 2,2-bis(*p*-aminophenoxy-*p*-phenylene)-hexafluoropropane. The blends were made by mixing the solutions of the corresponding polyamidic acids (**X**) and (**XI**) in NMP, one of the component being an ethylester, followed by solvent evaporation and thermal imidisation [24] (see **Figure 11.6**).

Mixtures of polyamic acids were made over a large range of composition at a total concentration of solids of 15%-20% and showed complete transparency. The films were

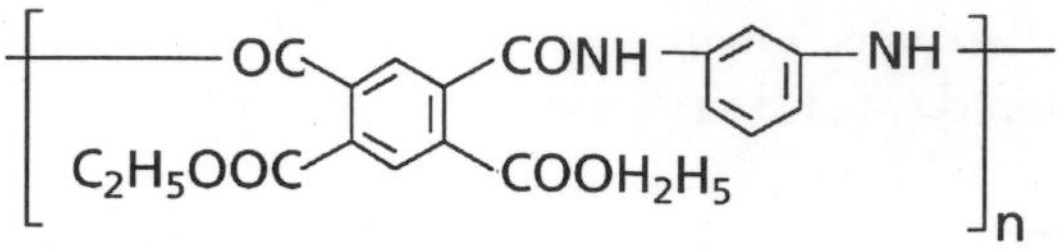

X

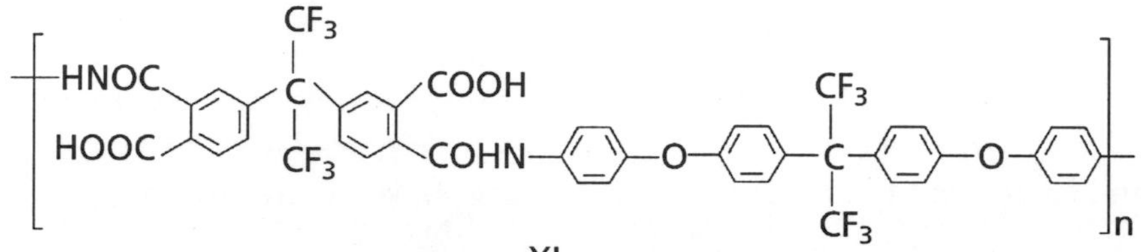

XI

Figure 11.6 Chemical structures of a polyamidic acid based on diethyl pyromellitate and *m*-phenylene (**X**) and of a polyamidic acid based on hexafluoroisopropylidene diphthalic anhydride and 2,2-bis(*p*-amino phenoxy)-*p*-phenylene-hexafluoropropane

prepared by spin-casting polyamic acid blends followed by drying and imidisation at 400 °C. The resulting films have a thickness in the range of 20-40 μm. The molecular weight of the polymer has a significant influence on the miscibility. Mixtures containing up to 15% low molecular weight 6F polyimide remained transparent and this was an indication of polymer miscibility. In any composition a phase-separation took place during the drying of the polyamic acid mixtures, when the solid content reached a value of approximately 65%. This morphology is then preserved throughout the thermal imidisation cycle at 400 °C. The particle size of the dispersed phase, approximately 1 μm or smaller, is determined by molecular weight (MW) and by the processing conditions. The mixtures containing low molecular weight 6F polyimide (MW = 10,000) gave more homogeneous films. The mixtures containing up to 25% low molecular weight 6F polyimide gave transparent films. The T_g of the neat 6F polyimide was found in all the mixtures regardless of their compositions, indicating complete demixing of rod-like and flexible polyimides. The mixture containing rod-like polyimide as a major component exhibited excellent dimensional stability up to 500 °C and low coefficient of thermal expansion comparable to the rigid-rod polyimide itself, and a lower dielectric constant of less than 3.0. The surface properties and adhesion strength of the mixtures are dominated by the 6F polyimide due to the segregation of this component to the surface region resulting in excellent self-adhesion properties.

A polyimide structure (**XII**) that is utilised in a large variety of blends is based on regular repeating units of ether and imide linkages and is prepared by the reaction of 2,2- bis [4-(3,4-dicarboxyphenoxy)phenyl] propane dianhydride with aromatic diamines [36] (see **Figure 11.7**).

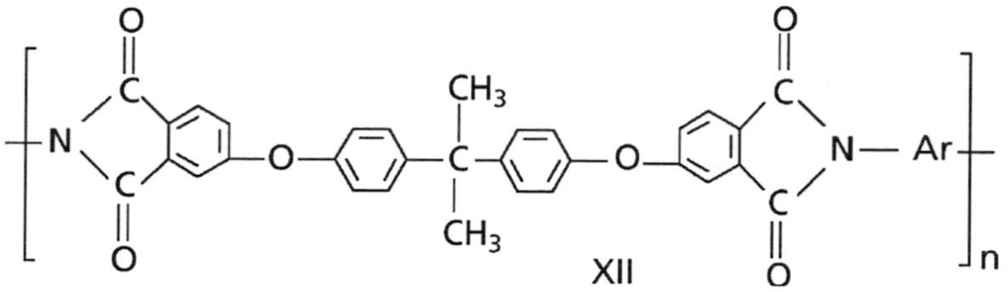

Figure 11.7 Chemical structure of a polyimide based on 2,2-bis [4-(3,4-dicarboxy phenoxy) phenyl] propane dianhydride and aromatic diamine (**XII**)

The dianhydride structure used in (**XII**) is the basis of a series of high performance engineering thermoplastics known under the trade name of ULTEM (General Electric). These polymers exhibit high heat resistance, excellent mechanical properties, inherent flame resistance and outstanding electrical properties. This is why this specific polyetherimide (PEI) structure has been utilised in blends with other polymers.

In the blends of PEI with polyimides based on traditional aromatic dianhydrides such as PMDA, the thermal stability was dependent on the PEI content while the α-relaxation occured at a slower rate and was broader and more cooperative in some blends compared to the neat polymers. Physical ageing of the blends in the glass state revealed changes in the relaxation rate. The dynamic-mechanical ageing behaviour was attributed to an increase in density on blending which decreased the free unoccupied volume and increased the constraints of molecular mobility. The adhesive properties and bond strength were found to vary with blend composition and test temperature [37]. Other blends based on PEI were made with polyamic-esters precursors of PI prepared from diphenyltetracarboxylic dianhydride and *p*-phenylene diamine, by dissolving them in DMAc and subsequently coagulating them in methanol. After thermal imidisation the blend showed a single T_g in both differential scanning calorimetry (DSC) and dynamic mechanical analyses (DMA) and the T_g increased with increasing PI content. These T_g values are reproducible in repeated heating cycles suggesting the true miscibility of the blend [38]. To improve the processability in thermoplastic PI, a method named the reactive plasticiser approach, was developed by using a small amount (5%-15% mol) of a less activated weak nucleophilic diamine comonomer as a reactive plasticiser to obtain resins which possess relatively low viscosity at low temperature and can be readily processed through autoclave cycle at low pressures. During a high temperature treatment the reactive plasticisers help the formation of high molecular weight polyimides and the desired properties are achieved. The most effective plasticisers are aromatic diamines such as 2,6-diaminopyridine. Such a reaction took place very efficiently in mixtures of PEI with a PI based on 1,4-bis(*p*-aminophenoxy) benzene [39]. Blends prepared from two different

PEI showed full miscibility when the synthesis of the two polymers was performed *in situ* in a melt of benzoic acid. Such blends showed a single T_g which took an intermediate position between the corresponding T_g of each of the two polymers. On the contrary, for the physical blend of the two PEI, the system showed a well-pronounced two-phase character with two T_g. The observed difference in the morphology is due to the fact that the *in situ* synthesis leads to the formation of some quantity of a block copolymer which functions as a compatibilising agent for the two PEI homopolymers [40]. Blends prepared from PEI and polyamide-imides based on trimellitic anhydride, in dimethylformamide (DMF) solution, were used to produce membranes having high water permeability: 1.5 cm^3/min-cm^2 by applying 1 kg/cm^2 pressure [41]. The thermoplasticity and flexibility of electrical insulating varnishes based on PI are increased by addition of unsaturated PEI based on maleic anhydride, trimellitic anhydride, diaminodiphenyl methane and ethylene glycol [42].

11.2.2 Blends of Polyimides with Poly-Ether-Ether-Ketones

A typical poly-ether-ether-ketone (PEEK; **XIII**) is poly(oxy-1,4-phenylene-oxy-1,4-phenylene-carbonyl–1,4-phenylene) which is an aromatic semi-stiff polymer possessing a fully *p*-aromatic backbone arrangement (see **Figure 11.8**). It is characterised by exceptional high performance properties, amongst which are thermal and chemical stability, the absence of polymorphic transition, and a large temperature interval of crystallisation. This polymer normally displayed common spherulitic features. In spherulites grown in bulk as well as in thin solution cast films, a lamellar structure was also observed [43].

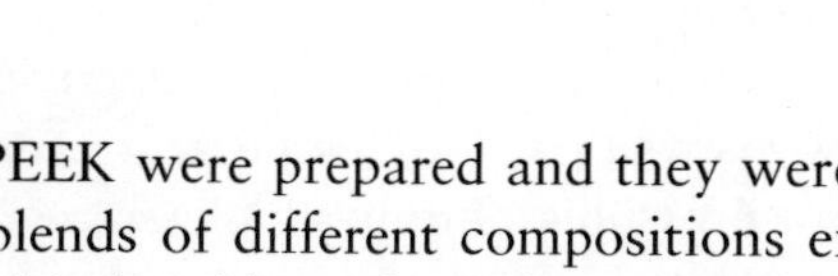

XIII

Figure 11.8 Chemical structure of poly(oxy-1,4-phenylene-oxy-phenylene-carbonyl-1,4-phenylene) (PEEK; **XIII**)

Blends of polyimides and PEEK were prepared and they were miscible over the entire composition range as the blends of different compositions exhibited a single T_g. The crystallisation of PEEK was hindered by polyimide, at which the dominant morphology was interlamellar [44]. Also, blends of PEEK with polyetherimide have been studied. Polyetherimide is fully miscible with PEEK in the amorphous state but it is rejected from PEEK crystals during crystallisation. For the blends with a low content of PEEK (up to

40%) a semicrystalline structure with neatly resolved sub-spherulitic details was shown by electron microscopy. At the finest scale, it was possible to see the stacks of lamellae as well as singular lamellae separated by zones of almost pure PEI. When the polyimide content is low, for example 40%, it becomes impossible to discern lamellar features in the spherulites, which clearly indicates that a large PEEK dilution is required in the blends to get lamellar structure, because these conditions ensure a sufficient openness of the spherulites. The values of T_g for pure polymers and blends are: PEI 218.5 °C; PEEK/PI 70/30 161 °C; PEEK/PI 60/40 167.7 °C and PEEK 143 °C. Upon crystallisation, the T_g of PEEK and blends PEEK/PI was found to increase significantly [43, 45].

The investigation of PEI/PEEK blends, by both DMA and dielectric behaviour, showed an α-relaxation process which was dependent on composition and the temperature of which increased with increasing the PEI content. The low temperature relaxation behaviour of PEI and the dielectric relaxation time in the T_g region was found to be affected by the addition of PEEK. The variation of loss and storage permitivities showed that the asymmetric broadening was strongly dependent on the blend composition, indicating a high degree of interaction between the components [46].

Blends of PEI with PEEK showed miscibility over the full range of compositions, with a single T_g varying almost linearly with composition between the two homopolymers. Enthalpic relaxation was a useful technique for studying the degree of compatibility of the system. Physical ageing was observed to embrittle the blends and there was a close correlation between the extent of enthalpy ageing and the change in mechanical and impact behaviour. The yield stress increased and the elongation to break decreased progressively with relaxation, in addition to a reduction in impact strength [47]. The crystallinity and mechanical properties of blends with different amounts of semicrystalline PEEK and amorphous PEI, prepared by melt mixing were studied by DSC to find the miscibility between the two components and to find proper conditions for the maximum development of crystallinity [48]. The sliding frictional behaviour of such blends follow a similar trend to that of the neat PEEK but the overall friction is higher for blends. There is a significant reduction in the failure potential of PEEK when PEI is incorporated. The potential for catastrophic thermally induced contact failure in the PEEK/PEI blend is less than that observed for unblended PEEK [49]. The dynamic-mechanical behaviour of PEEK/PEI blends was studied over a wide temperature range incorporating the α-transition and sub-T_g transition. The segmental motions responsible for the α-process were significantly slowed by blending and were dominated by PEI. Two sub-T_g transitions in PEI and the blends and one in PEEK were detected. The β-transition in PEI which occurred at around 100 °C was strongly affected by the presence of PEEK. The γ-transition of the blend shifts its temperature position with composition in a linear manner between the β-transition of the PEEK (–69 °C at 1 Hz) and the γ-transition in PEI (–81 °C at 1 Hz) [50]. The study of mechanical performance of an annealed PEEK/PEI blend prepared directly during injection moulding processes showed, at 23 °C, an overall synergistic behaviour in ductility and impact strength. The different crystallisation levels influenced the properties, but when there was no crystallinity effect, the intrinsic compatibility of PEEK and PEI was also clear and appeared as the main reason for the observed mechanical behaviour. At 125 °C, the modulus of elasticity and tensile strength of annealed blends and the modulus of elasticity of as-moulded blends were below the additivity values. However, in the as-moulded blends the tensile strength and ductility were synergistic [51].

The semicrystalline morphology in PEEK/PEI blends was investigated by dielectric relaxation spectroscopy as a function of a blend compositions and crystallisation conditions and it showed two glass-rubber relaxations for all samples corresponding to the coexistence of the mixed amorphous interlamellar phase and a pure PEI phase residing in interfibrillar/interspherulitic regions. Growth of PEEK spherulites from pure melt and its miscible blends with PEI have been studied by polarised optical microscopy which showed that the nucleation density of PEEK spherulites was depressed upon blending with PEI [52, 53]. Crystal morphology in the blend PEEK/PEI was also investigated by small angle X-ray scattering which showed that the lamellar thickness was constant (6 nm) and independent of blend compositions [54-56].

The deformation behaviour of thin films prepared from PEEK/PEI blends was studied over a wide temperature range by optical microscopy and tunnelling electron microscopy which showed that all the materials had localised shear deformation at temperatures well below T_g. In pure PEI and in blends with up to 60% PEEK content, a transition from shear deformation to disentanglement crazing occured as the temperature was raised. This transition was absent in PEEK which was deformed by shear over the whole temperature range, and similar behaviour was found for blends containing 80% PEEK. That means that at high PEEK content disentaglement crazing is supressed by strain-induced crystallisation and some evidence for crystallisation order in the deformed region of initially amorphous PEEK thin films was obtained by electron diffraction. The thin film deformation of the blend was also consistent with their bulk deformation behaviour, a high temperature ductile-brittle transition being observed at low PEEK contents in tensile tests [57].

From blends of PEEK/PEI, high performance foams were prepared by using water as blowing agent in the foaming process which took place at elevated temperatures in an extruder. The properties of the foams depend of the composition of the blend mixing, time and the temperature of the process [58].

Blends were made from PEEK with a polyesterimide prepared from *N,N*'-hexamethylene-bis(trimellitimide), 4,4'-dihydroxybenzophenone and *p*-hydroxybenzoic acid. The melt viscosity of such blends was significantly lowered which made them easily processable and mechanical properties were improved. When the content of polyesterimide in the blend was above 50%, a phase separation occured at 340 °C consisting of PEEK islands in a polyesterimide continuous phase. Optical textures observed at various temperatures support the phase separation phenomenon [59].

Sulfonated PEEK were prepared by sulfonating PEEK with 95% H_2SO_4 and were used in blends with PEI, followed by solution casting into films. A single T_g was observed by DSC in such blends which indicates phase miscibility [60]. The influence of doping with HCl and H_3PO_4 on the properties of sulfonated PEEK/PEI was studied in membranes made from such blends. Blending with PEI first results in an increase and then in a decrease in membrane swelling at concentrations higher then 5%. The electrical conductivity of blend membranes follows the same trend. Doping with the acids enhanced conductivity several fold and the effect of doping with HCl is more significant. PEI forms spherical particles dispersed in the PEEK matrix and at the same time partially dissolves the swelling of the matrix at higher PEI concentrations. The increase in the membrane capacity to absorb water at low PEI content is due to the formation of new water absorption sites along the interface between the dispersed particles and the matrix [61].

11.2.3 Blends of Polyimides with Polyamides

Blends of aromatic polyamides (PA), such as poly(*m*-phenyleneisophthalamide) (**XIV**), with aromatic PI based on aromatic dianhydrides, such as PMDA, and aromatic diamines, such as 4,4′-oxydianiline, were prepared by mixing the NMP-solution of precursor polyamic acid (**XV**) with NMP-solution of polyamide (see **Figure 11.9**).

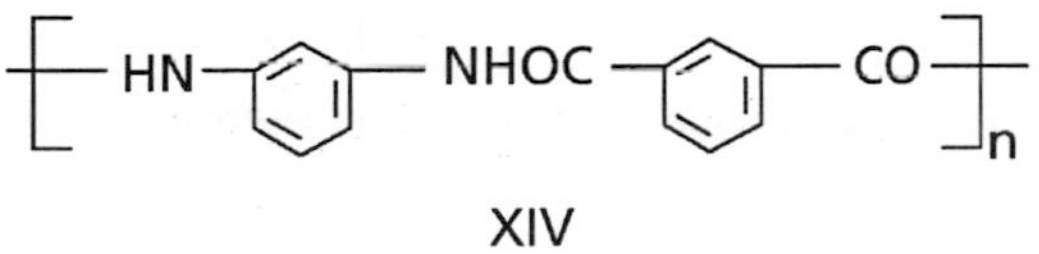

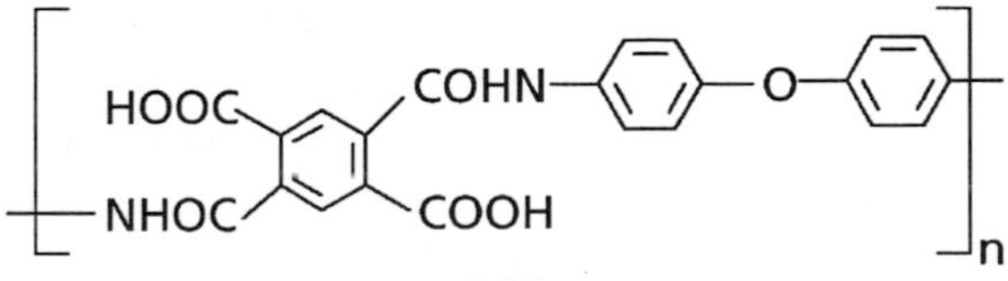

Figure 11.9 Chemical structures of poly(*m*-phenylene isophthalamide) (**XIV**) and of polyamic acid based on PMDA and ODA (**XV**)

Blend films were cast on glass plates and heated first at 80 °C to partially evaporate the solvent, then they were stretched and again heated at 300 °C. Such films showed high elasticity, strength and good productivity [62]. Very good mechanical characteristics and heat resistance resulted when the polyamic acid was prepared from a mixture of two dianhydrides, such as PMDA and benzophenonetetracarboxylic dianhydride (BTDA), with a mixture of two aromatic diamines, such as ODA and 4,4′-diaminodiphenylmethane, in *m*-cresol. The blend with PA showed a strong adhesion to copper foil, had a thermal decomposition above 450 °C, peel strength of 2.2 kg/cm, and no peeling in solder heat test at 260 °C for 10 seconds [63].

Blends of polyamic acid based on PMDA and ODA with aromatic polyamide Kevlar (Monsanto), poly(*p*-phenylene-terephthalamide) (**XVI**), were moulded to give sheets with low thermal expansion, flexural modulus of 300 kg/mm² and Izod impact strength of 8.7 kg cm/cm [64] (see **Figure 11.10**).

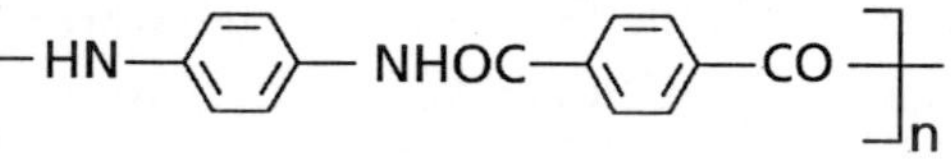

XVI

Figure 11.10 Chemical structure of poly(*p*-phenylene-terephthalalamide)

Aliphatic polyamides, such as Nylon 6, were also blended with aromatic PI based on dianhydrides, such as PMDA or BTDA, and diaminodiphenylether to give materials with good chemical stability. Thus, the stability of the blend films in concentrated H_2SO_4 and in 40% aqueous NaOH are equivalent to that of the control film and slightly better than commercially available Kapton PI film (DuPont de Nemours). The weight gains of the blend films in solvents like CH_2Cl_2, methylethylketone and water are dependent on the nature of the solvent and its interaction with the blend, but the BTDA based blend films showed better chemical resistance [65].

Polyetherimides were also blended with aromatic or aliphatic PA [66, 67]. In blends with aliphatic PA, such as Nylon 66, DSC analysis showed that crystallisation in such blends was retarded compared to that of pure Nylon 66 by incorporation of PEI with high T_g. The lowest growth rate of the spherulites was observed in the blends containing 10% or 15% fraction of PEI. The T_g increased with the content of PEI in the blends.

Aromatic-aliphatic PA prepared from terephthalic or isophthalic acid and aliphatic diamines were also blended with PEI and moulded to give materials with good stiffness and toughness even at 100 °C [68].

Ternary blends of PEI, PA and sulfonated PEEK were prepared and used in the manufacture of ionomer membranes for electrodialysis. DSC traces of the blend membranes showed only one T_g. In the blend membranes swelling was reduced by specific interactions due to hydrogen bonds between PA and PEI. The acid-base interactions also led to a decrease of ionic conductivity by partial blocking of SO_3^- groups for cation transport. The acid-base blends showed better ion permoselectivities, even at high electrolyte concentrations, and thus better performance in electrodialysis. The thermal stability of the blends was very good and even better than that of pure PEEK-SO_3H. The performance of these ionomer blend membranes is better than the performance of binary blend PEEK-PA, especially in the electrodialysis experiments with higher NaCl concentrations. This is mainly due to the lower swelling and thus better ion permoselectivity of the acid-base (ionomer) blend membrane, compared with the PEEK-polyamide blend [69].

11.2.4 Blends of Polyimides with Polyesters

Polyetherimides have been the most used PI-type polymers in blends with polyesters.

Blends of PEI and polyethyleneterephthalate (PET) were obtained by compression moulding and by direct injection moulding. Both procedures produced biphasic structures with similar homogeneity that showed a wide, single, T_g peak by DMA. The modulus of elasticity and the yield stress values were close to and above those predicted by the additivity rule. The ductility values *versus* composition were well below the additive values [70]. In another procedure, blends of PEI/PET were prepared in a twin-screw extruder and cast into transparent amorphous films and DSC showed a rapid rise in crystallinity followed by a much slower increase after the films reached the stress hardening point. The addition of PEI was found to hinder the formation of the crystalline lattice of PET and once formed, crystallites were highly distorted. Also, PEI caused the onset of stress hardening to move to lower draw ratios [71]. To improve the colour stability of PEI/PET blends phosphorus- or phenol-containing colour stabilisers were added to the blend, leading at the same time to better mechanical properties. Such materials showed a heat distortion temperature of 166 °C and a yellowness index of 78.89 [72].

Binary blends of PEI with PET were miscible in the melt, but showed simultaneous crystallisation and liquid-liquid demixing below the melting point. Small-angle X-ray scattering showed that the crystal thickness of PET was not perturbed upon blending with PEI. A larger amorphous layer thickness in the blend was identified, showing that some PEI was incorporated inside the interlamellar regions after crystallisation. Despite the swelling of the amorphous layer, the amorphous layer thickness was relatively independent of the blend composition. Significant extralamellar placement of PEI occurred when the PEI composition was higher than 20%. This morphology structure is interpreted in terms of simultaneous occurrence of liquid-liquid demixing and crystallisation [73]. The melting behaviour of PEI/PET blends was investigated by DSC which showed the multiple melting endotherms of PET. The study of the effects of blend composition, heating rate and crystallisation duration on the multiple melting of the blend shows that the extent of recrystallisation during DSC heating decreased with increasing of PEI composition and crystallisation duration [74].

Blends of PEI with polyarylates were miscible when 90/10 and 80/20 weight ratios of the two components were used. All other compositions were biphasic, one phase being almost pure polyarylate and the other being PEI containing a fairly constant polyarylate component of roughly 25%. This phase behaviour agreed with both the observed transparency and the fracture observed by scanning electron microscopy (SEM). The mechanical properties of the blends as a function of composition showed values close to additivity or even enhanced with an unexpected synergism in ductility [75]. Blends of aromatic polyesters with polyimides based on PMDA showed remarkable improved mouldability and crystallinity, and they were free from the bleeding of mouldings and excellent in weathering and delamination resistance. Such a test piece showed a volume change smaller than 0.1% in 75% hot water [76]. Other PI/polyester blends were prepared by mixing the polyamic acids with a liquid-crystalline aromatic polyester followed by injection moulding at 400 °C. Such a test piece showed tensile strength of 1349 kg/cm^2, elongation of 4.2% and Notched Izod strength of 13.87 kg.cm/cm [77]. In another procedure, polyester and PI are blended in the presence of low-molecular weight imide compounds having very low melting point for example an imide obtained from 1,2-cyclohexane dicaboxylic anhydride and butyl amine was used as a solvent for a blend of a PI based on PMDA and a polyester such as polyethylene-2,6-naphthalene dicarboxylate. The blend was prepared at high temperatures (300 °C) and then the solvent was distilled off and the resulting blend composition had a T_g of 122 °C and a melting point of 267 °C [78].

Polyester/PI blends with high adhesivity were prepared from PI, based on tetracarboxylic dianhydride and diisocyanate, and polyesters based on adipic acid and hexandiol, and they were used as adhesive layers or sealing materials for lead frames in semiconductor devices [79]. Blend films with high mechanical strength were manufactured by mixing a PI (**XVII**) having a T_g below 250 °C prepared from butantetracarboxylic acid and *m*-xylylendiamine with PET followed by extrusion and drawing to give biaxially oriented films with tensile moduli of 8.8 GPa and 7.4 GPa, in the machine and transverse direction, respectively [80] (see **Figure 11.11**).

Blends of polyesters with PI containing siloxane units which exhibited good antisoiling properties, were used to cover metal plates followed by baking at 150-300 °C [81].

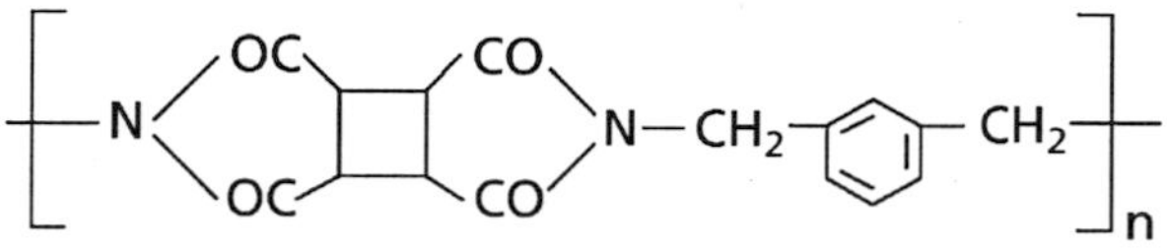

XVII

Figure 11.11 Chemical structure of a polyimide based on cyclobutanetetracarboxylic acid and *m*-xylylen-diamine (**XVII**)

A blend was prepared from a PI (**XVIII**) based on cyclobutane dianhydride and 2,2´-bis(*p*-aminophenoxyphenyl) propane and polyvinylcinnamate (**XIX**) (see **Figure 11.12**).

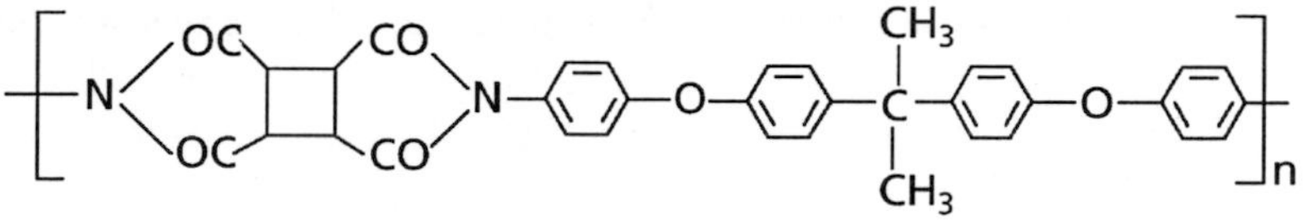

XVIII

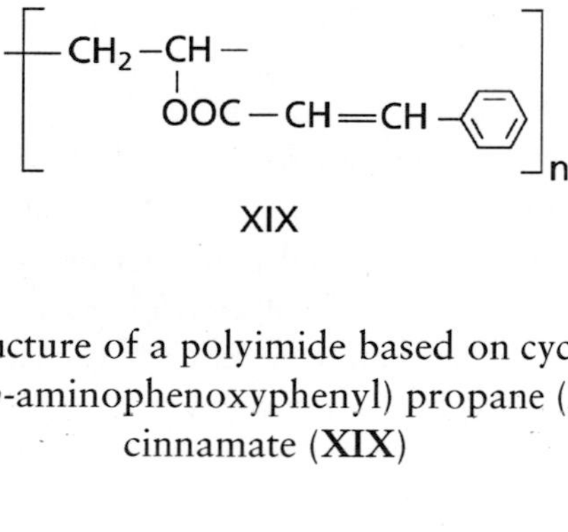

XIX

Figure 11.12 Chemical structure of a polyimide based on cyclobutane tetracarboxylic dianhydride and 2,2-bis(*p*-aminophenoxyphenyl) propane (**XVIII**) and of polyvinyl cinnamate (**XIX**)

Microscopic and thermal analysis revealed the miscibility of the two polymers in the thin layers. From UV spectroscopic investigations the existence of photo-dimerisation and the development of structural dichroism in the blend layers were confirmed. The photo-dimerisation reaction of the cinnamate was not significantly influenced by addition of PI, but the blend alignment layers provided uniform liquid-crystalline alignment, and better thermal stability of liquid-crystalline alignment compared to the pure polyvinylcinnamate alignment layer. The increase of the thermal imidisation temperature enhanced the thermal stability of the liquid-crystalline alignment, which indicates that the increase of T_g by introducing PI to polyvinylcinnamate is responsible for the improvement of the thermal stability of liquid-crystalline alignment [82].

Amorphous PEI and crystallisable polybutyleneterephthalate (PBT) were miscible in the melt over the entire composition range. Upon the crystallisation of PBT at temperatures below 200 °C, a strong segregation of PEI was observed. Investigations by optical microscopy of multiple melting behaviour show that the recrystallisation of PBT after the initial melting was hindered by the presence of PEI, and thus PEI/PBT exhibited quite similar phase behaviour to PEI/PET blends [83]. The DSC study of PEI/PBT blends indicates that they are partially miscible. The blending with PEI has a strong effect on the melting behaviour of PBT: first, as the content of PEI in the blend increases the multiple melting behaviour changes from three peaks (low, middle and high) to two peaks (middle and high); second, the size of the high peak relative to that of middle peak becomes smaller with the increasing of PEI content. Recrystallisation rate and heat of PEI/PBT blend decreased as the content of PEI increased. This suggests that remixing between PBT and PEI takes place after initial melting of the original PBT crystals and such a remixing makes the diffusion of PBT more difficult in subsequent recrystallisations [84]. The crystallisation behaviour in PEI/PBT blends was also studied by FTIR spectroscopic method and it was shown that the CO stretching peak of the blend shifted to a lower wavenumber as the crystallisation of the PBT proceeded. Crystallisation behaviour of PBT in the blend strongly depends on the composition: at high PBT content, crystallisation occurs without any phase separation [85].

Gas transport membranes were prepared from blends based on PEI with an aromatic polyester and CO_2 permeation and sorption capacity were compared with PEI films at different pressures and temperatures. In all cases permeability, diffusion and sorption coefficients decreased when the amount of polyester increased. The results were consistent with a polyester structure in the blend intermediate between an interfibrilar and a laminar morphology, according to electron microscopy evidence [86].

Blends were also prepared from PEI with polyethylene naphthalate (PEN) [87]. PEN is a relatively new polymer with good thermal, mechanical and barrier characteristics [88, 89]. PEN exhibits neck formation upon stretching from the amorphous state at temperatures between T_g and cold crystallisation. To enhance the optical and mechanical properties, blends were made with a PEI in a composition range varying from 100/0 to 70/30. PEN and PEI exhibit melt miscibility. Blends were melt casted into sheets or films, which were further biaxially stretched in the full compositional range. The increase of PEI fraction causes an increase of T_g while slightly decreasing the melting point. This is a result of increased inter-chain friction between the PEN chains in the presence of bulky PEI chains which disrupt the cooperative alignment of naphthalene groups parallel to the surface of the films. This was found to widen the processing window where films of uniform thickness can be obtained. The DSC studies of the amorphous blends show an increase of T_g with increasing PEI fraction, and a breadth of the T_g interval as well. This suggests the presence of a certain level of microphase segregation in these blends. The crystallisability of these blends is also influenced by composition as judged by the location of a cold crystallisation peak. The increased fraction of stiff and bulky PEI chains shifts the position of these peaks to higher temperatutres. This is mainly a result of dilution effect combined with the stiffening of the environment of crystallisable PEN chains at higher concentrations of PEI whose T_g (218 °C) is substantially higher than that of PEN (T_g = 120 °C). Below 30% weight PEI concentration the crystallisability is substantially diminished. To preserve the crystallisability of the blends the PEI concentration should not exceed 20% weight. The biaxial stretching of blend films was preformed at 20 °C above their T_g. When PEN films were stretched to low stretch ratios, neck regions appeared through the part. At high stretching ratios the films became uniform again after the disappearance of the necks. In 20% PEI blend films the cold crystallisation peak occurs immediately above the T_g and from the area below it can be seen that there is a larger percentage of oriented uncrystallised PEN chains in the blends whose crystallisation was hindered by the presence of PEI. In these films the melting endotherm also becomes broader. As for optical properties, the refractive indices increase with increased proportion of PEI and the values measured in all three directions are quite close to each other indicating that these films are nearly optically isotropic. When the films are stretched significantly, optical anisotropy rapidly develops. The refractive indices in the normal direction decrease with the stretching ratio. The addition of PEI to PEN, for the same stretching ratio, shows that the refractive indices in the normal direction increases with the addition of PEI. This increase is partially caused by the increased proportion of higher refractive indice in the blend, as a result of the disruption of preferential tendency of flat naphthalene group parallel to the surface, which was also, augmented with the data of X-ray analyses. Biaxially oriented PEN films show bimodal orientation, in the machine direction and in the transverse direction. Addition of a little PEI (10%) to PEN converts this bimodal orientation to in-plane isotropy in the equal biaxially stretched films. This conversion is the result of incresed interchain friction in the presence of stiff and bulky PEI chains that hinder the preferential orientation of the naphthalene planes paralel to one another and to the film surface. This orientation behaviour is also reflected in the mechanical properties. While PEN films exhibit bimodal orientation and show significantly lower moduli, the presence of PEI eliminates this minimum and the films exhibit a plane mechanical isotropy. A blend of 90/10 PEN/PEI showed a higher elongation to break than PEN films in all directions; PEN films exhibited very low elongation to break in the transverse direction.

Blends of PEI with PEN were also prepared by injection moulding, having a clear apperreance, Rockwell hardness of 104 L and flexural modulus of 2.4 GPa [90]. In another procedure the blends were prepared by melt polycondensation of naphthalenedicarboxylate with ethylene glycol in the presence of PEI, showing a T_g of 163 °C [91].

Ternary blends were prepared from PEI and two polyesters, PET and PBT, showing a rare case of thermodynamic miscibility in the amorphous state (quenched as well as molten state) with a single glass transition temperature observed in DSC and DMA studies [92]. Ternary blends of thermotropic liquid-crystalline polyesters with PEI and PEEK prepared by melt mixing exhibited a single T_g in DSC analyses. Although individual PEEK/PEI blends and the polyester undergo catastrophic tribologic failure at their respective T_g, the ternary blend was able to maintain sliding at temperatures in excess of the T_g of their constituent phases. This synergistic behaviour is explained by polyester reinforcement and PEEK cold crystallisation. PEI exhibits miscibility with both PEEK and polyester (PEN), these two ether polymers are known to be immiscible [93].

Binary blends of PEEK/PEN are immiscible but PEI acts as a solvent for both of them and ternary blends containing more than 40% weight PEI become miscible. The blending of these homopolymers improved the deformation behaviour of PEN films by eliminating necking which was observed in uni and biaxially stretched PEN films. While PEN and PEEK form separate crystal structure upon stretching and subsequent annealing, in blends containing more than 40% weight PEI, PEN crystallisation is hindered. Although PEEK can be crystallised at a concentration as low as 10 wt%, PEN remains amorphous [94].

Polyamide-imides prepared from an aromatic tricarboxylic anhydride with a diisocyanate were also used in blends with polyesters such as PET followed by injection moulding. Such blends showed appropriate tribological properties for use in sliding parts [95].

11.2.5 Blends of Polyimides with Polytetrafluoroethylene

Polyimides, for example based on benzophenone tetracarboxylic dianhydride and toluene diisocyanate, were blended with PTFE to give materials with a stable and improved friction coefficient compared to polyimide itself. This blend is useful for the manufacture of sliding parts such as piston rings, rod packings, bearings and others [96]. The blends are prepared by compression moulding of a mixture of polyimide resin powder and PTFE resin powder having M_w 500,000-1,000,000 and average particle diameter 5-20 μm, followed by sintering [97]. PTFE was selected as the abrasion-reduction and toughness-enhancement material for polyimide. The size of PTFE particles has an important effect on the abrasive property and microstructure of the blend. Research showed that PTFE particle size could be reduced to 10% of the original particles through air-current crush blending, which led to substantially improved friction and abrasive properties of the blends [98].

Ternary blends prepared from a polyetherimide (**XX**), polyarylketone and PTFE had good impact resistance and low kinematic coefficient [99] see **Figure 11.13**.

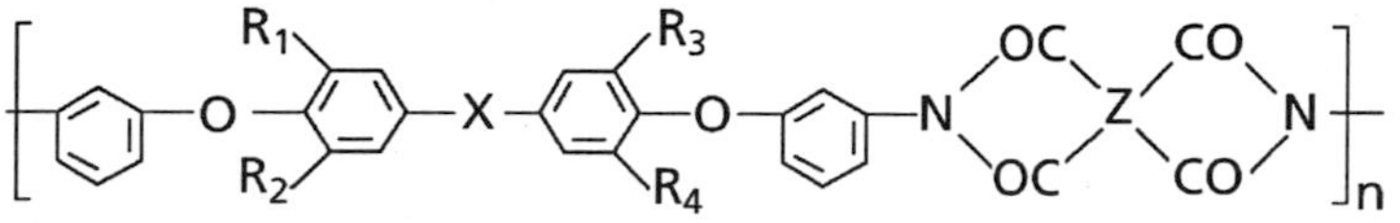

R₁, R₂, R₃, R₄ = H, alkyl, alkoxy, Cl, Br

X = -, S, C(CF₃)₂, CO, SO₂, O; Z = Tetravalent group

XX

Figure 11.13 General chemical structure of a family of polyetherimides (**XX**)

Ternary blends of a polyimide, PTFE and a polysiloxane were used to prepare antistick films for use with metals, ceramics, glass or enamelled surfaces. The adhesion of blend films to the substrate was very much influenced by the particle size of PTFE [100]. Polyimides prepared from PMDA and ODA were also used in ternary blends with PTFE and liquid crystalline polyesters. The blends were injection moulded to give materials with a coefficient of friction 0.25-0.32 [101]. Other ternary blends with good mechanical strength, sliding, and electric insulation properties were prepared by mixing polyimide powder, PTFE and novolak, followed by injection-moulding. Such a moulded part showed a bending strength of 110 MPa, charpy impact strength of 3.5 kJ/m², electric resistance of 1.2×10^{10} Ω, and a dynamic friction coefficient of 0.2 [102].

Ternary blends prepared from a polyamidimide, PTFE and polysiloxane are the main components in the bearings of slidable structures such as earthquake proof footings of buildings, where reduced friction and good heat resistance is required [103].

11.2.6 Blends of Polyimides with Polysulfones

Blends of PEI with polysulfones in various proportions were prepared by solution mixing followed by co-precipitation. Mechanical properties and morphology of the blends were studied by using tensile tests and SEM. Tensile moduli and ultimate strength exhibit positive deviations from simple additivity. The blends are not perfectly homogeneous, with fine dispersions, but the interphase between the two phases are well-bonded. These two polymers are partially miscible on the segment level [104]. The miscibility of polyimide-polysulfone blends was studied by DSC, rheological and X-ray scattering. The blends rich in polysulfone form a miscible blend when prepared by solution casting from a common solvent. Heating to temperatures above the T_g leads to phase separation into polysulfone rich domains and polyimide rich domains. Gas separation membranes

461

made from such blends showed that helium permeability is controlled by the polyimide component [105].

A polyimide (**XXI**), prepared from PMDA and 2,2´,6,6´-tetramethylbenzidine including chemical treatment with trifluoracetic anhydride-triethylamine in NMP, gave a miscible blend with polyethersulfone, poly(oxy-1,4-phenylensulfonyl-1,4-phenylene), (PEST; **XXII**) (see **Figure 11.14**). The study of compatibility of blend films was investigated by transmission electron microscopy (TEM) and it showed that the diameter of polyimide domains for 4:96 PI/polyether sulfone was approximately 0.2 µm, and the formation of the interpenetrating structure was confirmed by elemental analyses on the separated domain. Transparent films were obtained at polyimide content of 2-10% wt [106].

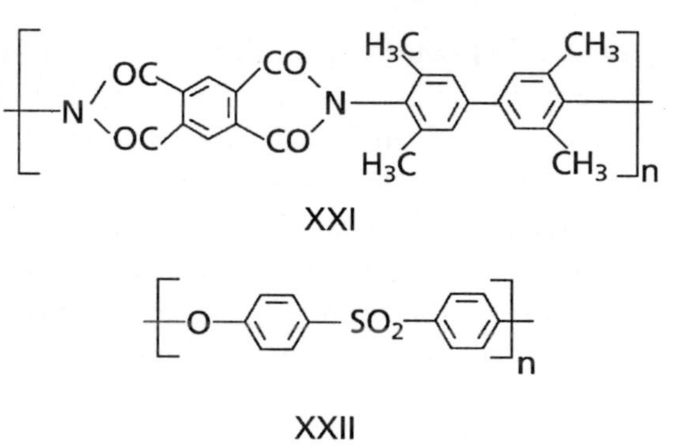

Figure 11.14 Chemical structures of a polyimide based on PMDA and 2,2´,6,6´-tetramethyl benzidine (**XXI**) and of poly(oxy-1,4-phenylenesulfonyl-1,4-phenylene) (**XXII**)

Blends of PEI and polyethersulfones were used in the preparation of hollow fibre membranes. The blend was dissolved in a mixture of NMP and ethanol followed by coagulation. The internal coagulant plays a very important role in this process and in this particular case it includes water, methanol, ethanol, 2-propanol and mixtures of these alcohols with water. Water is always used as the external coagulant [107].

Ternary blends containing an aromatic polyimide, polysulfone and polyamide-imide were used to prepare gas separation membranes with superior gas transport properties [108]. To increase the hydrophilicity and mechanical strength of gas membranes simultaneously, the polysulfone was sulfonated using H_2SO_4 and the products with different degrees of sulfonation were used to prepare blends with polyimides for hollow fibre membranes. Such membranes were used in air dehydration. When the degree of sulfonation and the degradation of sulfonated products were controlled appropriately, the membranes prepared from such blends displayed a very high ability in air dehydration, the ratio of water separation being above 95%, under low operating pressure (0.25 MPa). The property of gas dehydration membranes can be also improved when the sweep proportion and operating pressure were increased [109]. Such hollow fibre membranes made of polyimide/sulfonated polysulfone blends were also used for the separation of vapour methanol/methyl-*tert*-butyl ether mixtures. The separation coefficient of such hollow fibre membrane for vapour mixtures are extremely high, and therefore the application prospects are great [110].

Blends were prepared from PEI with polyphenylene sulfide by melt mixing followed by injection moulding to give products with improved flowability and good heat resistance and tensile strength. The spiral flow was 92 mm at 350 °C, melt index retention 70% and tensile strength 530 kgf/cm^2 [111].

11.2.7 Blends of Polyimides with Polycarbonates

Blends of polyetherimides and polycarbonates (PC) based on bisphenol A have been prepared and investigated by DSC, DMA, SEM and TEM. The blends were made by screw extrusion and solution casting, with weight fraction of PEI in the blends varying from 0.9 to 0.1. The maximum decrease of T_g of PEI is observed for the blends containing 90% wt PEI. In the study of the morphology, it was observed that when the minor component was PC, as in a PEI/PC blend (90/10), the size of the minor component domains was smaller (about 0.1 to 0.3 μm) than in the case when the minor component was PEI as in a PEI/PC blend (10/90), where the size of the minor component domains was 0.2 to 20 μm. This morphology behaviour is attributed mainly to the difference of viscosity ratio between the dispersed phase and continuous phase. The molecular weight of PC did not significantly influence the thermal behaviour, nor the morphology of the PEI/PC blends [112]. Mouldings made from PEI/PC blends retained their impact resistance after autoclave sterilisation [113].

Blends containing more than 80% wt PEI, less than 20% wt PC and up to 7% wt polyacrylate have been prepared for improving the environmental stresscrack resistance properties of polyetherimides. Such a blend was injection moulded to tensile bars which were mounted on testing devices with induced stress of 27 MPa and strain 0.83% and the bars showed no effect after immersion in trichloroethane for 15 minutes, compared to rupture for specimen made from PEI alone [114].

Other blends have been prepared by mixing polyetherimide with PC and polysiloxane and they exhibited high heat distortion temperature, improved room-temperature impact properties and improved impact strength and ductility at low temperature [115].

11.2.8 Blends of Polyimides with Polyurethanes

Polyurethanes (PU) have excellent abrasion resistance and properties of both rubber and plastics. They are becoming more important as engineering materials. Conventional PU exhibited small resistance to heat, for example the acceptable mechanical properties (strength, modulus) disappear from about 80-90 °C, and thermal degradation takes place above 200 °C [116]. Because of the poor heat resistance of pure PU their applications were limited. One possible method to improve the thermal stability of PU is by blending them with highly thermally stable heterocyclic polymers such as PI which have remarkable heat resistance and superior mechanical and electrical properties [117, 118]. Such blends were prepared by mixing a PU (**XXIII**) prepared by the reaction of a polyester polyol with 2,4-tolylene isocyanate and end-capped with phenol with poliamidic acid, which is a precursor of polyimide, prepared from PMDA and ODA end-capped with phthalic anhydride (see **Figure 11.15**). The two polymer solutions in NMP were blended at room temperature in various ratios and they were cast into films followed by thermal treatment at various temperatures. With the increase of PU component, the films changed from plastic to brittle and then to elastic. The elastomeric blend showed excellent mechanical properties and moderate thermal stability. The elongation of film was above 300%. TGA and DSC studies indicated that the thermal degradation was in the range of 250-270 °C which is significantly superior to PU itself. Films prepared from such blends showed good solvent resistance. The tensile modulus of the blend increased with increase of the imide component, but at the same time the elongation decreased greatly. At polyimide/polyurethane (50/50) the film was brittle and the elongation became the smallest. With further increase of polyimide/polyurethane ratio, elongation became larger and the tensile properties of films were close to those of polyimide. The tensile properties can be explained by the phase separation between PI and PU.

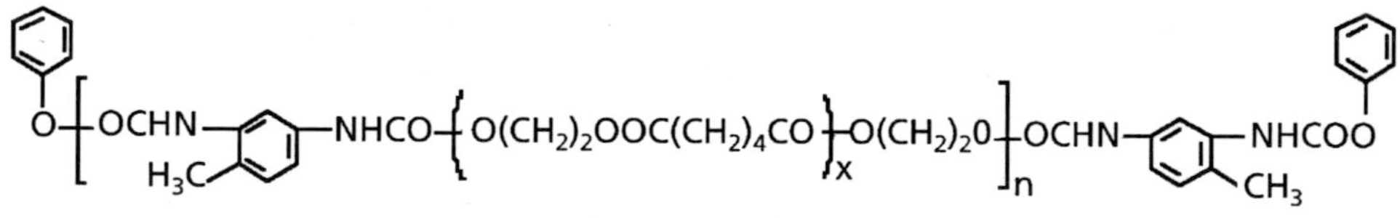

XXIII

Figure 11.15 Chemical structure of a polyurethane based on a polyesterpolyol and 2,4-tolulene diisocyanate end-capped with phenol (**XXIII**)

11.2.9 Blends of Polyimides with Silicones

Bisimide oligomers (**XXIV**) obtained from phthalic anhydride and a diamine containing polysiloxane groups were used in blends with a polyimide obtained from 1,3-bis(3-

aminophenoxy) benzene and benzophenonetetracarboxylic dianhydride (see **Figure 11.16**). The blend was extruded to give a film which was sandwiched with steel plates and pressed to give a highly heat resistant material showing good adhesion [119].

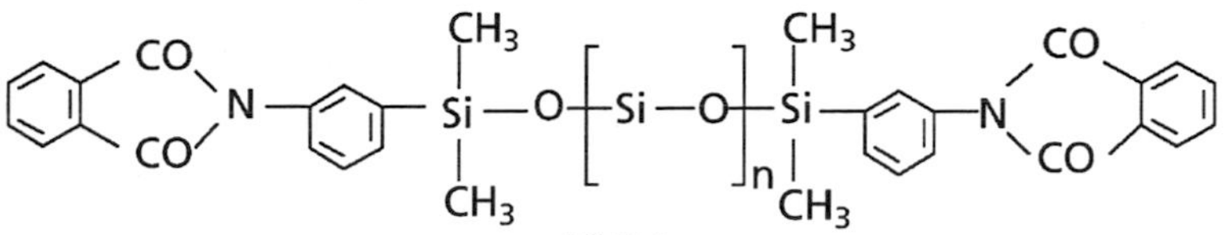

XXIV

Figure 11.16 Chemical structure of oligomers based on phthalic anhydride and a diamine containing polysiloxane groups (**XXIV**)

Polyamic acids, precursors to PI, based on hexafluoroisopropylidene diphthalic anhydride and ODA were blended with oligomeric polysiloxanes in a mixture of solvents prepared from NMP and tetrahydrofuran (THF). The thermal stability of such blends is good, but relatively lower when compared with that of a block-polyimide siloxane. Gas separation membranes made from such blends show high permeabilities and lower selectivities than those made from block polyimide-siloxane [120].

In another procedure two polyimides, one of which is based on a siloxane containing diamine, were dissolved in organic solvents and blended, followed by casting into films which were deposited onto polyester films. The resulting blend films showed good adhesion to stainless steel and bending resistance and are useful for electronic components [121]. The blends of aromatic polyimides and polyimides containing disiloxane units showed excellent heat resistance with 1% weight loss at 525 °C and very good mechanical properties and adhesivity with adhesive strength to steel laminate, 2.9 and 2.2 kg/mm^2 at 25 °C and 110 °C, respectively [122]. Blend films prepared from polyimides based on PMDA and polysiloxanes exhibited improved mechanical properties and heat resistance: for example a blend film containing a polysiloxane and a polyimide based on 4,4´-bis (3-aminophenoxy) biphenyl showed a tensile strength of 10.5 kg/mm^2, elongation of 52%, and a tensile modulus of 304 kg/mm^2 [123].

11.2.10 Blends of Polyimides with Polyaniline

Conducting polymers such as polyaniline have been widely investigated for both academic and industrial purposes over the last two decades. Their attractive electrical and optical properties have enabled their use in many potential applications. However, their poor mechanical properties and instability limit some industrial applications. Thus, considerable attention has been given to producing polyaniline/polyimide blends and blend films. Sheets useful for electrodes in batteries, electromagnetic shields and others have been prepared from a blend containing polyimide and polyaniline which had a modulus of elasticity of 200 kg/mm^2 and a volume resistivity of 10^7-10^{14} Ω cm. The sheets were prepared by casting the blend of polyamic acid with polyaniline, in NMP solution, followed by heating up to 250-400 °C [124, 125].

Also, blends of polyaniline and polyamic acid were processed into free standing, fully dense films showing improved thermal stability relative to polyaniline and they were used in gas separation membranes. These membranes showed greater gas selectivity and an increase in permeability. They were also used for the separation of liquid mixtures. The selectivity of such membranes for water - acetic acid mixtures was considerably improved compared to that of neat polyaniline and was comparable to that of HCl-doped polyaniline. This means that the polyamic acid acts as a polymeric dopant. Experiments with membranes based on blends of polyaniline with polyimide obtained from BTDA – ODA had permanent flux for acetic acid/water that was intermediate between polyaniline and polyimide membranes and selectivity that was intermediate between doped and non-doped polyaniline [126, 127].

Conducting films were prepared by casting, followed by thermal imidisation of blends prepared from polyimide and polyaniline doped with dodecylbenzene sulfonic acid. Study of electrical and physical properties of these films by IR, X-ray diffraction, thermal analyses, UV-Vis spectroscopy and conductivity measurements showed that the blends exhibited a relatively low percolation threshold of electrical conductivity at 5% polyaniline content and showed higher conductivity than that of pure polyaniline doped with dodecylbenzene sulfonic acid when the polyaniline content was higher than 20%. A lower percolation threshold and a lower compatibility was shown between the two components in the blends than those of polyaniline doped with camphor sulfonic acid blended with polyamic acid. A well-defined layered structure due to the alignment of the long alkyl chain dopant perpendicular to the polyaniline main chain was shown by X-ray spectra [128]. By using camphor sulfonic acid to protonate polyaniline, the counterion enabled the doped polyaniline to be processable as a solution. Thus blends of polyamic acid with polyaniline doped with camphor sulfonic acid were prepared in NMP and were cast into films and thermally imidised. The conductivity of polyaniline doped with camphor sulfonic acid/poliamic acid blend (50% wt polyaniline content) is greater than that of the pure polyaniline sample at room temperature. As the thermal imidisation proceeded, the molecular order of the polymer chain structure was improved in the resulting polyaniline/polyimide film due to the annealing effect of polyaniline chain and the film showed higher conductivity than pure polyaniline doped with camphor sulfonic acid and than polyamic acid/polyaniline doped with camphor sulfonic acid blend film.

Blend films of polyimide/polyaniline doped with camphor sulfonic acid had a good stability of conduction at high temperature [129].

Polyimide/polyaniline (emeraldine salt) blend was used as a hole transport layer in light emitting diodes (LED). The emissive layer was a blend containing oligophenylene vinylene, 1,4-distyryl benzene and 2-(4-biphenylyl)-5-4-(*tert*-butylphenyl) 1,3,4-(oxadiazole). These layers were sandwiched between indium and indium-tin oxide (ITO) electrodes. To increase the electron injection into emissive materials a thin magnesium-layer was inserted between the indium and the polymer blend. The electroluminescence spectra of these LED showed noticeable enhancement of the oscillator strength of the oligophenylene vinylene peak at 2.76 eV. This implies improved quantum efficiency of this blue LED resulting from the excitonic migration from oxadiazole to oligophenylene vinylene. The electroluminescence device with polyimide/polyaniline blend as host polymer displayed increasing device performance, lowering the turning point in intensity-voltage characteristics, compared to that of LED without polyaniline. Under normal illumination conditions, the devices with polyaniline showed visible blue-violet colour at room temperature after applying a bias exceeding 8 V [130].

11.2.11 Other Blends Containing Polyimides

PEI/polyvinylpyrrolidone (PVP) blends were prepared by adding PVP to a PEI solution, in the range of 5-30% wt and membranes were then made by casting such solutions. An effect of the added polymer upon selectivity could not be observed when non-condensable gases were applied. The influence of the PVP upon the membranes separation capability was obvious when a toluene/methane mixture was used. At 5% PVP, the selectivity compared to toluene reaches a value of 1000 that remains unaltered up to 200 °C. Besides the percentage of the PVP in the blend membrane, its molar mass has to be considered. Thus, lower molar masses increase the selectivity factor, but the stability decreases in the long run [131].

PI/polyacrylonitrile (PAN) blends were prepared by mixing a polyamic acid based on PMDA with PAN, followed by casting into films or spinning into fibres. The blend films showed improved tensile strength and thermal stability, and blend fibres have higher resistance to H_2SO_4 [132].

Blends of polyimides, for example based on PMDA and ODA, with elastomers such as nitrile rubber, in which PI is the major component or the minor one have been also reported in the literature [133]. A blend made from nitrile rubber as a major component and a polyether-polyester imide was moulded into a transparent test piece with a resistivity of 2×10^{13} Ω/cm^2 initially and after washing with detergents and having good mechanical strength [134].

11.3 Polybenzimidazole Blends

Poly [2,2'-(*m*-phenylene-5,5'-benzimidazole)] (PBI; **XXV**) is a commercially available material with a very high glass transition temperature (T_g ~430 °C). It has excellent mechanical properties, but is difficult to process into large parts and has high moisture absorption and poor thermo-oxidative stability at temperatures above 260 °C. On the other hand, thermoplastic PI such as PEI offer attractive thermo-oxidative stability and processability, but they lack the mechanical characteristics necessary to be used in applications such as matrices for use at temperatures above 300 °C in structural composites for aerospace use (for example, carbon fibre composites). Blends of PBI with PEI were made in order to combine the good mechanical properties and chemical stability of PBI with the good processability and thermo-oxidative stability of PEI. It was found that PBI was miscible with a wide range of PI including PEI, fluorinated PI and others. Blends made of polybenzimidazoles and PEI or of polybenzimidazoles and fluorinated PI exhibited excellent compressive properties (conveyed by polybenzimidazoles), and good solvent resistance being insoluble in those organic solvents which attack PEI and fluorinated polyimides [2]. PBI/PI blends were prepared by mixing poly[2,2'-(*m*-phenylene)-5,5'-dibenzimidazole] (**XXV**) with a polyamic acid (**XXVI**) synthesised from an aromatic dianhydride like 3,3',4,4'-benzophenontetracarboxylic dianhydride (BTDA) or 3,3',4,4'-diphenylsulfone tetracarboxylic dianhydride (DSDA) with aromatic diamines such as 4,4'-oxydianiline, 4,4'-diaminodiphenylsulfone, 4,4'-methylene dianiline or many others, in DMAc as solvent, followed by curing at a temperature higher than the T_g of the blend (see **Figure 11.17**). The blend systems were all miscible; evidence for miscibility were optically clear films, a synergistic single T_g at all compositions, and frequency shifts of the functional groups intermediate between those of the constituents.

This miscibility results from strong intermolecular interactions due to H-bonding between the NH of PBI and the CO of the PI. The strength of intermolecular interaction between PBI and various polyimides was measured by DSC, thermomechanical analysis (TMA) and FTIR and it was found to be higher in BTDA-based blends than DSDA systems. This difference was seen to arise from electron affinity in the carbonyl spacer of benzophenone in BTDA and the SO_2 spacer of diphenylsulfone in DSDA [135-139]. When the polar linkages CO, SO_2 were replaced by non-polar groups such as the hexafluoroisopropylidene group, the miscibility of PBI/PI blends was significantly reduced [140].

The study of films prepared from such solution blends showed that the solvent (DMAc) retained by the films significantly influenced their thermal stability. Vapour absorption

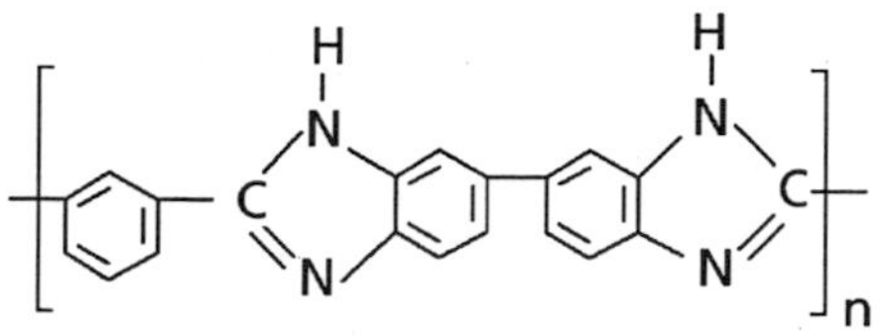

XXV

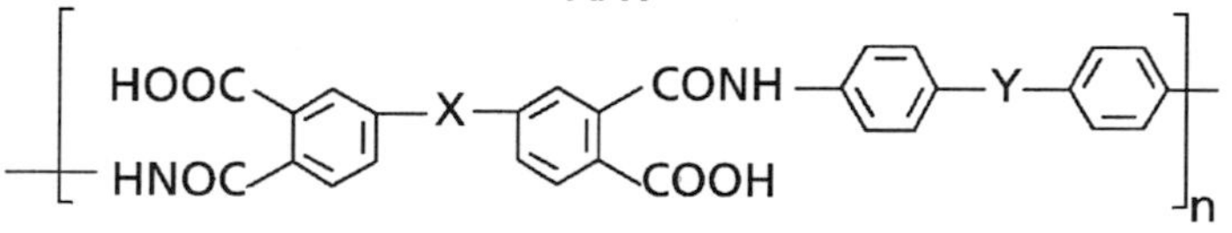

X - CO, SO₂; Y = O, CH₂, SO₂

XXVI

Figure 11.17 Chemical structure of poly[2,2´-*m*-phenylene-5,5´-benzimidazole)] (**XXV**) and of a polyamic acid based on an aromatic dianhydride and an aromatic diamine (**XXVI**)

and FTIR studies showed that solvents enter into strong interaction with the two polymers, especially with PBI component and therefore the thorough removal of solvents is an essential condition for the utilisation of films prepared from these blends [141, 142]. When PBI was blended with a polyimide containing siloxane groups the resulting blends could be melt processed [143].

Other blends were prepared from poly(arylene-ether-benzimidazole) (PAEBI), which is a good adhesion primer for copper, and polyamic acid (PAA) or polyamic acid di-ethylester (PAE) based on biphenyl-tetracarboxylic dianhydride and 4,4´-oxydianiline or *p*-phenylene diamine, precursors of representative PI being widely studied, in NMP and in the imidised state using light scattering technique. The PAA precursor was completely miscible with PAEBI, over the whole range of compositions. This miscibility results from strong interactions between carboxylic acid groups of the precursor PAA and imidazole groups of PAEBI via complex formation. However, during thermal imidisation of the PAA phase separation took place, leading to domains of 0.7-1.4 μm for the imidised blends containing 30%-70% wt polyimide. The other blends, containing less PI, were still optically transparent. In contrast the miscibility of the PAE precursor with PAEBI in NMP solution was limited to a concentration lower than 15%, leading to phase separation during thermal imidisation, producing domains of 0.8-3.0 μm. The immiscibility results from the relatively weak interactions of imidazole groups with both ester and amide linkages in PAE precursor. The differences in miscibility of PAEBI with the PI precursors was reflected in the adhesion strength to a copper joint. Usually the locus of failure is known to be within the PI layer. In the case of PAEBI/PAA blend, a relatively higher peel strength was found compared to the PAEBI/PAE blend, in the adhesion joint. Therefore, higher miscibility resulted in higher adhesion strength [144, 145].

The blends of polybenzimidazoles with PEI were widely investigated in the last decades because the miscibility of the two polymers occurs over a wide composition range, and thus the blends display single T_g which is intermediate. Thus, polybenzimidazole was mixed with a polyetherimide prepared by the reaction of 2,2´-bis (4-(3,4-dicarboxyphenoxy)phenyl) propane – tetracarboxylic dianhydride with *m*-phenylenediamine. The latter is a commercial polyetherimide named ULTEM 1000 (General Electric). The two polymers were dissolved in DMAc at a concentration of 2% each. The solutions were mixed and cast into films. Various mixtures were prepared in which the composition range was from 0 to 1 mass fraction of PI. Each blend composition gave optically clear films. FTIR studies showed that the strength of interaction between the two polymers is influenced by the traces of moisture remaining in the blend after washing and drying the films. There is a relatively weak interaction in the absence of moisture which can be explained by the stiffness of the chains composed of aromatic rings as well as self association of the component: electron-acceptor diimide groups and the electron-donor aromatic units of the polyimide forms charge transfer complexes resulting in aggregation of the solid polymer. Also, in polybenzimidazole the self-association is indicated by its high T_g. The blends exhibited single intermediate T_g. The water absorption by the blend is dependent on the composition. At high polybenzimidazole content the water uptake is diffusion controlled while at higher PI contents the absorption mechanism is changed and becomes a two-step process: it starts with a relatively high diffusion rate followed by a relaxation controlled mechanism. This study shows that the water absorption is increasing by increasing the PI content in the blend and it also showed that the strength of interaction between the polybenzimidazole and PI is lower than that of polybenzimidazole/water. However, in spite of the measured single intermediate T_g and evidence of intermolecular hydrogen bonding, polybenzimidazole and PI do not form thermodynamically homogeneous blends, in other words there is only partial miscibility between the two polymers imÙhe blend [146].

Blends of PBI/PEI in DMAc, at a concentration of 25% wt solids were wet-spun into hollow fibre membranes by using water as external coagulant, while either water or DMAc/water mixture was used as the bore fluid. TMA indicates that the blends are miscible, and the molecular interaction between PBI and PEI are so strong that their miscibility is independent of bore fluid chemistry and bore fluid flow rate. The T_g values of hollow fibre follow the theoretical prediction. Both SEM photographs and gas permeation data indicate that an increase of PBI percentages in the spinning solution resulted in hollow fibre with a tighter morphology, a lesser layer of finger-like voids and a significant lower gas permeability. The tensile strength of wet-spun PBI/PEI blend hollow fibre membranes is independent of PBI concentration while their elongation at break decreases with an increase of PBI concentration. A halo (ring) formation in the cross-section of asymmetric PBI/PEI membranes is a very interesting physical phenomenon of unique pore morphology. Uniform porosity was created in the middle of the hollow fibre cross-section area, which performs as a filter for light transmission. The addition of PBI to the solution of PEI in DMAc not only depresses the macrovoid formation, but also changes the precipitation path: nucleation growth *versus* spinoidal decomposition. The formation of a halo within the membrane is due to the fact that an uniform nucleation growth occurs in the ring region during the early stage of phase separation, because of high solution viscosity and diffusion controlled solvent-exchange process, and then separation grows in the mechanism of spinoidal decomposition from small amplitude composition fluctuations [147, 148].

Polynaphtimidazole (PNI; **XXVII**) which was prepared from 3,3′-diaminobenzidine and 2,6-naphthalene dimethylester was also blended with a polyimide prepared from BTDA and 1,4-phenylendiamine in methanesulfonic acid as solvent (see **Figure 11.18**).

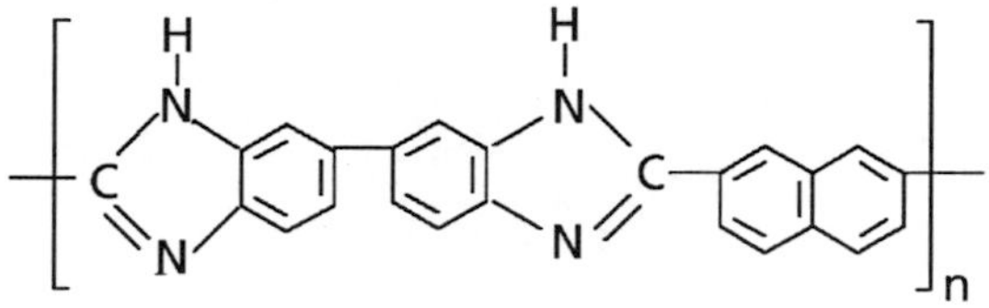

XXVII

Figure 11.18 Chemical structure of PNI based on 3,3′-diaminobenzidene and 2,6-naphthalene dimethylester (**XXVII**)

Blend films have been cast from such solutions and they showed single tan δ relaxation in DMA. The intermolecular interaction due to NH in imidazole and CO group in the imide ring played an important role in the miscibility of these two polymers. A blend film PNI/PI having the composition 80/20% wt has the highest tensile strength and modulus, in a series of blends having various compositions [149]. Blends were also prepared from polynaphtimidazole, obtained by the reaction of tetraaminodiphenylether with isophthaloyl bis(naphthalic anhydride), with polyetherketone [150].

PBI was used in blends with polyvinylacetate-vinyl alcohol copolymers. DSC, DMA, FTIR spectroscopy, SEM and optical measurement show that as the vinyl alcohol content of the copolymer component increases, the level of dispersion of the two components of the blends improves. The blends are not miscible at the molecular level, but the levels of dispersion in blends with polyvinylacetate/poly vinyl alcohol compositions of 56 and

73 mol% vinyl alcohol were good enough to allow the formation of optically clear films. Enhanced storage modulus for certain blends, which were larger than that of PBI alone, was also obtained. Evidence for significant hydrogen bonding between the blend component was found from FTIR measurements, but none of the blends exhibited a single T_g. Consequently it is estimated that the phase domain sizes for the well-dispersed blends lie in the range of 20-500 nm, which are larger than expected from single phase miscible blends [151].

11.4 Polyquinoxaline Blends

Blends of phenyl-substituted polyquinoxaline (PPQ; **XXVIII**) with other polymers were prepared in order to control specific properties (see **Figure 11.19**). Thus, in blends of PPQ with polycaprolactame it was shown that PPQ induced the formation of crystal grains distributed evenly over Nylon spherulites and strengthened the modified Nylon. The blend exhibited higher T_g than Nylon itself and showed typical reinforcing effects on mechanical properties [152].

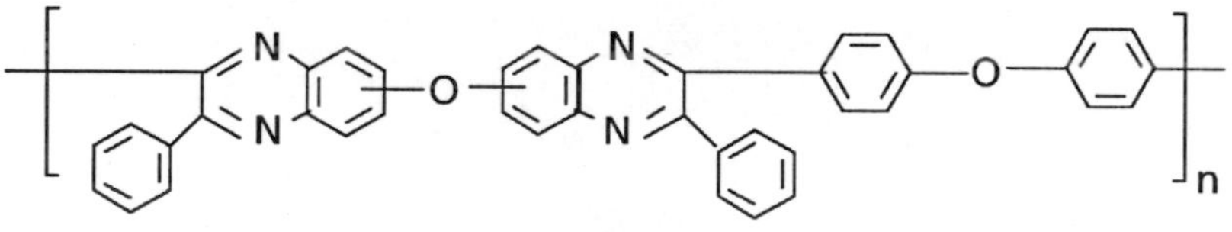

XXVIII

Figure 11.19 Chemical structure of a polyphenylquinoxaline (**XXVIII**)

Attempts were made to prepare blends of PPQ with PEI, but a compatibility study by a light scattering technique illustrates a significant self-agglomeration tendency of the PPQ resulting in the incompatibility of these polymers. The same self-agglomeration phenomenon was found in blends of PPQ with polymethylmethacrylate containing either electron-donor or electron-acceptor groups. Optical light- and atom force-microscopy (AFM) investigations reveal a better dispersion in these blends, but the self-agglomeration tendency is so intense that even the use of electron-donor components does not produce compatible blends [153].

Blends of substituted PPQ as n-type polymers and substituted polycarbazoles as p-type polymers have been prepared for use in light emitting electrochemical cells. These polymers are blue-green and blue emission polymer, respectively, both with high photoluminescent quantum efficiency. However, the photoluminescence of the polymer blend is completely quenched due to the charge transfer between the two polymers. A new and faint orange-yellow photoluminescence emission which has photonic energy consistent with the energy difference of the π band of polycarbazole and the π* band of polyquinoxaline has been observed. Light emitting electrochemical cells fabricated from these polymer blends show strong current injection and bright electroluminescence and new emission colour, which is due to the interpolymer radiative recombination of the electrons of the n-type polymers and holes from the p-type polymers. Such an independent p-doping of polycarbazole and n-doping of polyquinoxaline in the blend and interpolymer radiative recombination provide an interesting way of generation of new emission colour in the light emitting electrochemical cells systems [154].

Other heterocyclic conductive polymers, such as polypyrrole, have also been used in blends with insulating host polymers such as polycarbonate, polyamide. Several spectroscopic and thermal analyses of free standing blend films show an interaction between the host matrices and conducting polymer [155].

11.5 Polyoxadiazole Blends

Polymer blends in which a minor phase of rigid-rod macromolecules is dispersed in a more flexible matrix have received considerable scientific and practical interest due to their outstanding properties. Poly(ε-caprolactame), a conventional polyamide which is soluble in sulfuric acid without undergoing extensive degradation, was used to prepare blends with poly(*p*-phenylene-1,3,4-oxadiazole) (POD; **XXIX**) [156] (see **Figure 11.10**). The polymers were dissolved in sulfuric acid following a rigorous procedure and the resulting blend was coagulated in water and used for various studies. Wide-angle X-ray scattering (WAXS) and solid-state ^{13}C NMR analysis showed that the crystallinity of polyamide was strongly influenced by the presence of POD and was dependent on the blend composition. Below 30% polyamide content the crystallinity disappeared completely. Blends containing above 50% polyamide showed a clear melting process in the range of 210-220 °C which allows for easy processing. A blend containing 10% POD was compression moulded at 260 °C and the resulting sheets exhibited 30% increase of the modulus of elasticity and no change of the stress at rupture, compared to the polyamide itself. Transparent films were obtained by coagulation from polymer blends containing >30% polyamide. Thermogravimetric analysis (TGA) revealed interconnected two-phase structures, which are characteristic of spinoidal decomposition.

Poly(*p*-phenylene-1,3,4-oxadiazole) has been also used in blends with aromatic polyamides such as poly(*p*-phenylene-terephthalamide) [157, 158]. It was found that the modulus of elasticity of films and fibres drawn from these blends increased strongly when small amounts of aromatic polyamide were added to POD (up to 10%). At higher content of

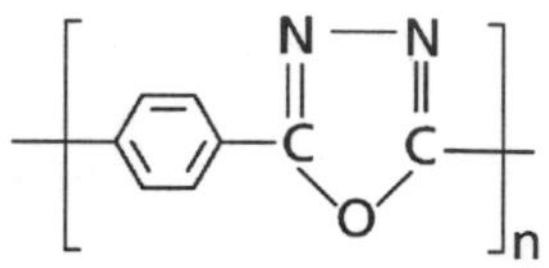

XXIX

Figure 11.20 Chemical structure of poly(*p*-phenylene-1,3,4-oxadiazole) (**XXIX**)

polyamide, a phase separation appeared and the POD was spherically dispersed in the polyamide, as shown by SEM. Also, blends of POD with polyoxymethylenes were prepared by mixing the two powdered components, followed by melting, cooling, extruding or moulding, to give products with good mechanical properties [159].

Acknowledgements

It is a pleasure to acknowledge the financial support provided by the National Agency of Science and Innovation (ANSTI) - Romania through the Contract No. 494/2000 (Program 'Orizont-2000') and the Grant No. 5052GR/99.

References

1.   J.M. Aducci in *Polyimides: Synthesis, Characterisation and Applications*, Volume 2, Ed., K.L. Mittal, Plenum Press, New York, NY, USA, 1984, 1023.

2.   M. Jaffe, P. Chen, E.W. Choe, T.S. Chung and S. Makhija, *Advanced Polymer Science*, 1994, **117**, 297.

3.   M.T. De Meuse, *Polymers for Advanced Technologies,* 1995, **6**, 1, 76.

4.   H. Tang, L. Dong and Z. Feng, *Cailiao Yanjiu Xuebao*, 1996, **10**, 449.

5.   J.W. Verbiki Jr., in *Concise Encyclopedia of Polymer Science and Engineering*, Ed., J.I. Kroschwitz, Wiley, New York, NY, USA, 1990, 826.

6.   C.E. Sroog, *Progress in Polymer Science,* 1991, **16**, 2, 561.

7.   P.E. Cassidy, T.M. Aminabhavi and V.S. Reddy *in Kirk-Othmer Encyclopedia of Chemical Technology*, Ed., J.I. Kroschwitz, Wiley, New York, NY, USA, 1994, Volume 12, 1045.

8. E. Hamciuc, M. Bruma and C.I. Simionescu, *Revue Roumaine de Chimie*, 1993, **38**, 11, 1311.

9. F.W. Mercer, M.T. McKenzie, M. Bruma and B. Schulz, *Polymer International*, 1994, **33**, 4, 399.

10. O. Petreus, F. Popescu, L. Rosescu and M. Bruma, *Revue Roumaine de Chimie*, 1994, **39**, 8, 971.

11. M. Bruma, I. Sava, F. Mercer, I. Negulescu, W. Daly, J. Fitch and P. Cassidy, *High Performance Polymers*, 1995, **7**, 4, 411.

12. C. Hamciuc, I. Sava, E. Hamciuc, M. Bruma, F. Mercer and N.M. Belomoina, *Revue Roumaine de Chimie*, 1996, **41**, 9-10, 815.

13. M. Bruma, J. Fitch and P. Cassidy in *Polymeric Materials Encyclopedia*, Ed., J.C. Salamone, CRC Press, Boca Raton, FL, USA, 1996, 4, 2456.

14. E. Hamciuc, C. Hamciuc, A. Airinei and M. Bruma, *Angewandte Makromolekulare Chemie*, 1997, **245**, 105.

15. C. Hamciuc, E. Hamciuc and M. Bruma, *Materiale Plastice (Bucharest)*, 1998, **32**, 2, 75.

16. C. Hamciuc, M. Bruma, F. Mercer, T. Kopnick and B. Schulz, *Macromolecular Materials Engineering*, 2000, **276/277**, 38.

17. C. Hamciuc, E. Hamciuc, I. Sava, I. Diaconu and M. Bruma, *High Performance Polymers*, 2000, **12**, 2, 265.

18. C. Hamciuc, M. Bruma, M. Szesztay and I. Ronova, *Journal of Macromolecular Science A*, 2000, **37**, 11, 1407.

19. M. Bruma, B. Schulz, T. Kopnick and J. Robison, *High Performance Polymers*, 2000, **12**, 3, 429.

20. M. Bruma and B. Schulz, *Journal of Macromolecular Science - Polymer Reviews*, 2001, **C41**, 1-2, 1.

21. M. Hasegawa, J.I. Ishii and Y. Shindo, *Journal of Polymer Science, Part B: Polymer Physics*, 1998, **36**, 5, 827.

22. M. Hasegawa, K. Okuda, M. Horimoto, Y. Shindo, R. Yokota and M. Kochi, *Macromolecules*, 1997, **30**, 19, 5745.

23. M. Hasegawa, J.I. Ishii and Y. Shindo, *Macromolecules*, 1999, **32**, 19, 6111.

24. S. Rojstaczer, M. Ree, D.Y. Yoon and W. Volksen, *Journal of Polymer Science, Part B: Polymer Physics*, 1992, **30**, 2, 133.

25. D.H. Lee, S.Y. Koo and S.T. Kim, *Polymer* (Korea), 1998, **22**, 1, 6.

26. R.F. Sutton and D.E . Coverdell, inventors; E.I. Du Pont de Nemours, assignee; US5,939,498, 1999.

27. A.P. Krasnov, I.A. Gribova, V.N. Adericha and Ya.V. Genin, *Trenie Iznos*, 1999, **20**, 2, 221.

28. H. Kido, M. Yoshimura, Y. Yoshida, K. Yanagihara, H. Oikawa, S. Tamai and T. Koba, inventors; Mitsui Chemical Inc., assignee; EP Patent 1,013,714, 2000.

29. N. Kido, S. Yoshimura, I. Yoshida, K. Yamagihara, H. Oikawa, S. Tamai and T. Koba, inventors; Mitsui Chemical Inc., assignee; JP Patent 191,907, 2000.

30. O. Oka and T. Sato, inventors; Tomoegawa Paper Company Ltd., assignee; JP Patent 11 100,565,1999.

31. I. Yoshida, K. Yanagihara and M. Yoshimura, inventors; Mitsui Chemical Inc., assignee; JP Patent 11 114,744,1999.

32. M. Bruma, J. Fitch and P. Cassidy, *Journal of Macromolecular Science C,* 1996, **36**, 1, 119.

33. P. Cassidy and J. Fitch in *Modern Fluoropolymers*, Ed., J. Scheirs, Wiley, New York, NY, USA, 1997, 173.

34. T-S. Chung, Proceedings of Antec '97, Toronto, Canada, 1997, Volume 2, 2658.

35. T-S. Chung, P. Foley and M. Jaffe, *Polymers for Advanced Technologies*, 1997, **8**, 9, 537.

36. I.W. Serfaty in *Polyimides: Synthesis, Characterisation and Applications*, Ed., K.L. Mittal, Plenum Press, New York, NY, USA, 1984, 1, 149.

37. J.A. Campbell, A.A. Goodwin, F.W. Mercer and V.N. Reddy, *High Performance Polymers*, 1997, **9**, 3, 263.

38. W. Huang, J. Xu and M. Ding, *Polymer,* 1997, 38, 16, 4261.

39. Y. Kim, M. Guo, L. Zhu, D. Kim, F.W. Harris and S.Z.D. Cheng, *Chinese Journal of Polymer Science,* 1999, **17,** 2, 171.

40. A.A. Kuznetsov, I.D. Egorov, G.K. Semenova, V.I. Berendyav, E.I. Akhmetieva and B.V. Kotov, *Vysokomolekulyarnye Soedineniya Seriya A and B,* 2000, **42,** 4, 683.

41. T. Okada, inventor; Nok Corp Japan, assignee; JP Patent 08 57,277, 1996.

42. G-D. Bidulescu, L. Tarko, I. Sitaru, F. Deak and T. Voinescu, inventors; Institutul de Cercetare Stiintifica si Inginerie Tehnologica Pentru Electrotehnica, Bucuresti, Romania, assignee; RO Patent 105,703, 1992.

43. D.A. Ivanov, P.D.M. Lipnik and A.M. Jonas, *Journal of Polymer Science, Part B: Polymer Physics,* 1997, **35,** 15, 2565.

44. X. Kong, H. Tang, L. Dong, F. Teng and Z. Feng, *Journal of Polymer Science, Part B: Polymer Physics,* 1998, **36,** 13, 2267.

45. D.A. Ivanov, and A.M. Jonas, *Journal of Polymer Science, Part B: Polymer Physics,* 1998, **36,** 5, 919.

46. M.J. Jenkins, *Polymer,* 2000, **41,** 18, 6803.

47. J. Hay and M. Jenkins, *Macromolecular Symposia,* 1999, **143,** 121.

48. M. Frigione, C. Naddeon and D. Acierno, *Polymer Engineering and Science,* 1996, **36,** 16, 2119.

49. B.H. Stuart and B.J. Briscoe, *High Performance Polymers,* 1996, **8,** 2, 275.

50. A.A. Goodwin and G.P. Simone, *Polymer,* 1997, **38,** 10, 2363.

51. A. Arzak, J.I. Equiazabal and J. Nazabal, *Journal of Macromolecular Science B,* 1997, **36,** 2, 233.

52. J.F. Bristow and D.S. Kalika, *Polymer,* 1997, **38,** 2, 287.

53. H.L. Chen and R.S. Porter, *Journal of Polymer Research,* 1999, **6,** 1, 21.

54. C.H. Lee, T. Okada, H. Saito and T. Inoue, *Polymer,* 1997, **38,** 1, 31.

55. J. Caspar, B. Dunges, R.G. Kirste, T. Heitz and A. Wiedenmann, *Physica B: Condensed Matter,* 1997, **234-236,** 240.

56. G. Giorgiev, P.S. Dai, E. Oyebode, P. Cebe and C. Malcom in *Applications of Synchrotron Radiation Techniques to Materials Science V*, Eds., S. R. Stock, D. L. Perry, and S. M. Mini, Materials Research Society Symposium Proceedings, No.590, 2000, 137.

57. R. Gensler, C.J.G. Plummer, H-H. Kausch and K. Münstedt, *Journal of Materials Science*, 1997, **32**, 11, 3037.

58. D.K. Brandom, J.P. Desouza, D.G. Baird and G.L. Wilkes, *Journal of Applied Polymer Science*, 1997, **66**, 8, 1543.

59. Y.S.Ni, J.I.Jin, S.J. Jeon and B.W. Jo, *Korean Polymer Journal*, 1998, **6**, 5, 349.

60. H.S. Kwon, Y.S. Chun, H.C. Jung, S.B. Kim and W.N. Kim, *Pollimo*, 1997, **21**, 5, 786.

61. S.D. Mikhailenko, S.M.J. Zaidi and S. Kaliaguine, *Journal of Polymer Science, Part B: Polymer Physics*, 2000, **38**, 10, 1386.

62. S. Demura and K. Haraguchi, inventors; Dainippon Inc and Chemicals Inc., assignee; JP Patent 09,143,282, 1997.

63. K. Goto and T. Aküke, inventors; Japan Synthetic Rubber Co., Ltd., JP Patent 09 77,972, 1997.

64. M. Furukawa, Y. Yamada and Y. Echigo, inventors; Unitika Ltd., assignee; JP Patent 10,292,112, 1998.

65. S. Niyogi, S. Maiti and B. Adhikari, *Polymer Degradation and Stability*, 2000, **68**, 3, 459.

66. A. Etxeberria, S. Guezala, J.J. Iruin, J.G. de la Campa and J. De Abajo, *Journal of Applied Polymer Science*, 1998, **68**, 13, 2141.

67. J.H. Lee, S.G. Lee, K.Y. Choi and J. Liu, *Polymer Journal*, (Tokyo), 1998, **30**, 7, 531.

68. M. Weber, H. Fisch, G. Pipper and A. Gottschalk, inventors; BASF AG, assignee; EP737,719, 1996.

69. W. Cui, J. Kerres and G. Eigenberger, *Separation and Purification Technology*, 1998, **14**, 1-3, 145.

70. J.M. Martinez, J.I. Equiazabal and J. Nazabal, *Journal of Applied Polymer Science,* 1996, **62**, 2, 385.

71. J. Choi and M. Cakmak, Proceedings of ANTEC '99, New York, NY, USA, 1999, Volume 2, 1656.

72. Y. Jin, M.J. Lindway, K. Xi and R.L. Utley, inventors; General Electric Company, assignee; WO Patent 12,604, 2000.

73. H.L. Chen and M.S. Hsiao, *Macromolecules,* 1998, **31**, 19, 6579.

74. H.L.Chen, J.C.Hwang and C.C. Chen, *Polymer,* 1996, **37**, 24, 5461.

75. S. Bastida, J.I. Equiazabal and J. Nazabal, *Polymer,* 1996, **37**, 12, 2317.

76. T. Matsuki, M. Tsukioka and J. Sadanobu, inventors; Teijin Ltd., assignee; WO Patent 98 23,682, 1998.

77. M. Kaku, inventor; E.I. Du Pont De Nemours and Co, USA, assignee; WO Patent 98 22,533, 1998.

78. T. Matsuki and J. Sadanobu, inventors; Teijin Ltd., Japan, asssignee; JP Patent 10 195,293, 1998.

79. S. Matsuura and S. Krishnamachari, inventors; Hitachi Chemicals Co. Ltd., assignee; JP Patent 10 219,109, 1998.

80. T. Higashioji, T. Tsunekawa and K. Amishima, inventors; Toray Industries Inc., assignee; JP 11 01,568, 1999.

81. Y. Kimata, R. Nishioka, H. Kanai and N. Furukawa, inventors; Nippon Steel Corporation, assignee; JP Patent 08,299,899, 1996.

82. H.T. Kim and J.K. Park, *Japanese Journal of Applied Physics Part 1,* 1999, **39**, 1A, 201.

83. H.S. Chen, J.C. Hwang, C.C. Chen, R.C. Wang, D.M. Fang and M.J. Tsai, *Polymer,* 1997, **38**, 11, 2747.

84. H.G. Kim and R.E. Robertson, *Han'guk Somyu Conghakhoechi,* 1999, **36**, 1, 17.

85. J. Jang and K. Sim, *Polymer Testing,* 1998, **17**, 7, 507.

86. C. Uriarte, J. Alfageme and J.J. Iruin, *European Polymer Journal,* 1998, **34**, 10, 1405.

87. J.C. Kim, M. Cakmak and X. Zhou, *Polymer,* 1998, **39**, 18, 4225.

88. M. Cakmak, Y.D. Wang and M. Simhambhatla, *Polymer Engineering and Science,* 1990, **30**, 6, 721.

89. Y. Ulcer and M. Cakmak, *Journal of Applied Polymer Science,* 1996, **62**, 10, 1661.

90. C.E. Scott, inventor; Eastman Chemical Co., assignee; US5,648,433, 1997.

91. T. Ishiwatari and J. Sadanobu, inventors; Teijin Ltd. Japan, assignee; JP Patent 11 343,334, 1999.

92. S.N. Yau and E.M. Woo, *Macromolecular Rapid Communications,* 1996, **17**, 9, 615.

93. J. Hanehi and N.S. Eiss, *Tribology Transactions,* 1997, **40**, 1, 102.

94. E. Bicakci, X. Zhou and M. Cakmak, Proceedings of ANTEC '97, Toronto, Canada, 1997, Volume 2, 1593.

95. M. Miura, T. Yamada, T. Mioshy and H. Ban, inventors; Mitsubishi Gas Chemical Co. Ltd., assignee; JP Patent 10 298,428, 1998.

96. N. Tanaka, inventor; Nippon Pillar Packing, Japan, assignee; JP Patent 08,319,391, 1996.

97. D.E. George, A. Yokoyama and S. Nakagawa, inventors; E.I. du Pont de Nemours and Co., USA, assignee; JP Patent 47,493, 2000.

98. H. Li, D. Xu and H. Cheng, *Gaofenzi Cailiao Kexue Yu Gongcheng,* 1999, **15**, 3, 81.

99. H. Furukawa, T. Koba, A. Morita and M. Yamaki, inventors; Mitsui Toatsu Chemcals, Inc. Japan, assignee; JP Patent 09 87,517, 1999.

100. J. Herber, inventor; Weilburger Lackfabrik Jacob Grebe Gmbh, assignee; DE Patent 19,902,078, 2000.

101. J.S. Bloom, inventor; E.I. DuPont de Nemours and Co., US5,700,863, 1997.

102. H. Akimoto, K. Asami and Y. Nakajima, inventors; Toshiba Chemical Corp., assignee; JP Patent 226,495, 2000.

103. T. Nakamaru and Y. Yamamoto, inventors; Oiles Industries Co. Ltd., Japan, assignee; JP Patent 74,136, 2000.

104. J. Huang and Q. Guo, *Journal of Macromolecular Science B*, 1997, **36**, 3, 423.

105. G.C. Kapantaidakis, S.P. Kaldis, G.P. Sakellaropoulos, E. Chira, B. Loppinet and G. Floudas, *Journal of Polymer Science, Part B: Polymer Physics*, 1999, **37**, 19, 2788.

106. A. Mochizuki, K. Yamada, M. Ueda and R. Yokota, *Polymer Journal (Japan)*, 1997, **29**, 4, 339.

107. D. Wang, K. Li and W.K.Teo, *Membrane Formation and Modification*, Eds., I. Pinnau and B. D. Freeman, ACS Symposium Series No.744, 2000, 96.

108. O.M. Ekiner, inventor; L'air Liquide S.A., France, assignee; US5,608,014, 1997.

109. X. Peng, Y. Wu, J. Liu and J. Zheng, *Gaofenzi Cailiao Kexue Yu Gongcheng*, 1998, **14**, 6, 79.

110. B. Shi, Y. Wu, J. Liu, Q. Kong and X. Peng, *Mo Kexue Yu Jishu*, 1999, **19**, 6, 48.

111. S. Hiruta, M. Yoshimura, T. Kido, T. Sato and H. Taichi, inventors; Mitsui Chemicals Inc., Japan, assignee; JP 11 80,545, 1999.

112. Y.S. Chun, H.S. Lee and W.N. Kim, *Polymer Engineering and Science*, 1996, **36**, 22, 2694.

113. S.M. Cooper, D. Nazareth and R.A. Greenberg, inventors; General Electric Co, assignee; EP Patent 711,810, 1996.

114. M.J. El-Hibri, inventor; General Electric Co, assignee; US5,905,120, 1999.

115. R. Puyenbroek, inventor; General Electric Co, assignee; EP Patent 926,204, 1999.

116. B. Masiulanis and R. Zielinski, *Journal of Applied Polymer Science*, 1985, **30**, 2731.

117. J.A. Cella in *Polyimides Fundamentals and Applications*, Eds., M.K. Ghosh and K.L. Mittal, Marcel Dekker, New York, NY, USA, 1986, 343.

118. M. Zuo and T. Takeichi, *Journal of Polymer Science, Part A: Polymer Chemistry,* 1997, **35**, 17, 3745.

119. T. Yoshimura, A. Shibuya, Y. Sakata, W. Yamashita, H. Oikawa and M. Ota, inventors; Mitsui Petrochemicals Inc., assignee; JP Patent 09,328,546, 1997.

120. Y.B. Lee, H.B. Park, J.K. Shim and Y.M. Lee, *Journal of Applied Polymer Science,* 1999, **74**, 4, 965.

121. S. Okaaki and K. Takahama, inventors; Sumitumo Bakelite Company Ltd., Japan, assignee; JP Patent 10 231,425, 1998.

122. A. Shibue, T. Yoshimura, Y. Sakata, W. Yamashita, H. Oikawa and M. Ota, inventors; Mitsui Toatsu Chemicals Inc., assignee; JP 1017,769, 1998.

123. I. Ookawa, M. Tamai and T. Yamaguchi, inventors; Mitsui Toatsu Chemicals, assignee; JP Patent 09 12,882, 1997.

124. G. Min, *Synthetic Metals,* 1999, **102**, 1-3, 1163.

125. Y. Uetani, M. Nakamura and M. Abe, inventors; Nitto Denko Corp Japan, assignee; JP Patent 08,259,709, 1996.

126. T.M. Su, I.J. Ball, J.A. Conklin, S.C. Huang, R. K. Larson S.L. Nguyen, B.M. Lew and R.B. Kaner, *Synthetic Metals,* 1997, **84**, 1-3, 801.

127. I.J. Ball, S-C. Huang and R.B. Kaner, Proceedings of Antec '98, Atlanta, GA, USA, 1998, Volume 2, 1301.

128. M.G. Han and S.S. Im, *Journal of Applied Polymer Science,* 1999, **71**, 13, 2169.

129. M.G. Han and S.S. Im, *Journal of Applied Polymer Science,* 1998, **67**, 11, 1863.

130. J.G. Lee, B.C. Park, H.S. Woo, Y. Kim, C.S. Ha, C.M. Lee, K. Jeong, J.H. Ha and Y.R. Kim, *Solid State Communications,* 1997, **102**, 12, 895.

131. C. Hying and E. Stande, *Journal of Membrane Science,* 1998, **144**, 1-2, 251.

132. Yu.N. Sazanov and A.N. Gribanov, *Zhurnal Prikladnoi Khimii* (St. Petersburg), 1999, **72**, 11, 1896.

133. J.F. McNamara, inventor; E.I. Du Pont De Nemours and Company, USA, assignee; WO Patent 99 00,453, 1999.

134. M. Someta, M. Kaneko, M. Kawada, K. Kuroda, H. Narusawa and S. Asai, inventors; Mitsui Chemicals Inc, assignee; JP 11 60,880, 1999.

135. T.K. Ahn, M. Kim and S. Khoe, ACS *Polymeric Materials Science and Engimneering*, 1996, **75**, 1, 22.

136. T.K. Ahn and M. Kim and, *Macromolecules*, 1997, **30**, 11, 3369.

137. S. Lee, S.H. Chun, J.G. Lee, H. Lee and S. Choe, *Polymer Preprints*, 1997, **38**, 1, 284.

138. T.K. Ahn, *Kongop Hwahak*, 1998, **9**, 2, 185.

139. S. Lee, J.G. Lee, H. Lee and S. Choe, *Macromolecules*, 1999, **32**, 18, 5961.

140. Y.J. Kim, *Han'guak Somyu Konghakhoechi*, 1996, **33**, 5, 420.

141. E. Fekete, Z. Peredy, E. Foldes, F.E. Karasz and B. Pukanszky, *Polymer Bulletin*, 1997, **39**, 1, 93.

142. Z. Peredy, *Muanyag es Gumi*, 1997, **34**, 6, 191.

143. W.J. MacKnight, S.W. Kantor and H.Zhu, Proceedings of Antec '96, Indianapolis, IN, USA, 1996, Volume 2, 1594.

144. J.Y.M. Ree, T.J. Shin, X. Wang, W. Cai, D. Zhou and K.W. Lee, *Journal of Polymer Science, Part B: Polymer Physics*, 1999, **37**, 19, 2806.

145. J.Y.M. Ree, T.J. Shin, X. Wang, W. Cai, D. Zhou and K.W. Lee, *Polymer*, 2000, **41**, 1, 169.

146. E. Földes, E. Fekete, F.E. Karasz and B. Pukansky, *Polymer*, 2000, **41**, 4, 975.

147. T.S.Chung and Z.L. Xu, *Journal of Membrane Science*, 1998, **147**, 1, 35.

148. T.S.Chung, Z.L. Xu and C.H.A.Huan, *Journal of Polymer Science, Part B: Polymer Physics*, 1999, **37**, 14, 1575.

149. Y.H. Kao and L.W. Chen, *Materials Chemistry and Physics*, 1997, **47**, 1, 51.

150. A.L. Rusanov, A.P. Kiselev, L.D. Frolova, E.G. Bulycheva, Z.B. Shifrina and I.V. Yaroshenko, *Plasticheskie Massy*, 1995, **5**, 1, 38.

151. G.V. Adams and J.M.G. Cowie, *Polymer*, 1998, **40**, 8, 1993.

152. G. Yang and F. Lu, *Chinese Journal of Polymer Science*, 1992, **8**, 1, 87.

153. K. Andress, C. Hackmann, H. A. Schneider, A.L. Rusanov and N.M. Belomoina, *New Polymeric Materials*, 1994, **4**, 1, 97.

154. Y. Yang and Q. Pei, *Applied Physics Letters*, 1997, **70**, 15, 1926.

155. L. Toppare, *Turkish Journal of Chemistry*, 1997, **21**, 1, 30.

156. B. Immirzi, M. Malinconico, E. Martuscelli, M. Paci and A. Segre, *Makromolekulare Chemie, Rapid Communications*, 1992, **13**, 2, 231.

157. E. Schulz, E. Leibnitz, A. Bauer and R. Schmolke, *Angewandte Makromolekulare Chemie*, 1992, **195**, 79.

158. C. Kummerlöwe, H.W. Kammer, M. Malinconico and E. Martuscelli, *Polymer*, 1993, **34**, 7, 1677.

159. H-J. Sterzel and M. Meyer, inventors; BASF AG, assignee; EP Patent 517025A1, 1992.

12 Blends and Interpenetrating Networks Based on Polyurethanes

Aurelian Stanciu and Leonard Ignat

12.1 Introduction

It is well known that polyurethane (PU) chemistry opened the way to a new class of high performance materials such as coatings, adhesives, elastomers, fibres, foams, etc. [1]. Segmented polyurethanes are often used in multicomponent polymeric systems, such as polymer compounds, blends and interpenetrating polymer networks (IPN) [2-4]. These multicomponent systems provide the possibility of combining properties of different polymeric materials.

Based on a simple polyaddition reaction, the PU proved to be very versatile polymers, with tailor-made properties [5]. The advantages of these compounds are related to their high hardness for a given modulus, high abrasion and chemical resistance, excellent mechanical and elastic properties, blood and tissue compatibility, and also some other specific properties [6].

Incorporating a PU elastomer in a blend [7-12] or IPN [13, 14] increases the toughness and impact resistance. The resulting polymer blend is immiscible, but compatible because of the intermolecular hydrogen bonds [15].

Recently, IPN based on PU have gained widespread acceptance in industrial applications and new IPN showing better performance are emerging day-by-day [14]. The PU added to a rubber, plastic, resin, fibre or other polymeric conventional compounds, as a minor replacement, generally improve physical properties, abrasion resistance, and oil and solvent resistance. In the same manner, replacing a part of a urethane rubber with a conventional polymer can give improved compression set and resistance to moisture and glycols, and lower compound cost. For some particular chemicals, these materials offer even better protection with enhanced vapour water transmission than other currently used materials [16].

12.2 Polyurethane Blends

Blends of rubbers are fairly common in rubber compounding because of two factors: the ability to improve the properties of one rubber by the addition of a second, and the ability to lower the cost of the more expensive rubber by blending with a less expensive rubber, and hopefully keeping most of the benefits of the expensive one [17].

From this point of view, urethane compounds can generally improve the properties of polymers it is added to and when small portions of other polymers are added to it, a lower cost compound with many of the good properties of urethane can be the result. Optimising the cure system and polymer blend ratio, in most cases can result in compounds that perform in the most cost effective manner [16].

To overcome the immiscibility problem, compatibilisers are used. The effect of compatibiliser or emulsifiers on the structure of immiscible polyurethane blends has been investigated for many years [10, 11, 17-22]. An efficient compatibiliser should reduce the interfacial energy between phases leading to a finer dispersion, provide good stability against large-scale segregation and result in an improved interfacial adhesion.

12.2.1 Segmented Polyurethane Elastomers

PU are block and graft copolymers with many advantageous properties, classified as 'thermoplastic elastomers', which have lead to their use in a wide range of applications. The PU elastomer morphology may be described as resulting from the occurrence of phase-segregated regions [6]: a region with low glass transition temperature (T_g), where the dominant chemical species do not undergo considerable hydrogen-bonding; a region with high T_g, where strong hydrogen-bonding is the rule and a region of interface, populated with both species. The mechanical properties, which are the consequence of this morphology, make these block copolymers behave as thermoplastic elastomers.

Their basic chemical structure is the urethane group (R-NHCOO-R´) and they are synthesised from the reactions of polyether-based or polyester-based diols and diisocyanates, followed by the introduction of chain extenders to form the macromolecules [6, 23]. One appropriate mechanism for elastomeric PU synthesis has been proposed by Pappas [24]. In the first step, dibutyltin dilaurate (Bu_2SnL_2), a catalyst of polymerisation, reacts with the alcohol and then with the isocyanate, which results in an intermediate (I). In the second step an intramolecular transfer occurs, the intermediate (I) becomes very unstable and reacts with another alcohol molecule giving the urethane bond and active species that can reinitiate the process of polymerisation (**Scheme 12.1**).

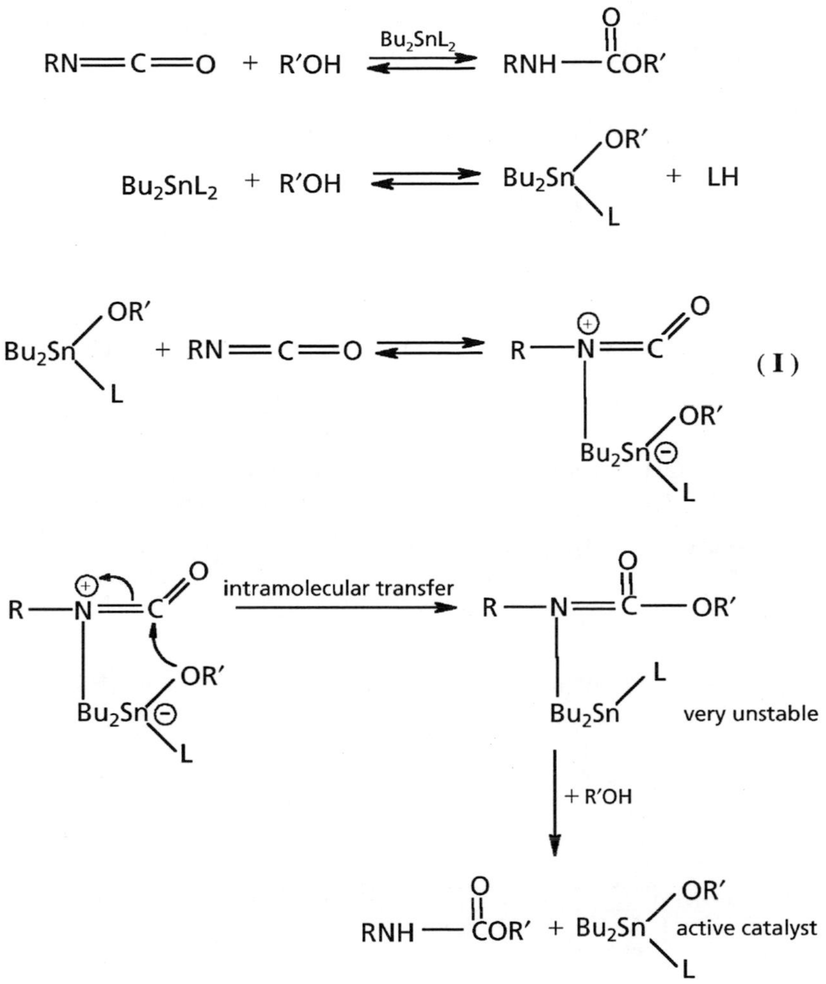

$$RN{=}C{=}O \ + \ R'OH \ \underset{}{\overset{Bu_2SnL_2}{\rightleftharpoons}} \ RNH{-}\overset{\overset{\displaystyle O}{\|}}{C}OR'$$

$$Bu_2SnL_2 \ + \ R'OH \ \rightleftharpoons \ Bu_2Sn{\overset{\displaystyle OR'}{\underset{\displaystyle L}{<}}} \ + \ LH$$

$$Bu_2Sn{\overset{\displaystyle OR'}{\underset{\displaystyle L}{<}}} \ + \ RN{=}C{=}O \ \rightleftharpoons \ R{-}\overset{\oplus}{N}{=}C{\overset{\displaystyle O}{<}} \qquad (\textbf{I})$$

intramolecular transfer

very unstable

+ R'OH

$$RNH{-}\overset{\overset{\displaystyle O}{\|}}{C}OR' \ + \ Bu_2Sn{\overset{\displaystyle OR'}{\underset{\displaystyle L}{<}}} \quad \text{active catalyst}$$

Scheme 12.1 Polyurethane elastomer synthesis mechanism

Polyurethane-based multicomponent systems, which contain a thermoplastic hard component to improve their mechanical properties, are made by two basic procedures: the prepolymer technique (**Scheme 12.2**) and/or the 'one shot' method (**Scheme 12.3**).

Due to the different polarity and chemical nature of both blocks they separate into two phases designated as 'soft' or 'hard' phases (**Figure 12.1**).

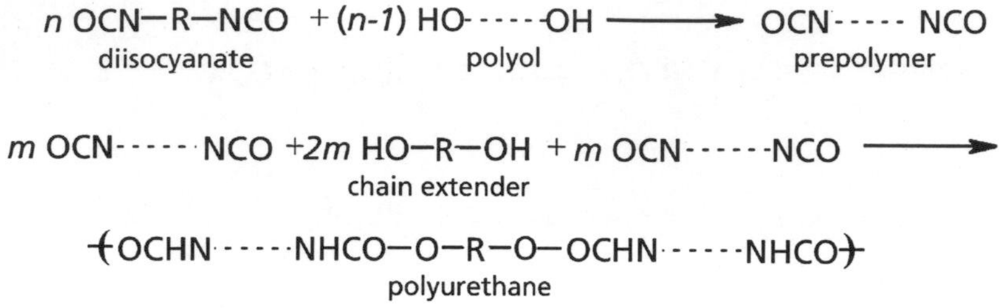

n OCN—R—NCO + (n-1) HO------OH ⟶ OCN----- NCO
diisocyanate
polyol
prepolymer
m OCN-----NCO +2m HO—R—OH + m OCN-----NCO ⟶
chain extender
(OCHN-----NHCO—O—R—O—OCHN-----NHCO)
polyurethane

Scheme 12.2 Schematic representation of 'prepolymer technique' method

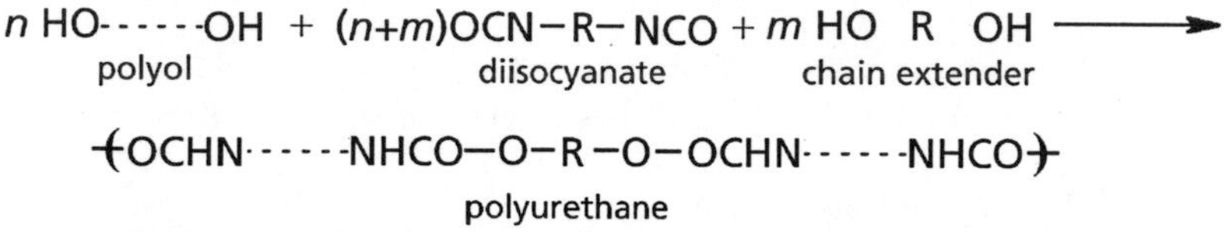

n HO-----OH + (n+m)OCN—R—NCO + m HO R OH ⟶
polyol
diisocyanate
chain extender
(OCHN-----NHCO—O—R—O—OCHN-----NHCO)
polyurethane

Scheme 12.3 Schematic representation of 'one shot' method

Soft segments
Hard segments

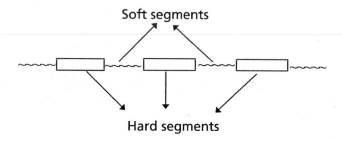

Figure 12.1 Microphase separations of segmented polyurethanes

Indeed, segmented polyurethanes are typical block copolymers of the type $(A\text{-}B)_n$, where the block A consists of relatively long and flexible soft segment (SS) and the block B is a highly polar hard segment (HS). Very often the SS is a polyether or polyester and the HS is diisocyanate extended with a low molecular weight diol. Microphase separation at mesoscopic length scales, resulting in a morphology consisting of microdomains rich in HS (HS microdomains) and a microphase rich in SS (SS microphase), is a fundamental property of segmented polyurethanes, arising from the incompatibility of HS and SS [25-28].

The hard blocks of the molecules are connected with each other by intermolecular hydrogen bonding, generating domains that act as junctions of physical crosslinking for the soft blocks. Hydrogen bond distribution influences the extent of microsegregation in PU and a set of properties typical for these materials. As consequence, the properties of PU blended systems with other polymers would depend on how the added polymer influences the extent of microsegregation of PU [29].

In fact, many of the versatile properties of segmented PU, which make these systems useful for diverse technological applications, are related to their incomplete microphase separation. Thus, it is not surprising that several studies have been devoted to the investigation of the factors controlling microphase separation and to the development of methodologies to quantitatively study microphase separation in PU [25-28, 30-34]. In addition, methods have been developed to calculate the microphase composition and to characterise the degree of microphase separation [35].

The concentration of 'soft' and 'hard' segments is usually expressed by Equations 12.1 and 12.2:

$$\text{SSC} = (M_{POL} - 2M_{OH}) \times \frac{100}{M_0} = \frac{100 \times (M_{POL} - 34)}{M_{POL} + M_{ISO} + r(M_{EX} + M_{ISO})} \tag{12.1}$$

$$\text{HSC} = 100 - \text{SSC} \tag{12.2}$$

where SSC is soft segment concentration;

HSC – hard segment concentration;

M_{POL} – molecular weight of the polyol;

M_{OH} – molecular weight of hydroxyl groups;

M_{ISO} - molecular weight of the isocyanate;

M_0 – molecular weight of the repetitive unit consist in one soft and one hard segment

M_{EX} - molecular weight of chain extender

r – chain extender/polyol molar ratio

Some types of PU used in blends based on polyurethanes are summarised in **Table 12.1**.

12.2.2 Blends Based on Polyurethane Elastomers

Solid urethane rubbers are commonly used due to their excellent properties, especially tensile and tear strength, and very high abrasion resistance. PU also have excellent

Table 12.1 Polyurethane elastomers used in blend formulations					
Code	Polyurethane elastomers	Hard segments	Soft segments	Trade name	Refs.
PU-1	TDI/Cl-PD/POEOD	TDI/Cl-PD	POEOD		[7]
PU-2	Polyester urethane rubber			Vibrathane 5008	[16]
PU-3	Polyether urethane rubber Milled			Adiprene CM	[16]
PU-4	Polyether urethane rubber Unmilled			Adiprene C	[16]
PU-5	MDI/BD/PPO-EO	MDI/BD	PPO-EO	Polimersintez Corp. (Russia)	[29]
PU-6	MDI/BD/PEBA	MDI/BD	PBA-2000	Polimersintez Corp. (Russia)	[29]
PU-7	MMDI/BD/PBA	MMDI/BD	PBA		[36]
PU-8	Polyether urethane rubber			Estane 58245	[37]
PU-9	TDI-prepolymer/ polyether polyols				[38]
PU-10	MDI/BD/PPO	MDI/BD	PPO-4000		[39]
PU-11	MDI/BD/PTMO	MDI/BD	PTMO-2000		[39]
PU-12	MDI/BD/PTMA	MDI/BD	PTMA	Bayer Co.	[40]
PU-13	MDI/BD/PTMO	MDI/BD	PTMO	Pel55D	[41]
PU-14	MDI/BD/PTMO	MDI/BD	PTMO	Ela85A	[41]
PU-15	MDI/N,N-bis (hydroxyethyl) Isonicotinamide/ PTMO		PTMO-1000		[42] [43]
PU-16	TDI/MDEA/PTMO	TDI/MDEA	PTMO-1000		[44]
PU-17	MDI/BD/PTMA	MDI/BD	PTMA		[45, 50]

	Table 12.1 Continued				
Code	Polyurethane elastomers	Hard segments	Soft segments	Trade name	Refs.
PU-18	MDI/BD/PTMO	MDI/BD	PTMO		[23, 46]
PU-19	MDI/BD/ PPO-PC	MDI/BD	PPO-PC		[45]
PU-20	MDI/BD/PTMA	MDI/BD	PTMA-2000	Desmopan 359	[47]
PU-21	MDI/BD/PEGA-PDMS	MDI/BD	PEGA-PDMS		[48]
PU-22	TDI/BD/PEGA-PDMS	TDI/BD	PEGA-PDMS		[49]
PU-23	HDI/BD/PEGA-PDMS	HDI/BD	PEGA-PDMS		[50]
PU-24				Vibrathane 5004, Uniroyal Co., USA.	[8]
PU-25	MDI/BD/PCL	MDI/BD	PCL		[46]
PU-26	MDI/BD/PPO	MDI/BD	PPO		[46]
PU-27	MDI/BD/PTMO	MDI/BD	PTMO	P-21T	[51]
PU-28	Polyurethane 58311			Chemplast Ltd, India	[52]

BD: 1,4-butanediol
Cl-PD: 3-chloro-1,2-propanediol
HDI: 1,6 hexamethyl diisocyanate
MDEA: N-methyldietanol amine
MDI: 4,4´-methylenebis(phenyl isocyanate)
MMDI: modified 4,4´ diphenylmethane diisocyanate
PBA: poly(butylene adipate)
PBA-2000: poly(butylene adipate) with M_n = 2000
PCL: poly ε–caprolactone
PDMS. polydimethylsiloxanes
PEBA: poly(ethylene butylene adipate)
PEGA: poly(ethylene glycol adipate)
POEOD: poly(oxyethylene oxide diane)
PPO-EO: poly(propylene ethylene) oxide
PTMA: poly(tetramethylene adipate)
PTMG: polytetramethylene glycol
PTMO: poly(tetramethylene oxide)
PPO-PC: poly(propylene oxide)-polycarbonate
TDI: 2,4-toluene diisocyanate

resistance to oils and similar fluids because of their polar nature. These special properties make urethane rubbers the ideal base polymers for compounding [53]. Thus PU are blendable with other polar rubbers such as acrylonitrile butadiene rubber (NBR), acrylonitrile rubber/polyvinyl chloride (NBR/PVC) and hydrogenated nitrile rubber (HNBR), or metal ionomers, typically giving intermediate properties for the respective blends. They may also be blended with less polar polymers such as polydimethylsiloxanes (PDMS), butadiene rubber (BR), styrene butadiene rubber (SBR) and Kraft lignin, or with the nonpolar ethylene propylene diene rubber (EPDM) [16, 38]. The interactions that could appear in such blends generally include hydrogen bonding, Coulombian attraction forces between ions and ion-dipole interactions.

The benefits of adding small portions of urethane rubber to other conventional polymers as NBR/PVC will improve abrasion resistance, tensile and tear strength, retain high temperature, oil and solvent resistance properties and low temperature properties. On the other hand, adding several portions of conventional polymers to urethane improve reversion resistance, compression set (NBR), resistance to water, glycol, methyl ethyl ketone (MEK) and lower the cost of compounds.

Although there are many types of PU blends, those containing PVC, are among the most important from both a scientific and a commercial point of view [7, 54, 55]. There are contradictory results concerning miscibility and the properties of PVC/PU blends, in that PU plays the role of permanent plasticiser for PVC [56]. However, there is no established systematic variation of properties, since various authors have used different PU systems.

Polyurethanes are reasonably compatible with chlorinated polyethylene (CPE), styrene-acrylonitrile copolymers (SAN), acrylonitrile-butadiene-styrene copolymers (ABS) and butadiene-styrene block copolymers (BS).

Blending PU with SAN results in many property variations due to the numerous combinations of different PU types (hard and soft segment character, domain separation, etc.), or with different SAN copolymers (weight% of acrylonitrile and different distributions of styrene and acrylonitrile) that are possible [3, 57].

Incorporation of the polybispropoxy-phosphazene (PPO-PhZ), with high phosphorus and nitrogen content, into the PU during the synthesis [23, 46] gives a blend with lower flammability.

In recent years PDMS, due to their good performance, has received considerable interest as a component in PU alloys and blends aiming at functional polymeric materials [10, 11, 41, 58, 59].

The main types of blend formulations based on PU are reviewed in **Table 12.2**.

No.	Polyurethane blends	Refs.
	**Table 12.2 Blends based on polyurethane elastomers**	
1	PU-1/PVC	[7]
2	PU-2/NBR 40; PU-2/NBR 50; PU-2/HNBR; PU-2/SBR; PU-2/EPDM	[16]
3	PU-3/BR	[16]
4	PU-4/NBR/PVC	[16]
5	PU-5/PA12	[29]
6	PU-6/PA12	[29]
7	PU-7/PVC	[36]
8	PU-8/poly α-olefin; PU-8/polyester PU	[37]
9	PU-9/Kraft lignin	[38]
10	PU-12/SAN; PU-12/EVA; PU-12/EVA/SAN	[40]
11	PU-13/PDMS	[41]
12	PU-14/PDMS	[41]
13	PU-15/metal acetates	[42, 43]
14	PU-16/PVC/PSMA	[44]
15	PU-17/SAN; PU-17/EVA; PU-17/EVA/SAN; PU-17/PVC	[45, 46]
16	PU-18/PPO-PhZ; PU-18/PVC	[23, 46]
17	PU-20/metal ionomers/HDPE	[47]
18	PU-24/CPE; PU-24/VAMAC	[8]
19	PU-27/Li salt complexes/PDMS-PMPhS	[51]
20	PU-28/CA	[52]

PA12: polyamide 12
EVA: ethylene-vinyl acetate copolymer
PSMA: poly(styrene-co-maleic anhydride)
HDPE: high-density polyethylene
VAMAC: ethylene acrylic elastomer
PDMS-co-PMPhS: poly(dimethylsiloxane-co-methylphenylsiloxane)
CA: cellulose acetate

12.2.3 Polyurethane Blending

The mixing operation is the most important step in blend preparation and it is well known that certain properties depend on it [60, 61]. Two different reactive polymers can chemically interact at high temperature to form interchain crosslinked polyblends. Das and co-workers have developed such interchain, crosslinkable, elastomer-elastomer and elastomer-plastic blends [62-65].

Two commonly used techniques for obtaining PU blends are:

1. preblending technique (the two elastomers are first blended in given ratios, allowed to equilibrate, and then curatives are incorporated into the blend at high temperature),

2. preheating/preblending technique (the preblend is subjected to heat treatment at high temperature and then the curatives are incorporated into the blend at room temperature).

Recent studies also showed that PU are able to crosslink with polyethylene co-EVA [40, 45, 46, 66] at high temperature without any curatives. De and co-workers have reported that rubbers having appropriate functional groups interact with each other when blended and thus crosslink at high temperature in the absence of any curatives [67-72]. The temperature profile can be adjusted based on the processing temperatures of both individual materials and their weight ratios in the blends. Also, speed of mixing can be adjusted to obtain different ratios to accomplish the final thickness requirements.

Polyurethane blend preparation can be conducting by solution, melt or reactive mixing (**Tables 12.3** and **12.4**).

Table 12.3 Polyurethane blends obtained by solution mixing					
Blend type	Temperature	Solvent	Additive	Apparatus	Refs
PU-2/NBR	160 °C/press cured		Sulfur or peroxide	Farrell BR1600 Banbury internal mixer	[16]
PU-16/PVC/ PSMA	Room temperature	DMF, THF		Stirrer	[44]
PU-28/CA	Room temperature	DMF	PVP	Brabender Plasticoder 814400 single screw extruder	[52]
CA: *cellulose acetate* DMF: *dimethylformamide* PVP: *poly(vinylpyrrolidone)* THF: *tetrahydrofuran*					

Table 12.4 Polyurethane blends obtained by melting or mechanical mixing			
Blend type	Cure conditions	Apparatus	Refs
PU-2/ Zn^{2+}/HDPE	Melting/240 °C/inert gas (Ar); Compression moulding, 240 °C/10 MPa	Home-made and-cup type mixer	[16]
PU-12/SAN	Melting/180-190 °C; Compression moulding, 160 °C/6 MPa	Two-roll mill	[40]
PU-19/EVA	Melting/180-190 °C; Compression moulding, 160 °C/6 MPa	Twin-roll machine	[45]
PU-7/PA12	Mechanical extrusion/195 °C	Werner and Pfleiderer ZSK-40 extruder	[29]
PU-8/poly α-olefin	Mechanical extrusion/165-195 °C	Brabender Plasticoder 814400 single screw extruder	[37]

The main techniques for obtaining highly homogenous PU blends, such as those presented in **Table 12.3** (solution or press cured mixing) and **Table 12.4** (melting or mechanically mixing), allow mixing of PU with almost all conventional plastics, rubbers, fibres, etc.

12.2.4 Morphology of Elastomeric Polyurethane Blends

The structural and morphological properties of a polymeric material are significantly determined by the presence of hydrogen bonding. The dynamic behaviour of hydrogen bonding systems, depends on the working conditions, particularly the temperature and the presence of other components that are able to interact in a similar manner.

Macroscopic properties of polymer blends largely depend on their microscopic morphologies. At present scanning electron microscopy (SEM) and transmission electron microscopy (TEM) are generally used to inspect the quality of polymer blends [73]. Atomic force microscopy (AFM) [74] can be used for routine measurement of organic surfaces [75-78]. Furthermore, topography, elasticity, friction, and conductivity of the surface can be studied locally by using different imaging techniques [79, 80]. Thus AFM techniques are useful tools for the routine investigation of polymer blends and represent complementary techniques to conventional electron microscopys (SEM, TEM) as well as X-ray and energy loss spectroscopy techniques [81].

Differential scanning calorimetry (DSC) was often used to study the thermal properties of various samples. A T_g at low temperatures and a melting peak at high temperatures were observed for every PU sample associated with the non-crystalline region of soft segments and strong hydrogen-bonding/crystalline region of the hard segments caused them.

The morphology of the chlorinated PU-1/PVC (**Table 12.2**) blends, studied by SEM, is of a uniform type, but with differing degrees of roughness [7]. So, a melting endotherm for the PU-1 hard segments occurs in the temperature range of 230-250 °C. This observation agrees well with morphogical models proposed for different PU-1/PVC systems [82-84], which suggest that the PVC molecules may disperse in the soft segment phase with little influence on the microphase separation and mixing of the hard and soft segments.

The ring-like voids surround the dispersed spheres in polytetramethylene adipate (PTMA)-PU-17/SAN/EVA ternary blend, was obtained. This interesting morphology clearly indicated that the most PU-17 - PTMA is located between the dispersed EVA phase and SAN matrix acting as a compatibiliser. Although PU-17 - PTMA and SAN are immiscible, there are moderate specific interactions between them [45].

The urethane and ether groups of the PU are of particular importance, especially when blending with polyamides [85, 86]. It is well known that PU contains polar functionality that can also interact with the polar component of polyamide 6 (PA6). The interaction between PU and PA6 in the corresponding blends can enhance the toughness of polyamide due to its elastic properties. Hydrogen bonding has been observed between molecules of either PU-PA6 and PA6-PA6, and it is sufficiently strong to cause the polymer chains to distort rather than disrupt the hydrogen bonds. When groups of like polarity, such as carbonyl groups, come into proximity, the polymer chains again distort from their regular conformation because of mutual repulsion. The hydrogen bonding interaction is what determines the compatibility of PA and PU and the interfacial adhesion in their blends [87].

The addition of polyamide 12 (PA12) into segmented PU causes rearrangement in the system of labile hydrogen bonds, the nature of which depends on the polyurethane origin. For ester-based PU-6 (**Table 12.1**) the extent of domain microsegregation decreased during blending. On the contrary, in oligoether-based PU-5, the microphase separation was more pronounced when PA12 was added. Infra red-spectroscopy (IR) showed the two PU versions blended with PA12 to form more high-energy hydrogen bonds with polyamide macromolecules than in homopolyurethanes [29].

Micrographs were made of the top surface of CA, PU-28, CA/PU-28 blend and the cross sections of blend membranes [52]. The micrographs showed the existence of phase separation and compatibility in the blend membranes (CA/PU-28). The sponge like structures in the PU membranes was similar to those observed by Koenhen and co-workers

[88] in PU/dimethylformamide (DMF)/H$_2$O system. Further, when three such spheres (CA/PU-28/PVP contact each other a void is formed. The space between aggregates revealed small pores in the membrane since CA/PU-28 blend is highly hydrophilic.

The structure of MDI-PTMO/BD-PPO-PhZ blend flame retardant was investigated by DSC, Fourier-transform infrared (FTIR), SEM and nuclear magnetic resonance spectroscopy [23]. It was found that the addition of PPO-PhZ would accelerate the urethane reactions and introduce a formation of micro-crosslinks. Consequently, the T$_g$ of PU-17 was increased. Using SEM, PPO-PhZ additive was observed to phase-separate from the PU-17 matrix with good interfacial bonding.

Addition of small concentrations of PDMS to PU-13 or PU-14 promoted more efficient packing of PTMO, the soft domain of PU. On the other hand, PDMS inhibited crystallisation of PTMO and evidenced that it may work in synergy with improved packing to promote fatigue resistance and flexibility and elongation properties of blends [41].

12.3 Properties of Polyurethane Blends

12.3.1 Glass Transition

The compatibility of a new polymer blend is often the first property studied. There are many experimental techniques to characterise the compatibility level of the polymer blends, but DSC and dynamic mechanical analysis (DMA) are the most widely used. These techniques, by measuring the T$_g$ of each phase, indicate the number of phases present. The miscibility level, i.e., the amount of the B component miscible in the A phase may be deduced by means of the shift of T$_g$ of each phase compared to that of the pure component.

Since PU can be easily modified by gradual replacement of soft segments, such as PTMA, PTMO, PCL, PPO, etc., the relationship between chemical structure of the PU and its miscibility with PVC can be readily investigated. Wu and co-workers [46] studied the miscibility of the PU blended with chlorinated polyvinyl chloride (CPVC) by DSC and FTIR spectroscopy. DSC results showed that the polyester polyurethane blends, PCL-PU-25/CPVC and PTMA-PU-17/CPVC (**Table 12.2**), are miscible over the entire composition as they exhibit only one major T$_g$. The T$_g$ of the blends regularly increased with increasing CPVC content. Increasing the molecular weight of the glycol can decrease the miscibility due to the crystallisability of the soft segments. Two distinct T$_g$ for all polyether polyurethane blends (PPO-PU-26/CPVC and PTMO-PU-18/CPVC) were obtained, which did not change with the changing of the blend's composition, suggesting that the blends of are immiscible in all cases. On the other hand, the FTIR spectroscopic results showed a strong interaction between CPVC and the polyester soft segments. In addition, it was found that the miscibility of PU and CPVC decreased upon increasing the hard segment content. Finally, they found that polyester polyurethane blends with CPVC form miscible systems while polyether polyurethane-CPVC blends are immiscible.

The DSC plots of PVC/chlorinated PU-1 systems revealed a single T_g in the region of 55-65 °C, showing the blend's miscibility [7]. The compatibility of these blends can be due to hydrogen-bonding between the C=O groups of the polyester and the α-hydrogen of chlorinated polymer [89, 90] or a dipole-dipole –C=O....Cl-C- interaction [91, 92].

Hernandez and co-workers [93], using PVC/copolyester-urethane blends and DMA, concluded that PU compatibility with PVC is due to the presence of 32% PCL and 38% PTMA in the main chain of the copolyester-urethane polymer.

The degree of compatibility of the polyurethane, Vibrathane-5004/chlorinated polyethylene (PU-24/CPE) blends with different preblending and preheating techniques was studied by low temperature DSC [8]. For a PU/CPE preblended sample a single T_g was observed at –44 °C, whereas for the PU-24/CPE preheated preblended sample, the T_g value shifted to a higher temperature. This is probably because of the interchain crosslinking reaction occurring between the two elastomer phases during the heat treatment. For PU-24/VAMAC (ethylene acrylic elastomer) preblended and preheated preblended samples, two T_g values were observed, indicating their incompatibility. The preblended sample showed the T_g values at –32 °C and at –10 °C, whereas the preheated preblended sample showed the T_g values at –27 °C and at –8.2 °C.

Morphology stabilisation and diminution of dispersed phase in PU-20/polyolefin blends can be obtained by using a compatibiliser, usually in the melt mixing preparation processes. Thus could be improved impact strength and elasticity of high crystalline polyolefin and also paintability and moisture inertness of the resulting blends. Maleated polypropylene (PP-MA), maleated polypropylene grafted with polyethylene oxide [(PP-MA)-g-PEO], polyethylene-*co*-acrylic acid (PEAA), polystyrene-*co*-maleic anhydride (PSMA), polybutyl acrylate-*co*-acrylic acid (PBAA) [94-96] were used as compatibilisers for PU-20/ polypropylene (PP) blends. A modified ethylene Zn^{2+} ionomer in moderate concentrations (10-15% wt) in PU-20/HDPE was used [47, 97]. In all cases compatibilisation is due to the strong adhesion between PU or polyolefin component and compatibiliser through specific forces, mainly hydrogen bonding.

In the ester-PU/PA12 blends, miscibility can be achieved owing to close solubility parameters of soft segment of PU-6 and that of PA12. These differ by 1.2 $(MJ/m^3)^{0.5}$, while for oligoester-PU-5/PA12 blends, the solubility parameters differ by 4.7 $(MJ/m^3)^{0.5}$ [29].

The T_g of PDMS-*co*-methylphenylsiloxane (PDMS-*co*-PMPhS) blended with PU was measured by DSC to evaluate the compatibility between PU-27 and PDMS-*co*-PMPhS. Each single polymer (PU and PDMS-*co*-PMPhS) showed T_g at –72 °C and –118 °C, respectively. Two T_g corresponding to two components was observed for the PU-27/PDMS-*co*-PMPhS blends, which are almost the same values as those of single polymers. In contrast to this observation, regarding the PU-20/polyether-modified polysiloxane (PES) blends, only one T_g was observed, which decreased with increasing content of PES [51].

12.3.2 Degradation

There is a definite correlation between the chemical structure of a polymer and its degradation temperatures. Blends prepared through the different blending techniques, having the same compounding formulation and same blending ratio differ in thermal behaviour.

Thermal analysis of the preblended polyurethane PU-24 type Vibrathane-5004/CPE showed that the initial degradation (T_1) started at 254 °C and continued up to 376 °C where the second degradation step (T_2) started. The degradation of the preheated sample started at 303 °C and led to the second degradation at 411 °C. Whereas the T_1 of PU-24/VAMAC preblended samples occurred at 246 °C and continued up to 366 °C where the T_2 started. For the blend obtained by heating the preblend, followed by curative addition, the degradation occurred at relatively higher temperature. The T_1 started at 252 °C and continued up to 377 °C where T_2 occurred [8]. The early degradation of the preblended sample compared to the preheated preblended sample with the same elastomer ratio, indicated the presence of some physical linkages between the two phases. An interchain crosslink reaction occurred in the blend of the two elastomers having reactive functional groups as a result of blending before addition of curatives. This extent of crosslinking increased when the preblend was subjected to heat treatment before curative addition. Thus, the thermal stability and the degradation temperature of the preheated preblended sample increased significantly.

Pielichowski and Hamerton [7] examined, using DSC, SEM and thermogravimetric analysis (TGA), either alone or coupled with FTIR, the thermal behaviour of a series of solution-cast blends of PVC)/chlorinated TDI-based polyurethane (PVC/PU-1) polymers. They found that the decomposition proceeds through a two-step route; the main, decisive degradation stage in the 200-320 °C temperature range was found to be a result of parallel reactions of PVC and PU-1 decomposition. This was also confirmed by Ozawa-Flynn-Wall kinetic analysis, the activation energy remained constant for degrees of conversion greater than 0.3 [98-102]. The reasons for the better thermal stability of some PVC/PU-1 blends can be explained by analysis of specific interactions between the

C=O groups of the urethane segments and the α-hydrogen of the chlorinated polymer or a dipole-dipole –C=O.....Cl-C- interactions. On the other hand, the rate of diffusion of volatile products through microphase domain structure may differ due to changes in morphology arrangement thus considerably affecting the overall decomposition route.

12.3.3 Mechanical Behaviour

The stress-strain behaviour of the three binary blends [PTMA-PU-17/SAN, PTMO-PU-17/SAN and polypropylene oxide-polycarbonate (PPO-PC)-PU-17/SAN] showed that, in terms of yield strength, fracture strength and elongation, the order was PU-PTMA>PU-PTMO>PU-PPO-PC. Also, a remarkable improvement of the mechanical properties of the SAN/EVA blends was realised by incorporation of some PU-17-PTMA [45].

Addition of PA12 to PU provides wide variations in the PU mechanical strength and hardness. At low concentrations of PA12 (up to 20-30 wt% PU), mechanical strength values for the blends were higher than additive ones. Blends containing 30-50 wt% have good impact resistance; oligoester-PU-5/PA12 compositions showed higher impact resistance at low temperature (–40 °C) testing, than ester-PU-6/PA12. The lower T_g of the oligoester-PU-6 soft segment compared to the ester-PU can explain these properties [29].

Blending low concentrations of PDMS with PU-13 or PU-14 led to improvement in mechanical properties (tensile strength up 10-40%, elongation at break up 20-50% and Young's modulus up 10%), while at higher levels of PDMS (>3%), physical properties begin to be affected [41, 103].

12.3.4 Electrical Properties

There have been extensive investigations on the electronic transport properties of pseudo semiconducting PU [51, 104-108].

Seki and co-workers [109] reported that the complexes of ether-based PU prepared from MDI/BD/PTMO with lithium salts, exhibit ionic conductivity of approximately 10^{-6}/S/cm at room temperature. Also, the improvement of the ionic conductivity of the PU-27/lithium salt complexes by blending polysiloxanes and polyether-modified polysiloxane was studied [51].

Ionic conductivity (σ) of the PU-27 electrolyte can be expressed by Equation 12.3:

$$\sigma = n \cdot q(\mu_+ + \mu_-) \qquad (12.3)$$

where, n is the ion density, q is charge and μ_+ and μ_- are the cationic and anionic mobility, respectively.

The blending of PDMS-*co*-PMPhS with PU possessing oligobutylene oxide (PBO) segments was investigated to evaluate the effect of the flexible polysiloxane component on the ionic conductivity [51]. The ionic conductivity of PU-27/PDMS-*co*-PMPhS electrolyte was higher than that of PU electrolyte. When the weight percent of PDMS-*co*-PMPhS in the polymer was 10 wt%, the highest conductivity (4.1×10^{-7} S/cm) was obtained at a LiClO$_4$ concentration of 1.5 mmol/g of polymer. Further addition of PDMS-*co*-PMPhS caused the decrease of the conductivity. On the other hand, PU/PES electrolytes showed the highest conductivity at higher LiClO$_4$ concentration (2.5 mmol/g of polymer) than the PU-27 or PU-27/PDMS-*co*-PMPhS electrolyte (1.5 mmol/g of polymer). This suggests that PU/PES has higher ion solvation ability than PU-27 or PU-27/PDMS-*co*-PMPhS. To verify these assumptions, ion mobility of the PU-27/PDMS-*co*-PMPhS and PU-27/PES blends containing 1.5 mmol LiClO$_4$/g of polymer was measured with the voltage polarity reversing method [110]. The drifting velocity was obtained from the time (t) when the current peak occurred after reversing the polarity, and from this time, the applied voltage (E) and the distance between the electrodes (d), the mobility (μ) of the cation and anion were calculated using the Equation 12.4:

$$\mu = d^2 /(E \times t) \tag{12.4}$$

Both the cationic and anionic mobility (μ_+ and μ_-) of both the electrolytes increased with increasing content of PDMS-*co*-PMPhS and PES. In conclusion, ionic conductivity of PU-27-based electrolyte increased by blending PDMS-*co*-PMPhS or PES. The improvement of the conductivity for the PU-27/PES electrolyte is attributed to the increase of both ion mobility and ion density. On the other hand, it is thought that the increase of ion mobility, which compensates for the decrease of ion density, contributes to the improvement of the conductivity for the PU-27/PES electrolyte. It is interesting that ion mobility increases by the addition of polysiloxane even in the phase-separated PU-27/PDMS-*co*-PMPhS blend.

12.4 Applications of Polyurethane Blends

Polymer electrolytes based on blends of polyether urethane and polysiloxanes have attracted considerable interest in the last few years because of their potential application as electrolytic membranes in primary and secondary high energy density lithium batteries.

Since polyurethanes show high porosity, low weight to volume ratio, good resilience character, abrasion resistance and oil resistance, they can be incorporated into CA to introduce balanced hydrophilicity in the resultant blend membrane and hence achieve optimum membrane performance in terms of better rejection and flux.

Blends of PP/PU are widely used in the automobile and electronic industry as well as for household and sport supplies, and in medical components.

12.5 Polyurethane Interpenetrating Networks

Polymer network systems composed of least two kinds of components with different physicochemical properties are expected to be promising functional polymeric materials [111-113]. The attempts to connect some different chemical components by chemical bonds provide a convenient route for the modification of properties to meet specific needs and were first made for the copolymers synthesised from multi-components with block or random sequences [10, 11]. Among these methods, those leading to IPN, which could be considered as a subgroup of the broad class of polymer blends, represent a new, essential approach to solve the problem of mutual incompatibility of polymers [13, 14, 114]. IPN generally possess enhanced physical and engineering properties compared to the normal polyblends of their components because of the synergistic effect of individual polymers.

IPN are traditionally defined as a combination of two or more chemically distinct polymers in network form, held together ideally and solely by their permanent mutual entanglements, in which at least one is synthesised and/or crosslinked in the immediate presence of the other. IPN could also be called polymer alloys and have been intensively studied [6, 14, 111, 113, 115-135].

The most commercially important PU elastomers used in IPN are phase-separated systems consisting of a MDI or TDI based hard segment and a polyether or polyester soft segment. Frequently synthesised and studied series of polyurethane IPN consists of crosslinking their networks with acrylic [129-131, 133, 136-140], styrene [141-148], polysiloxanes [149-155], unsaturated polyester [126, 156-162] and epoxy polymers [120, 163- 164]. Some representative examples are illustrated in **Table 12.5**.

12.5.1 Preparation of Polyurethane IPN

IPN synthesis is a method of blending two or more polymers to produce a mixture in which phase separation is not as extensive as it would be otherwise. It is the only way of combining crosslinked polymers, conventional blending or mixing results in multiphase morphologies due to the thermodynamic incompatibility of polymers. Permanent entanglements prevent total phase separation and stabilise the morphology.

No.	PU	PU/IPN type	Hard segments	Soft segments	Refs.
1	TDI/TEA/castor oil	PU/PMMA	TDI/TEA	Castor oil	[119]
2	TDI/BD/PPO	PU/epoxy	TDI/BD	PPO	[14]
3	TDI/TMP/PTMGO	PU/PEA	TDI/TMP	PTMGO	[165]
4	H_{12}MDI/PPG/BEPD	PU/PMMA	H_{12}MDI/ TMP	BEPD	[138]
5	H_{12}MDI/TMP/PPG	PU/PMMA	H_{12}MDI/TMP	PPG	[129]
6	TDI/TMP/PPOG/PhTS	PU/PBMA-PDMA	TDI/TMP	PPOG/PhTS	[131]
7	MDI/TMP(BD)/PTMO	PU/ANR	MDI/TMP(BD)	PTMO	[132]
8	TDI/BD/PPO	PU/VER	TDI/BD	PPO	[128]

Table 12.5 Polyurethanes in IPN formulations

ANR: *allyl novolac resin*
BEPD: *2-butyl-2-ethyl-1,3-propanediol*
H12MDI: *hydrogenated 4,4´-methylenebis(phenyl isocyanate)*
PBMA: *poly(butyl methacrylate)*
PDMA: *polydimethacrylate*
PEA: *polyester acrylate*
PMMA: *poly (methyl methacrylate)*
PPG: *poly(oxypropylene)glycol*
PTMG: *polytetramethylene glycol*
TEA: *triethylamine*
TMP: *trimethylpropane triol*
VER: *vinyl ester resin*

Two basic methods are currently used for making IPN: when both polymers are formed and crosslinked simultaneously, the network is known as a simultaneous interpenetrating network (SIN), whereas if the crosslinking of polymer II follows that of polymer I, the material is a sequential IPN. When only one polymer is crosslinked, a semi-IPN is produced, and when all polymers are in network form it is a full-IPN (**Scheme 12.4**).

There are also classes of thermoplastic IPN (involving physical crosslinks rather than chemical crosslinks so that they could be considered as hybrids between polymer blends and IPN), latex IPN, which often exhibit core-shell morphologies, and gradient IPN, when the crosslink density or composition varies from one network location to another. In fact, most IPN, including polyurethane IPN, do not interpenetrate on a molecular

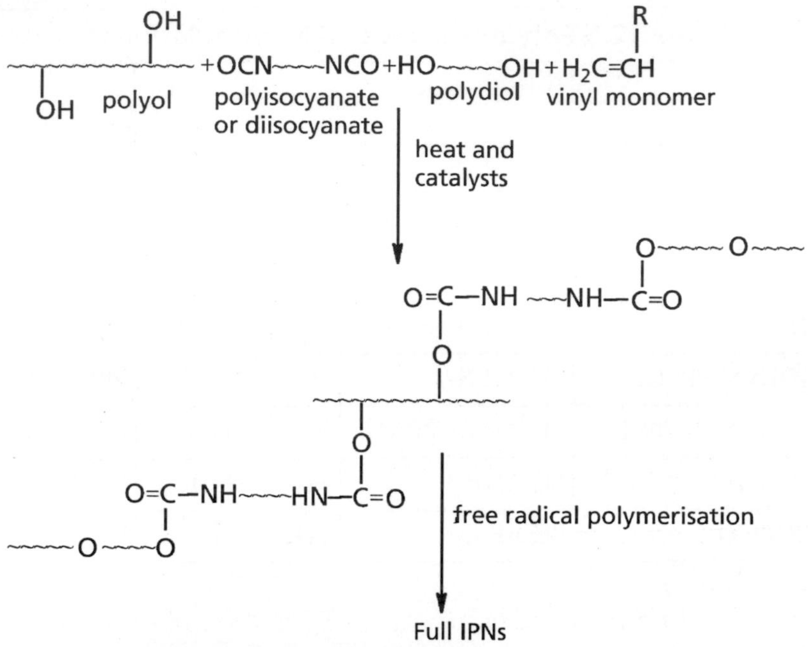

Scheme 12.4 Reactions occurring during polyurethane IPN synthesis

scale. Polymers in the system may form finely divided phases of only hundreds of Ångström units in size. However, these phases may be continuous on a macroscopic scale ('dual phase continuity'). In some cases, true molecular interpenetration is thought to take place at the phase boundaries [118].

Many methods of polyurethane IPN preparation have been reported in the literature (**Table 12.6**).

The preparation of PU/PS IPN could be done by polymerisation of PU and PS through independent condensation and free radical polymerisation. Sperling and co-workers [166, 167] have carried out investigations on PU/PS IPN systems based on castor oil polyurethane and castor oil polyester networks. Other PU/PS simultaneous IPN obtained from castor oil, TDI, styrene and DVB were synthesised under conditions where the free radical polymerisation of styrene and crosslinking reaction of PU progress at comparable rates [168]. Castor oil and MDI- based PU/PS IPN with a fixed styrene level with varying amounts of DVB were also synthesised [169].

Table 12.6 Methods of preparation for polyurethane IPN				
PU IPN	Methods of synthesis	Catalysts	Post-cure treatments	Refs.
PU/PS	One-shot procedure, 60 °C/30 min/vacuum	DBTDL, BPO,	Mould, 80 °C/2 h; 100 °C/24 h	[143]
PU/PS	Simultaneous polymerisation, 60 °C	BEPD, DBTDL, DVB, LPO	Mould, 100 °C/4 h; 120 °C/10 h/vacuum	[138]
PU/PMMA	Sequential polymerisation	BPO, EGDMA	Air circulated oven, 70 °C/24 h; 120 °C/4 h	[119]
PU/PMMA	Simultaneous polymerisation, 60 °C	BEPD, DBTDL, TEGDM, LPO	Mould, 100 °C/4 h; 120 °C/10 h/vacuum	[138]
PU/PBMA-PDMA	Sequential polymerisation	BIE	UV	[131]
PU/epoxy	Simultaneous polymerisation	BPO, DBTDL,	Glass mould, 25 °C/10 h; 80 °C/1 h; 110 °C/3 h	[14]
PU/epoxy E51	Simultaneous polymerisation	AIBN, DETA	Glass mould, 55 °C/2h; 70 °C/5 h; 120 °C/2 h	[120, 163]
PU/ANR	Simultaneous polymerisation, 70-75 °C	BPO	Oven, on Petri dishes, 60 °C/24 h; 70 °C/12 h/vacuum	[132]
PU/PADC	Simultaneous polymerisation, 80 °C	BPO, DBTDL	Mould, 50 °C to 85 °C/35 h	[135]

PS: polystyrene
DBTDL: dibutyl tin dilaurate
BPO: benzoyl peroxide
BEPD: 2-butyl-2-ethyl-1,3-propanediol
TEGDM: tetraethylene glycol dimethacrylate
LPO: lauroil peroxide
DETA: diethylene triamine
AIBN: 2,2 ́azobis-isobutyronitrile
EGDMA: ethylene glycol dimethacrylate
DVB: divinyl benzene
BIE: benzoin isobutyl ether
PADC: polyallyl diglycol carbonate

The interconnected IPN could be prepared by first forming a PU network based on an unsaturated polyester polyol with toluene TDI in the presence of a triol followed by interconnecting the PU chain segments by reaction of styrene with sites of unsaturation [170]. Since a great difference in solubility parameters between PU and PS causes phase separation, several other approaches to the preparation of IPN were investigated, such as introduction of oppositely charged groups in the polymeric components [171-173], synthesising IPN under high pressure [174, 175], introduction of a crosslinking agent having multiple hydroxyl and vinyl groups [143] and preparation of PU/PS-grafted IPN using a microgel process [176].

A lot of information is available on oil-based PU and the sequential or simultaneous interpenetration with vinyl monomers to get the polymers of desired properties [177-182]. Athawale and co-workers [183-185] report the synthesis of IPN from different kinds of elastomeric polyurethanes and plastic PMMA and IPN based on uralkyd resin as the elastomeric component and PS as the plastic component. IPN based on crosslinked polyurethane and crosslinked PEA were synthesised by the simultaneous curing of both networks [187].

Semi-IPN were prepared by crosslinking PDMS prepolymer with 3-aminopropyltriethoxysilane in the presence of linear polyurethane [149].

Latex IPN PU/polyhydroxyethyl acrylate (PHEA) were synthesised by first polymerising the 2-hydroxy ethyl acrylate-terminated prepolymers in the presence of a peroxide in aqueous media and then adding a prepolymer dispersion of a polyurethane ionomer, subsequently UV-cured [139].

It is well known that introducing filler into the reaction system at the stage of formation of crosslinked polymer, specifically PU, essentially affects the reaction, the conditions and the degree of microphase separation.

At the simultaneous formation of PU/polybutyl methacrylate (PBMA) semi IPN, Lipatov and co-workers [134, 187] used talc and polymeric fine-disperse TEGDM as particulate fillers. Also, use of alumina and especially finely dispersed PEA and Aerosil as fillers at synthesis of PU/PEA IPN leads to a very rapid rate of reaction [165].

12.5.2 Properties of Polyurethane IPN

Like blends, graft or block copolymers, IPN are multicomponent polymer materials. However, IPN exhibit specific characteristics: in solvents, flow and creep are suppressed, and they swell but do not dissolve; also, the domain size can be tailored [118]. The multipolymeric systems thus obtained display a broad range of properties from toughened elastomers to high impact plastics, based on the selected composition ratio and synthetic details.

It was found that the chemical bonds between the two networks have a great effect on the kinetic behaviours of SIN formation as well as the morphology development and the physical properties of such SIN.

The degree of segregation (DS) in polyurethane IPN can be calculated from the parameters of the relaxation maxima [187]:

$$DS = \left(h_1 + h_2 = -\frac{h_1l_1 + h_2l_2 + h_ml_m}{L}\right)\bigg/\left(h_1^0 + h_2^0\right)$$

where h_1^0 and h_2^0 are the maximum values of mechanical loss for the pure components (case of complete phase separation); h_1 and h_2 are the values for each component at different degrees of segregation; h_m is the maximum loss for the relaxation transition, responsible for the interfacial layer appearance; L is the interval between the T_g of the pure components; l_1, l_2 and l_m are the shifts of the respective maxima in temperature scale.

Most of the materials considered are chemically crosslinked. So, the structure of polyurethane IPN can be discussed on the basis of molecular theory of rubber elasticity using the stress-strain data [154-155, 188-189]. The stress-strain data were interpreted in terms of the reduced stress |f*| and the Mooney-Rivlin semi-empirical equations:

$$\left|f^*\right| = \frac{f}{A(\alpha - \alpha^{-2})} \tag{12.5}$$

and/or

$$\left|f^*\right| = 2C_1 + 2C_2/\alpha$$

where, f is the applied force, α is the elongation $\alpha = \frac{L}{L_i}$, L and L_i are the stretched and unstretched lengths, respectively), A is the cross-sectional area of the unstressed sample, C_1 and C_2 are constants.

The reduced stress can be plotted as a function of reciprocal elongation to determine the crosslink density of the sample. The empirical Equation 12.5 provides a linear relation between |f*| and $1/\alpha$.

If one compares reduced stress for $\alpha \to 1$ according to Equation 12.5 and Flory expression for the modulus [189] the following equation is obtained:

$$2C_1 + 2C_2 = \frac{\rho}{M}[(1 - 2/\phi)(1 + f_c/f_{ph})]_{\alpha \to 1}RTV_{2c}{}^{3/2} = A'_\phi \frac{\rho}{M}RTV_{2c}{}^{2/3}$$

where, A'_f designates the term in square brackets, M means the molecular weight between crosslinks, ρ is the sample density, ϕ is the crosslink functionality, f_c is the contribution due to the constraints imposed by the junction fluctuations in the polymer coils, f_{ph} is the contribution due to the assimilation of the polymer network with a 'phantom' network, $V_{2c} = V/V_0$ (V_0 is the initial volume of the sample, and V is the volume of the sample strained with a force f).

However, the synthesis of IPN-like materials having a thermoplastic nature can involve physically crosslinked systems. These bonds of a physical nature may arise from ionic portion of an ionomer, glassy portion of a block copolymer, crystalline portion of a semi-crystalline polymer, or hydrogen bonding.

In the preparation of simultaneous IPN, different formation rates of the two networks have been shown to give a material with a range of morphologies and mechanical properties [190-192].

The most popular and well-studied polyurethane IPN is PU/PMMA, which is a good system for modelling and yields to a variety of materials with useful properties [133, 138]. The highly immiscible PU/PMMA systems could reveal different morphologies as function of polymerisation routes, more precisely as function by the events order of PU and PMMA gelation and phase separation of these two polymers. These also explain the differences in morphology and modulus between PU/PMMA 50/50, for that first gel PMMA, and PU/PS 50/50 IPN, for that first gel PU.

For PU/PADC, when formation of the PU network occurs well ahead of allyl diglcol carbonate (ADC) gelation, extensive phase separation resulted, while the simultaneous formation of both networks or ADC first gelation give a fine continuous or very fine dispersed morphology and better mechanical properties [135]. The dispersed domains of PU/PS IPN (7/3) was significantly reduced by introducing crosslinks in PS domains using DVB [139]. From the applications point of view it is desired that phase separation occurs after at least one component gels, as is the case of PU/PMMA 50/50 IPN, where phase separation appears after 8 minutes of reaction and gelation after 50 minutes of reaction.

Comparative studies of IPN resulting from the transesterified castor oil PU and PMMA with the corresponding unmodified castor oil PU and PMMA IPN could reveal the effect of polyol modification of castor oil on PU prepolymer, the effect of NCO/OH ratio of the PU prepolymer on mechanical, chemical and thermal properties of resultant IPN, and the interpenetration in heterogeneous IPN system, as well [119].

The PU/PBMA semi IPN exhibited microphase separation that is increased by slowing the reaction rates; each phase should be treated as an independent IPN. The maximum compatibility of the two polymers was observed at the blend composition of 50% PU-50% PBMA, since this proportion of the blend exhibits maximum interpenetration [123]. Lipatov and co-workers [187, 192-194] found that reaction kinetics and viscoelastic properties of PU/PBMA semi IPN are interrelated, so that the PBMA formation rate affect the PU formation and higher rates will give higher degrees of segregation and higher differences between the two corresponding T_g. Kinetics and microphase separation for PU/PBMA system could be adjusted by introducing a finely dispersed filler, either inorganic (talc), or polymeric (TEGDM). These fillers bring a restriction into the molecular mobility of PU and PBMA, initiator efficiency and microphase separation. At low initiator concentrations, use of filler leads to an increase of PU and PBMA formation rate for any filler concentration, whereas for high initiator and low filler concentrations an opposite effect takes place. Moreover, high initiator coupled with high filler amounts (>20%) results in decrease of both PU and PBMA network formation.

For PU/unsaturated polyester (UPE) IPN, the driving force of phase separation was the formation of the PU phase and is due to the incompatibility between the resulting PU polymers and the unreacted monomers of PU and UPE resin. Final morphology is strongly dependent of curing temperature and reaction rate of PU and UPE [130].

During the formation of PU/vinyl ester simultaneous IPN it was found that the kinetics of formation for the two networks significantly depends on the morphology development and chemical binding between them [195-199].

The IPN formation occurs by two simultaneous processes, i.e., curing and phase separation arising from the thermodynamic incompatibility of growing polymeric chains. This largely determines the structure of IPN and therefore their end properties. There have been a number of studies on IPN formation kinetics [200-202] but only few of these deal with the connection between IPN formation kinetics and microphase separation.

Based on an analysis of the shift, form and size of the carbonyl group stretching vibration band, an assumption was made that, with a low content of one of the polyurethane IPN components (PEA) in the initial reaction mixture, contacts between growing chains of heterogeneous networks are preferable, and so in the initial stages of the reaction, a retardation of the PU formation takes place [192]. As curing proceeds, the composition of the system and the intermolecular interactions continuously change and at a certain stage thermodynamic incompatibility appears. It has been shown that after the onset of microphase separation, the influence of the admixture networks diminishes markedly. As the concentration of PEA in the IPN increased, the time before the onset of microphase separation reduced sharply. The formation of three-dimensional structures and microphase separation proceeding simultaneously leads to a continuous alteration in phase composition and in interaction forces. The influence of the PEA networks on the urethane formation rate in IPN of the same composition depends strongly on the PEA curing rate, determined in the present case by the initiator concentration. The PEA micro-regions having been heterogeneous in the reaction initial stages are now uniformly distributed in the PU matrix as fillers [165].

In this way, the PU IPN formation kinetics and the ratio of the constituent network curing rates determine the rate and degree of microphase separation. This in turn determines the boundary layer composition and structure and hence the structure of IPN based materials.

Formation of polymers both in bulk and on various hard surfaces is often accompanied by the appearance of internal shrinkage stress, which affects the polymeric material properties. Therefore, it is advisable for polymeric composite production to be carried out under conditions such that the stress relief time is comparable with the time of polymer formation. In this respect, the production of multicomponent polymeric compositions based on the IPN principle, where the individual component formation processes can be separated for an appropriate time, appears to be the most promising.

The internal shrinkage stress in a multicomponent system containing components differing largely in elastic modulus is closely connected with the formation kinetics of these systems. The decrease of the formation rate of the high-modulus component in the IPN brings about a pronounced decrease in internal shrinkage stress in the system, which improves the mechanical properties of the material obtained.

All the fillers, regardless of their nature, shorten the time of appearance and increase the internal shrinkage stress values in the filled IPN with predominantly high-modulus component concentration as compared with unfilled systems. However, the onset of the internal shrinkage stress and the absolute value are determined by the chemical nature of the filler.

According to the method suggested in [203] the value of the polymer-filled interaction Δg^*_{P-F} in the above IPN was estimated. The calculation was carried out based on the analysis of the vapour sorption by an unfilled polymer, the filler and filled polymer according to the equation:

$$\Delta g^*_{P-F} = \Delta g_1 + n\Delta g_{11} - \Delta g_{111}$$

where Δg_1 and Δg_{111} are the free energies of the filled and unfilled polymers interaction with a large amount of solvent; Δg_{11} is the free energy of filler with a large amount solvent.

For filled IPN the sign of the value of Δg^*_{P-F} depends on the ratio of components, on introducing the filler into the composition: for small ionomer contents the value of Δg^*_{P-F} is negative, i.e., the system is thermodynamically stable, at an ionomer content of 30% or greater, the value of Δg^*_{P-F} is positive, i.e., the system is destabilised.

The thermodynamic stability of multicomponent polymeric systems (on introducing fillers) depends on the system components' degree of interaction with a hard surface.

When fillers are introduced into an IPN consisting of components which are thermodynamically unstable under certain conditions, e.g., filler content, components ratio, affinity of only one component, negative Δg^*_{P-F} values are observed, i.e., in presence of a filler, systems with higher thermodynamic stability are formed as compared with unfilled systems. In this case the filler favours component compatibility in the IPN.

The introduction of filler implies pronounced structural changes in the polyurethane IPN under the influence of a hard surface, leading to the expansion of vitrification region.

12.5.3 Applications of Polyurethane IPN

Because of their phase separated elastomeric nature, polyurethanes opened the way to a new class of high impact IPN materials such as coatings, adhesives, elastomers, fibres and foams. Numerous applications have been patented in the fields of electrical insulation, coatings and encapsulants, noise and vibration damping, adhesives and membranes, materials for optics.

Urethane alkyds are usually used to make conventional paints that cure by air oxidation of the unsaturated groups in the presence of metallic driers to give tough coatings with high abrasion resistance. Uralkyds have superior adhesion, hardness, abrasion resistance, and durability and chemical resistance as compared to traditional alkyds. Moreover, cured uralkyd resin is elastomeric in nature. This property of the uralkyd resin is used to develop elastomers.

Polysiloxane crosslinked in presence of aromatic urethanes gave IPN with wear rates much lower than pure polyurethanes or pure silicones. This type of IPN can be applied as biomedical materials, synthetic fabrics, or composite foams with adhesive layers.

Many plastics, such as PS, are brittle and may be toughened by incorporation of small quantities of elastomeric PU [13].

Pernice and co-workers [204] pointed out that SIN composed of polyurethane in combination with epoxy resin, polyacrylates, or unsaturated polyester resin could be applied in reaction injection moulding (RIM) processes. Also, SIN consisting of different polyurethane acrylates and vinyl ester resin in which the main backbone of epoxy is maintained have been synthesised and studied for a RIM process to prepare toughened epoxy materials [120, 205-207].

Concluding Remarks

The generation of new or modified polyurethane blends and polyurethane interpenetrating networks becomes more and more an interesting sphere of activity. Some well-known types of commercial thermoplastic polyurethane elastomers used in blend and IPN syntheses are: Elastane and PELLETHANE (Dow Chemical Company), Elastollan (BASF), Pearlthane (Merquinsa), Desmopan 445 and Texin 245 (Bayer), PurSil and CarboSil (Polymer Technology Group).

Because plant design is not expensive, a broad spectrum of development projects are envisaged in this area, which will enrich continuously the existing range of PU materials. In the some cases PU products, which are not reproducible even with a sophisticated technology, can be obtained by blending.

References

1. Z. Wirpsza, *Polyurethanes: Chemistry, Technology and Applications*, Ellis Horwood/Prentice-Hall, London, 1993.

2. E. Mitzner, H. Goering and R. Becker, *Die Angewandte Makromolekulare Chemie*, 1994, **220**, 177.

3. A. Kanapitsas, P. Pissis and A. Garcia Estrella, *European Polymer Journal*, 1999, 35, 923.

4. S. Dabdin, R.P. Burford and R.P. Chaplin, *Polymer*, 1996, 37, 785.

5. A. Eceiza, J. Zabala, J.L. Egiburu, M.A. Corcuera, I. Mondragon and J.P. Pascault, *European Polymer Journal*, 1999, 35, 1949.

6. Z.S. Petrovic and J. Ferguson, *Progress in Polymer Science*, 1991, **16**, 695.

7. K. Pielichowski and I. Hamerton, *European Polymer Journal*, 2000, **36**, 171.

8. M. Maity, B.B. Khatua and C.K. Das, *Polymer Degradation and Stability*, 2000, **70**, 263.

9. O. Olabisi, L.M. Robeson and M.T. Shaw, *Polymer-Polymer Miscibility*, Academic Press, New York, NY, USA, 1979.

10. L.A. Utracki, *Polymer Alloys and Blends: Thermodynamics and Rheology*, Hanser Publishers, Munich, 1989.

11. *Polymer Blends*, Eds., D.R. Paul and S. Newman, Academic Press, New York, NY, USA, 1978.

12. D. Colombini, G. Merle, J.J. Martinez-Vega, E. Girard-Reydet, J.P. Pascault and J.F. Gerard, *Polymer*, 1999, **40**, 935.

13. Siddaramaiah, P. Mallu and A. Varadarajulu, *Polymer Degradation and Stability*, 1999, **63**, 305.

14. G.Y. Wang, Y.L. Wang and C.P. Hu, *European Polymer Journal*, 2000, **36**, 735.

15. M.M. Coleman, J.F. Graf and P.C. Painter, *Specific Interactions and the Miscibility of Polymer Blends. Practical Guides for Predicting and Designing Miscible Polymer Mixtures*, Technomic, Lancaster, PA, USA, 1991.

16. T.L. Jablonowski, Proceedings of the 155th ACS Rubber Division Meeting, Chicago, IL, USA, Spring 1999, Paper No. 46.

17. C. Auschra and R. Stadler, *Macromolecules*, 1993, **26**, 6364.

18. L. Leibler, *Makromolekulare Symposia*, 1988, **16**, 1.

19. S. Thomas and R.E. Prud'homme, *Polymer*, 1992, **33**, 4260.

20. D.G. Bucknall and J.S. Higgins, *Polymer*, 1992, **33**, 4419

21. J. Heuschen, J. Vion, R. Jerome and P. Teyssie, *Polymer*, 1990, **31**, 1473.

22. L. Leibler, *Physica A*, 1991, **175**, 258.

23. W.Y Chiu, P.S Wang, and T.M. Don, *Polymer Degradation and Stability*, 1999, **66**, 233.

24. S.P. Pappas, *Notes from Coatings Science Course*, North Dakota State University, USA, 1988.

25. L.M. Leung and J.T. Koberstein, *Macromolecules*, 1986, **19**, 706.

26. Y.V. Savelyev, E.R. Akhranovich, A.P. Grekov, E.G. Privalko, V.V. Korskanov, V.I. Shtompel, V.P. Privalko, P. Pissis and A. Kanapitsas, *Polymer*, 1998, **39**, 3425.

27. J.T. Koberstein, A.F. Galambos and L.M. Leung, *Macromolecules*, 1992, **25**, 6195.

28. B. Chu, T. Gao, Y. Li, J. Wang, C.R. Desper and C.A. Byrne, *Macromolecules*, 1992, **25**, 5724.

29. S.S. Pesetskii, V.D. Fedorov, B. Jurkowski and N.D. Polosmak, *Journal of Applied Polymer Science*, 1999, **74**, 1054.

30. V.P. Privalko, E.S. Khaenko, A.P. Grekov and Y.V. Savelyev, *Polymer*, 1994, **35**, 1730.

31. L. Apekis, P. Pissis, C. Christodoulides, M. Spathis, E. Niaounakis, E. Kontou, E. Schlosser, A. Schoenhals and H. Goering, *Progress in Colloid & Polymer Science*, 1992, **90**, 144.

32. Y. Li, J. Kang, O. Stoffer and B. Chu, *Macromolecules*, 1994, **27**, 612.

33. G. Pompe, P. Pohlers, P. Poetschke and J. Piontek, *Polymer*, 1998, **39**, 5147.

34. L. Cuve, J.P. Pascault, G. Boiteux and G. Seytre, *Polymer*, 1991, **32**, 343.

35. J.T. Koberstein and L.M. Leung, *Macromolecules*, 1992, **25**, 6205.

36. S. Parnell, G. Suppiah and K. Min, Proceedings of Antec '99, New York, NY, USA, 1999, Volume 3, 3625.

37. R. Xu, J.L. Mead, S.A. Orroth, R.G. Stacer and Q.T. Truong, Presented at the 155th ACS Rubber Division Meeting, Chicago, IL, USA, Spring 1999, Paper No. 78.

38. D. Feldman and M.A. LaCasse, *Journal of Applied Polymer Science*, 1994, **51**, 701.

39. S.D. Seneker, N. Barksby and B.D. Lawrey, Proceedings of Polyurethanes Expo '96, Las Vegas, NV, USA, 1996, 305.

40. J. Xie, H. Ye, Q. Zhang, L. Li and M. Jiang, *Polymer International,* 1997, **44**, 35.

41. T. Bremner, D.J.T. Hill, M.I. Killeen, J.H. O'Donell, P.J. Pomery, D. St. John and A.K. Whittaker, *Journal of Applied Polymer Science,* 1997, **65**, 939.

42. B.P. Grady, E.M. O'Connell, C.Z. Yang, and S.L. Cooper, *Journal of Polymer Science: Polymer Physics Edition,* 1994, **32**, 2357.

43. E.M. O'Connell, C.Z. Yang, T.W. Root and S.L. Cooper, *Macromolecules,* 1996, **29**, 6002.

44. H.A. Al-Salah, *Polymer Bulletin,* 1998, **40**, 477.

45. J. Xie, Q. Zhang, H. Ye and M. Jiang, *Journal of Polymer Research,* 1995, **2**, 203.

46. W. Wu, X.L. Luo and D.Z. Ma, *European Polymer Journal,* 1999, **35**, 985.

47. C.P. Papadopoulou and N.K. Kalfoglou, *Polymer,* 1999, **40**, 905.

48. A. Stanciu, A. Airinei, D. Timpu, A. Ioanid, C. Ioan and V. Bulacovschi, *European Polymer Journal,* 1999, **38**, 1959.

49. S. Ioan, G. Grigorescu and A. Stanciu, *Polymer Plastics Technology and Engineering,* 2000, **39**, 807.

50. A. Stanciu, V. Bulacovschi, V. Condratov, C. Fadei, A. Stoleriu and S. Balint, *Polymer Degradation and Stability,* 1999, **64**, 259.

51. M. Shibata, T. Kobayashi, R. Yosomiya and M. Seki, *European Polymer Journal,* 2000, **36**, 485.

52. *Plastics Additives and Modifiers Handbook,* Ed., J. Edenbaum, Chapman and Hall, London, UK, 1996.

53. M. Sivakumar, R. Malaisamy, C.J. Sajitha, D. Mohan, V. Mohan and R. Rangarajan, *European Polymer Journal,* 1999, **35**, 1647.

54. M. Sivakumar, M. Carmel, L.A. Utracki, J.P. Szabo, I.A. Keough and B.D. Favis, *Polymer Engineering and Science,* 1992, **32**, 1716.

55. J.T. Haponiuk and A. Balas, *Journal of Thermal Analysis,* 1995, **43**, 215.

56. A.J. Varma, S.V. Deshpande and P. Kondapalli, *Polymer Degradation and Stability,* 1999, **63**, 1.

57. B. Zerjal, V. Musil, I. Smit, Z. Jelcic and T. Malavasic, *Journal of Applied Polymer Science,* 1993, **50**, 719.

58. *Siloxane Polymers,* Eds., S.J. Clarson and J.A. Semlyen, PTR Prentice Hall, Englewood Cliffs, NJ, USA, 1993.

59. S. Kohjiya, H. Tsubata and K. Urayama, *Bulletin of the Chemical Society of Japan,* 1998, **71**, 961.

60. J.A. Manson and L.H. Sperling, *Polymer Blends and Composites,* Plenum Press, New York, 1976.

61. F.P. La Mantia, A. Valenza and D. Acierno, *Polymer Degradation and Stability,* 1985, **13**, 1.

62. S.K. Singha Roy and C.K. Das, *Polymer Composites,* 1995, **3**, 5.

63. A.R. Tripathy and C.K. Das, *Journal of Applied Polymer Science,* 1994, **51**, 245.

64. A.R. Tripathy and C.K. Das, *Plastics & Rubber & Composites Processing & Applications,* 1994, **21**, 5.

65. S.K. Singha Roy and C.K. Das, *Journal of Polymer Engineering,* 1995, **14**, 175.

66. M. Maity and C.K. Das, *International Journal of Polymeric Materials,* 1998, **45**, 123.

67. R. Alex, P.P. De and S.K. De, *Polymer Communications,* 1990, **31**, 67.

68. S. Mukhopadhyay and S.K. De, *Journal of Applied Polymer Science,* 1992, **45**, 181.

69. S. Mukhopadhyay, P.P. De and S.K. De, *Journal of Applied Polymer Science,* 1991, **43**, 347.

70. T. Bhattacharya and S.K. De, *European Polymer Journal,* 1991, **27**, 1065.

71. S. Mukhopadhyay, T.K. Chaki and S.K. De, *Journal of Polymer Science: Polymer Letters Edition,* 1990, **28**, 25.

72. S. Mukhopadhyay and S.K. De, *Journal of Applied Polymer Science,* 1991, **42**, 2773.

73. L.C. Sawyer and D.T. Grubb, *Polymer Microscopy*, Chapman and Hall Ltd., London, UK, 1987.

74. G. Binnig, C.F. Quate and C. Gerber, *Physics Review Letters*, 1986, **56**, 930.

75. J. Frommer, *Angewandte Chemie - International Edition*, 1992, **31**, 1298.

76. H. Fuchs, *Journal of Molecular Structure*, 1993, **29**, 292.

77. L.M. Eng, H. Fuchs, K.D. Jandt and J. Petermann, *AIP Conference Proceedings*, 1992, **241**, 262.

78. L.M. Eng, K.D. Jandt, H. Fuchs and J. Petermann, *Applied Physics A*, 1994, **59**, 145.

79. P. Maivald, H.J. Butt, S.A.C. Gould, C.B. Prater, B. Drake, J.A. Gurley, V.B. Elings and P.K. Hansma, *Nanotechnology*, 1991, **2**, 103.

80. H. Fuchs, *Physics Bulletin*, 1994, **50**, 837.

81. D. Reifer, R. Windeit, R.J. Kumpf, A. Karbach and H. Fuchs, *Thin Solid Films*, 1995, **264**, 148.

82. G.M. Estes, R.W. Seymour and S.L. Cooper, *Macromolecules*, 1971, **4**, 452.

83. C.B. Wang and S.L. Cooper, *Journal of Applied Polymer Science*, 1981, **26**, 2989.

84. F. Xiao, D. Shen, X. Zhang, S. Hu and M. Xu, *Polymer*, 1987, **28**, 2335.

85. S. Cimmino, L. D'Orazio, R. Greco, G. Maglio, M. Malinconico, C. Mancarella, E. Martuscelli, R. Palumbo and G. Ragosta, *Polymer Engineering and Science*, 1984, **24**, 48.

86. T.K. Kang, Y. Kim, W.J. Cho and C.S. Ha, *Polymer Testing*, 1997, **16**, 391.

87. A. Genovese and R.A. Shanks, *Computational and Theoretical Polymer Science*, 2001, **11**, 57.

88. D.M. Koenhen, M.H.V. Mulder and C.A. Smolders, *Journal of Applied Polymer Science*, 1977, **21**, 199.

89. A. Garton, *Infrared Spectroscopy of Polymer Blends, Composites and Surfaces*, Hanser Publishers, Munich, Germany, 1992.

90. A. Koscielecka, *European Polymer Journal,* 1993, **29**, 23.

91. D. Allard and R.E. Prud'homme, *Journal of Applied Polymer Science,* 1982, **27**, 559.

92. M. Aubin, Y. Bedard, M.F. Morrissette and R.E. Prud'homme, *Journal of Polymer Science, Polymer Physics Edition,* 1983, **21**, 233.

93. R. Hernandez, J.J. Pena, L. Irusta and A. Santamaria, *European Polymer Journal,* 2000, **36**, 1011.

94. T. Tang, X. Jing and B. Huang, *Journal of Macromolecular Science B,* 1994, **33**, 287.

95. H. Stutz, P. Potschke and U. Mierau, *Macromolecular Symposia,* 1996, **112**, 151.

96. K. Wallheinke, P. Potschke and H. Stutz, *Journal of Applied Polymer Science,* 1997, **65**, 2217.

97. C.P. Papadopoulou and N.K. Kalfoglou, *Polymer,* 1998, **26**, 7015.

98. J. Sestak, *Thermophysical Properties of Solids, Their Measurements and Theoretical Thermal Analysis,* Elsevier, Amsterdam, The Netherlands, 1984.

99. J.M. Criado, M. Gonzales, A. Ortega and C. Real, *Journal of Thermal Analysis,* 1988, **34**, 1387.

100. C.D. Doyle, *Journal of Applied Polymer Science,* 1962, **6**, 39.

101. T. Ozawa, *Bulletin of the Chemical Society of Japan,* 1965, **38**, 1881.

102. J.H Flynn and L.A Wall, *Polymer Letters,* 1966, **4**, 232.

103. D.J.T. Hill, M.I. Killeen, J.H. O'Donnell, P. Pomery, D. St. John and A.K. Whittaker, *Journal of Applied Polymer Science,* 1996, **61**, 1757.

104. *Electronic Properties of Polymers,* Eds., J. Mort and G. Pfister, Wiley, New York, NY, USA, 1982.

105. *Conjugated Polymeric Materials. Opportunities in Electronics, Optoelectronics and Molecular Electronics,* Eds., J.L. Bredas and R.R. Chance, Kluwer Academic Press, Dordrecht, 1990.

106. *Organic Semiconducting Polymers,* Ed., J.E. Katon, Marcel Dekker, New York, NY, USA, 1968.

107. *Electronic Properties of Polymers and Related Compounds*, Eds., H. Kuzmany, M. Mehring and S. Roth, Springer-Verlag, Berlin, 1995.

108. M. Rusu, A. Stanciu, V. Bulacovschi, G.G. Rusu, M. Bucescu and G. I. Rusu, *Thin Solid Films*, 1998, **326**, 256.

109. M. Seki, K. Sato and R. Yosomiya, *Die Makromolekulare Chemie*, 1992, **193**, 2971.

110. M. Watanabe, M. Rikukawa, K. Sanui and N. Ogata, *Journal of Applied Physics*, 1985, **58**, 736.

111. L.H. Sperling, *Interpenetrating Polymer Networks and Related Materials*, Plenum Press, New York, NY, USA, 1981.

112. *Advances in Interpenetrating Polymer Networks*, Eds., D. Klempner and K.C. Frisch, Technomic, Lancaster, PA, USA, 1994.

113. *Interpenetrating Polymer Networks*, Eds., D. Klempner, L.H. Sperling and L.A. Utracki, Advances in Chemistry Series No. 239, ACS, Washington, DC, USA, 1994.

114. S.C. Tang, C.P. Hu and S.K. Ying, *Polymer Journal*, 1990, **22**, 70.

115. J.R. Millar, *Journal of the Chemical Society*, 1960, **263**, 1311.

116. *IPNs Around the World*, Science and Engineering, Eds., S.C. Kim and L.H. Sperling, Wiley, New York, NY, USA, 1997.

117. Y.S. Lipatov and L.M. Sergeeva, *Interpenetrating Polymer Networks*, Naukova Dumka, Kiev, 1979

118. R. Bischoff and S.E. Cray, *Progress in Polymer Science*, 1999, **24**, 185.

119. V. Athawale and S. Kolekar, *European Polymer Journal*, 1998, **34**, 1447

120. F.J. Hua and C.P. Hu, *European Polymer Journal*, 1999, **35**, 103.

121. L.W. Barret and L.H. Sperling, *Trends in Polymer Science*, 1993, **1**, 45.

122. Y.S. Lipatov, *Journal of Macromolecular Science C*, 1990, **30**, 209.

123. S. Raut and V. Athawale, *European Polymer Journal*, 2000, **36**, 1379.

124. E.E.C. Monteiro and J.L.C. Fonseca, *Polymer Testing*, 1999, **18**, 281.

125. D. Puett, *Journal of Polymer Science, Part A*, 1967, 5, 839.

126. S. Ajithkumar, N.K. Patel and S.S. Kansara, *European Polymer Journal*, 2000, **36**, 2387.

127. G.C. Marti, E. Enssani and M. Shen, *Journal of Applied Polymer Science*, 1981, **26**, 1465.

128. L.H. Fan, C.P. Hu and S.K. Ying, *Polymer Engineering and Science*, 1997, 37, 338.

129. L.H. Sperling and V. Mishra, *Macromolecular Symposia*, 1997, **118**, 336.

130. Y.C. Chou and L.J. Lee, *Polymer Engineering and Science*, 1994, **34**, 1239.

131. A. Kanapitsas, P. Pissis, L. Karabanova, L. Sergeeva and L. Apekis, *Polymer Gels and Networks*, 1998, **6**, 83.

132. W.Y. Chiang and D.M. Chang, Proceedings of Antec '94, San Francisco, CA, USA, 1994, Volume 1, 959.

133. V. Mishra and L.H. Sperling, *Journal of Polymer Science, Polymer Physics Edition*, 1996, **34**, 883.

134. Y.S. Lipatov, L.V. Karabanova and L.M. Sergeeva, *Polymer International*, 1994, **34**, 7.

135. S. Dadbin, R.P. Burford and R.P. Chapin, *Polymer Gels and Networks*, 1995, 3, 179.

136. S.C. Kim, D. Klempner, K.C. Frisch, W. Radigan and H.L. Frisch, *Macromolecules*, 1976, **9**, 258.

137. H. Djomo, A. Morin, M. Damyanidu and G.C. Meyer, *Polymer*, 1983, **24**, 65.

138. V. Mishra, F.E. Du Prez, E. Gosen, E.J. Goethals and L.H. Sperling, *Journal of Applied Polymer Science*, 1995, 58, 331.

139. B.K. Kim, *Macromolecular Symposia*, 1997, **118**, 195.

140. Y. Rharbi, A. Yekta, M.A. Winnik, R.J. DeVoe and D. Barrera, *Macromolecules*, 1999, **32**, 3241.

141. S.C. Kim, D. Klempner, K.C. Frisch, H.L. Frisch and H. Ghirardela, *Polymer Engineering and Science*, 1975, **15**, 339.

142. K. Kaplan and N.W. Tschoegl, *Polymer Engineering and Science,* 1975, **15**, 343.

143. N. Chantarasiri, N. Farkrachang, K. Kawnaramit and A. Eurpermkiati, *European Polymer Journal,* 2000, **36**, 95.

144. N. Devia, J.A. Manson, L.H. Sperling and A. Conde, *Polymer Engineering and Science,* 1979, **19**, 12.

145. W. Ku, J. Liang, K. Wei, H. Liu, C. Huang, S. Fang and W. Wu, *Macromolecules,* 1991, **24**, 4605.

146. X. He, J. Widmaies and G.C. Meyer, *Polymer International,* 1993, **32**, 289.

147. L.H. Sperling, *Polymer Engineering and Science,* 1985, **25**, 517.

148. N. Devia, J.A Manson, L.H. Sperling and A. Conde, *Macromolecules,* 1979, **12**, 360.

149. H. Xiao, Z.H. Ping, J.W. Xie and T.Y. Yu, *Journal of Polymer Science, Polymer Chemistry Edition,* 1990, **28**, 585.

150. J.R. Ebdon, D.J. Hourston and P.G. Klein, *Polymer,* 1984, **25**, 1633.

151. J.R. Ebdon, D.J. Hourston and P.G. Klein, *Polymer,* 1988, **29**, 1079.

152. D.J. Hourston and M. Zarandouz in *Advances in Interpenetrating Polymer Networks,* Volume 2, Eds., D. Klempner and K.C. Frisch, Technomic Publishing Co., Lancaster, PA, USA, 1990.

153. J. Geetha, *Polymer Science Symposium, Polymer Proceedings,* 1991, **2**, 882.

154. A. Stanciu, A. Airinei and S. Oprea, *Polymer,* 2001, **42**, 7257.

155. A. Stanciu, V. Bulacovschi, M. Lungu, S. Vlad, S. Balint and S. Oprea, *European Polymer Journal,* 1999, **64**, 259.

156. H.L. Frisch, K.C. Frisch and D. Klempner, *Polymer Engineering and Science,* 1974, **14**, 646.

157. L.T. Nguyen and N.P. Suh, *Polymer Engineering and Science,* 1986, **26**, 799.

158. S.S. Lee and S.C. Kim, *Polymer Engineering and Science,* 1991, **31**, 182.

159. T.J. Hsu and L.J. Lee, *Journal of Applied Polymer Science,* 1988, **36**, 1157.

160. T.J. Hsu and L.J. Lee, *Polymer Engineering and Science*, 1988, **28**, 956.

161. Y.S. Yang and L.J. Lee, *Macromolecules*, 1987, **20**, 1490.

162. J.H. Kim and S.C. Kim, *Polymer Engineering and Science*, 1987, **27**, 1243.

163. F.J. Hua and C.P. Hu, *European Polymer Journal*, 2000, **36**, 27.

164. H.L. Frisch, R. Foreman, R. Schwartz, H. Yoon, D. Klempner and K.C. Frisch, *Polymer Engineering and Science*, 1979, **19**, 284.

165. L.M. Sergeeva, S.I. Skiba and L.V. Karabanova, *Polymer International*, 1996, **39**, 317.

166. G.M. Yenwo, L.H. Sperling, J. Pulido, J.A. Manson and A. Conde *Polymer Engineering and Science*, 1977, **17**, 251.

167. N. Devia, J.A. Manson, L.H. Sperling and A.J. Conde, *Journal of Applied Polymer Science*, 1979, **24**, 569.

168. V.G. Kumar, M.R. Rao, T.R. Guruprasad and K.V.C. Rao, *Journal of Applied Polymer Science*, 1987, **34**, 1803.

169. S. Bai, D.V. Khakhar and V.M. Nadkarni, *Polymer*, 1997, **38**, 4319.

170. S.B. Pandit, S.S. Kulkani and V.M. Nadkarni, *Macromolecules*, 1994, **27**, 4595.

171. M. Rutkowska and A. Eisenberg, *Journal of Applied Polymer Science*, 1984, **29**, 755.

172. K.C. Frisch, D. Klempner, H.X. Xio, E. Cassidy and H.L. Frisch, *Polymer Engineering and Science*, 1985, **25**, 758.

173. K.H. Hsieh, L.M. Chou and Y.C. Chiang, *Polymer Journal*, 1989, **21**, 1.

174. J.H. Lee and S.C. Kim, *Macromolecules*, 1984, **17**, 2193.

175. J.H. Lee and S.C. Kim, *Macromolecules*, 1986, **19**, 644.

176. H. Chen and J.M. Chen, *Journal of Applied Polymer Science*, 1993, 50, 495.

177. G.M. Yenwo, J.A. Manson, J. Pulido, L.H. Sperling, A. Conde and N. Devia, *Journal of Applied Polymer Science*, 1977, **21**, 531.

178. M. Patel and B. Suthar, *Journal of Polymer Science, Polymer Chemistry Edition,* 1987, **25**, 2251.

179. P. Patel and B. Suthar, *Polymer,* 1990, **31**, 339.

180. P.L. Nayak, S. Lenka, S.K. Panda and T. Patnaik, *Journal of Applied Polymer Science,* 1993, **47**, 1089.

181. P. Chakrabarty and B. Das, *Journal of Applied Polymer Science,* 1996, **60**, 2125.

182. P. Nayak, D.K. Mishra, D. Parida, K.C. Sahoo, M. Nanda and S. Lenka, *Journal of Applied Polymer Science,* 1997, **63**, 671.

183. V. Athawale and S. Kolekar, *Polymer Journal,* 1998, **30**, 813.

184. V. Athawale and S. Raut, *Polymer Journal,* 1998, **30**, 963.

185. V. Athawale and S. Kolekar, *Journal of Macromolecular Science, Pure and Applied Science, Part A,* 1998, **35**, 1929.

186. V. Athawale and S. Kolekar, *European Polymer Journal,* 1998, **34**, 1451.

187. Y.S. Lipatov, T.T. Alekseeva, V.F. Rosovitsky and N.V. Babkina, *Polymer International,* 1995, **37**, 97.

188. L.R.G. Treloar, *The Physics of Rubber Elasticity,* 3rd Edition, Clarendon Press, Oxford, UK, 1975.

189. M. Pegararo, L. Di Landro, F. Severini, N. Cao and P.J. Donzelli, *Journal of Polymer Science, Part B: Polymer Physics Edition,* 1991, **29**, 365.

190. D. Jia, L. Chen, B. Wu and M. Wang in *Advances in Interpenetrating Polymer Networks,* Volume 1, Eds., D. Klempner and K.C. Frisch, Technomic Publishing Co., Paris, 1989.

191. D.S. Lee, J.H. An and S.H. Kim in *Interpenetrating Polymer Networks,* Eds., D. Klempner, L.H. Sperling and L.A. Utracki, Advances in Chemistry Series No. 239, ACS, Washington, DC, USA, 1994.

192. Y.S. Lipatov, G.M. Semenovitich, S.I. Skiba and L.V. Karabanova, *Polymer,* 1992, **33**, 361.

193. Y.S. Lipatov, T.T. Alekseeva, V.F. Rosovitsky and N.V. Babkina, *Polymer,* 1992, **33**, 610.

194. Y.S. Lipatov, and T.T. Alekseeva, *Polymer Communications*, 1991, **32**, 254.

195. N.P. Chen, Y.L. Chen, D.N. Wang, C.P. Hu and S.K. Ying, *Journal of Applied Polymer Science*, 1992, **46**, 2075.

196. L.H. Fan, C.P. Hu, Z.P. Zhang and S.K. Ying, *Journal of Applied Polymer Science*, 1996, **59**, 1417.

197. L.H. Fan, C.P. Hu and S.K. Ying, *Polymer*, 1996, **37**, 975.

198. L.H. Fan, C.P. Hu, Z.Q. Pan, Z.P. Zhang and S.K. Ying, *Polymer*, 1997, **38**, 3609.

199. L.H. Fan, C.P. Hu and S.K. Ying, *Polymer Engineering and Science*, 1997, **37**, 338.

200. L. Hermant and G. Meyer, *European Polymer Journal*, 1984, **20**, 85.

201. S. Jin and G. Meyer, *Polymer*, 1986, **27**, 592.

202. N.J. Harrick, *Internal Reflection Spectroscopy*, Wiley, New York, NY, USA, 1967

203. C. Clark-Monks and B. Ellis, *Journal of Polymer Science, Polymer Physics Edition*, 1973, **11**, 2089.

204. R. Pernice, K.C. Frisch and R. Navare, *Journal of Cellular Plastics*, 1982, **3**, 121.

205. N.P. Chen, Y.L. Chen, D.N. Wang, C.P. Hu and S.K Ying, *Journal of Applied Polymer Science*, 1992, **46**, 2075.

206. L.H. Fan, C.P. Hu, Z.Q. Pan, Z.P. Zhang and S.K. Ying, *Polymer*, 1997, **38**, 3609.

207. L.H. Fan, C.P. Hu and S.K. Ying, *Journal of Polymer Engineering and Science*, 1997, **37**, 338.

13 Blends and Networks Containing Silicon-Based Polymers

Valeria Harabagiu, Mariana Pinteala and Bogdan C. Simionescu

13.1 Introduction

The positioning of the silicon atom in the third period and fourth main group of the periodic table accounts for its chemical behaviour being similar to that of its neighbour, carbon. However, the higher atomic value and the presence of unoccupied $3d$ orbitals give marked differences between the properties of silicon atoms and silicon derivatives as compared to their carbon analogues. The higher electropositivity and the vacant $3d$ orbitals of the silicon atom diminish its tendency to self-linking and induce a strong affinity for electronegative elements possessing unshared p electrons, especially oxygen. Thus, the common silicon-based polymers are not the polysilanes but rather the polysiloxanes. However, Si-Si and Si-C incatenation in polysilanes and polycarbosilanes, respectively, is also possible.

The properties of silicon-based polymers are strongly dependent on both the structure of chain backbone (Si-Si, Si-C or Si-O-Si) and the nature of the organic substituents attached to the silicon atom. Polysilanes are characterised by interesting photo- and electrochemical behaviour related to the conjugation of σ electrons into the Si-Si skeleton. Polycarbosilanes, especially those containing aromatic groups, are generally thermally stable compounds. In polysiloxanes, the double inorganic (Si-O backbone) - organic (organic radicals linked to the silicon atom) nature of the macromolecular chains determines a quite unique combination of properties. Polydimethylsiloxane (PDMS), the most simple and used siloxane polymer, is characterised by high flexibility, hydrophobicity and physico-chemical stability, low surface tension and surface energy, low solubility parameter and dielectric constant and UV transparency. In addition, these properties show only relatively small variation over a wide range of temperatures. Polysiloxanes also possess film forming ability, high gas permeability, release action, and chemical and physiological inertness. This is why Rochow wrote some years ago 'For decades I avoided mixtures or copolymers of organic resins with silicone, even abhorred them, because I did not want the pristine properties of my silicon brain children to be degraded or ruined by adulteration' [1]. However, the poor mechanical properties of polysiloxanes impelled the development of two- and multi-component polymeric materials. Several reasons encouraged the preparation of such materials containing chemically linked or physically mixed silicon-based polymers:

a. the wish to impart their useful properties to the newly obtained material and to diminish the shortcoming of their poor mechanical properties,

b. cost reasons (silicon polymers are relatively expensive),

c. their capacity to provide model blend systems appropriate to theoretical studies. In principle, three types of bicomponent systems, differing in their structure and ultimate properties could be distinguished: crosslinked copolymers, blend systems or interpenetrating networks (IPN) (**Scheme 13.1**).

In blend systems of silicon-based polymers the low morphological stability of the polymeric material is a consequence of the immiscibility of blend components that could induce a macrophase separation. Copolymer networks are composed from two different polymer sequences chemically linked into a single network. In such systems the crosslinking prevents the macrophase separation and only a microphase separation is possible. The IPN contain two different polymer networks (full-IPN) or a linear polymer and a polymer network (semi-IPN); the structure of the polymeric material is stabilised by crosslinking and chain entanglement.

This chapter considers the recent developments in blend systems, copolymer networks and IPN, containing as one of the partners, polysiloxanes, polycarbosiloxanes, polysilanes or polycarbosilanes possessing the properties presented in **Table 13.1**.

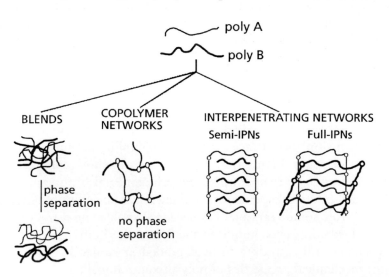

Scheme 13.1 Schematic representation of two-component polymeric materials

Table 13.1 Silicon-based polymers – components of polymer blend systems			
Code	Structure	Properties	Refs.
POLYSILOXANES			
PDMS	CH_3 / $-Si-O-$ / CH_3	$T_g = -123$; $T_m = 47.5$ $T_{dec} = 300$; $\delta = 7.5$ $\gamma = 21.0$ Permeability: O_2: 353; N_2: 695	[2, 3]
PDPhS	C_6H_5 / $-Si-O-$ / C_6H_5	$T_g = 35$ $\delta = 9.5$	[4]
PPhMS	CH_3 / $-Si-O-$ / C_6H_5	$T_g = -85$ $\delta = 9.0$; $\gamma = 26.1$ Permeability: O_2: 24	[5]
POHMS	CH_3 / $-Si-O-$ / OH	$T_g = -128$ (freshly prepared)	[6]
PDMS-PDPhS	C_6H_5 CH_3 / $-(Si-O)_n(Si-O)_m-$ / C_6H_5 C_6H_5	$T_g = -115$ to -35 for copolymers with 10%-80% DPhS units	[4]
PDMS-NH_2	CH_3 CH_3 / $-(Si-O)_n Si-R-NH_2$ / CH_3 CH_3	$T_g = -123$ to -115; increases with decreasing MW $\gamma = 19.9 - 21.0$; decreasing with increasing MW	[7]
POLYCARBOSILOXANES			
PTMPS	CH_3 CH_3 / $-Si-⬡-Si-O-$ / CH_3 CH_3	$T_g = -15$ $T_m = 129$ Crystallinity = 45	[8]

Code	Structure	Properties	Refs.
Table 13.1 Continued			
POLYSILANES			
PDMDPhSi	CH_3 CH_3 $-Si-Si-$ CH_3 C_6H_5	no T_g $T_m = 55-75$	[9]
PR¹R²Si	R^1 $-Si-$ R^2	Polymethylpropylsilane exhibits columnar liquid crystalline phase	[10, 11]
POLYCARBOSILANES			
PDMSiM	CH_3 $-Si-CH_2-$ CH_3	$T_g = -88$	[12]
PSiαMS	CH_3 $-Si-CH_2-$ C_6H_5	$T_g = 22$	[13]
PDMSiPh	CH_3 $-Si-\bigcirc-$ CH_3	$T_g = 15$ $T_m = 100-160$ Crystallinity = 45	[8]

PDMS = polydimethylsiloxane; PDPhS = polydiphenylsiloxane; PPhMS = polyphenylmethylsiloxane; POHMS = polyhydroxymethylsiloxane; PDMS-PDPhS = poly(dimethyl-co-diphenyl)siloxane; PDMS-NH₂ = aminoalkyl-terminated PDMS; PTMPS = polytetramethylsil-phenylenesiloxane; PDMDPhSi = poly(dimethyl-co-diphenyl)silylene; PR¹R²Si = differently substituted polysilylenes (R¹=R²=n-C₅H₁₁; R¹=R²=n-hexyl; R¹=CH₃ and R²=C₃H₇; R¹=CH₃ and R²=C₁₈H₃₇; R¹=R²=n-butyl); PDMSiM = polydimethylsilylenemethylene; PSiαMS = polysila-α-methylstyrene; PDMSiPh = polydimethylsilphenylene;

T_g = glass transition temperature (°C); T_m = melting point (°C); T_{dec} = decomposition temperature (°C); crystallinity (%); δ= solubility parameter (cal/cm³)¹/²; γ= surface tension (mN/m); perm. = permeability (cm³·cm/cm²·s·Pa) at 273 K and 1013 x 10⁵ Pa

13.2 Blend Systems of Silicon-based Polymers

It is well established that the morphology of polymer blends is controlled by thermodynamics, kinetics and viscosity, and is highly influenced by processing conditions [14-17]. As the final properties of polymer blends are fundamentally determined by their morphology, the investigation of polymer blend thermodynamics is of a great theoretical and practical interest.

Polysiloxanes, especially PDMS, are characterised by an extremely low solubility parameter (δ = 7.5 $(cal/cm^3)^{1/2}$) [2] and surface tension (γ = 21 mN/m) [2]. As a consequence, they proved to be highly incompatible and, moreover, immiscible with almost all other polymers. This is why only few blends containing pure PDMS homopolymer [18] or dimethylsiloxane containing copolymers [19] were described in the older literature. As the practical interest is directed to miscible blends, recent studies on blends containing silicon-based polymers focus on the fundamental approaches on mixtures containing carbon-based polymers and their carbosilane or carbosiloxane analogues [8, 9, 12, 13], on binary (functional polysiloxanes/solvent) [20-22], ternary (organic polymer/PDMS/solvent) and quaternary (organic polymer/PDMS/solvent/solvent) solutions [23]. The engineering of polymer miscibility through the use of polymer structures able to provide specific interactions such as hydrogen bonding [24, 25] and of copolymers [26, 27] was also considered.

13.2.1 Thermodynamic Aspects

13.2.1.1 Interaction Parameter

The simplest relationship describing the miscibility of polymer blends is the well-known Flory-Huggins theory [28, 29] giving the free energy of mixing ΔG_m as a function of interaction parameter, χ:

$$\Delta G_m = RTV \, (\phi_1/V_1 \ln\phi_1 + \phi_2/V_2 \ln\phi_2 + \chi/V_r \, \phi_1\phi_2) \qquad (13.1)$$

where R = gas constant; T = absolute temperature; V and V_r = total and reference volumes; V_i (i = 1 or 2) = volume of a chain polymer of component i and ϕ_i = volume fraction of component i.

The first two terms in Equation 13.1 represent the combinatorial entropy of mixing. In most cases, where both blend components are high molecular weight polymers, the entropy of mixing is negligible and the phase behaviour of the blend are governed by the sign and temperature dependence of the interaction parameter. The stability condition of the mixture is expressed in Equation 13.2; polymers are miscible only in those rare instances when χ is negative or close to zero:

$$2\chi < V_r \, (1/V_1\phi_1 + 1/V_2\phi_2) \tag{13.2}$$

A deeper insight into polymer blend thermodynamics is offered by equation of state theories (EOS) [30] which describe a polymer liquid in terms of reduction parameters p^*, v_{sp}^*, and T^*, with different numerical values for different theories. These parameters are used to evaluate the corresponding reduced parameters, $\bar{p}, \bar{v}, \bar{T}$:

$$\bar{p} = p/p^*, \quad \bar{v} = v_{sp}/v^*, \quad \bar{T} = T/T^*$$

where p = pressure; v_{sp} = specific volume and T = temperature.

The EOS reads:

$$\bar{p}\bar{v}/\bar{T} = \bar{v}^{1/3}/\bar{v}^{(1/3-1)} - 1/\bar{v}\bar{T}$$

and the free-energy of mixing also contains free-volume contributions, which were completely neglected by Flory-Huggins theory. For EOS theories based on the Prigogine statement [31] it is possible to derive an expression for χ which *a priori* contains entropical contributions by using the more simple Flory-Huggins theory to extract the χ parameter from EOS theories. However, there are different ways to determine a composition dependent χ parameter from these theories and the results may be different. The most used EOS theories are Flory-Orwoll-Vrij [32-34], Patterson [35, 36] or modified cell model of Dee and Walsh [37]. All these theories provide the expression of a total interaction parameter χ_{tot} as a sum of χ_{int} (enthalpic contribution accounting for the exchange of energy of the two blend components) and χ_{fv} (free-volume contributions):

$$\chi_{tot} = \chi_{int} + \chi_{fv} \tag{13.3}$$

This theoretical treatment offers a logical interpretation of the complicated phase behaviour of polymer blend systems. It has been verified on several model blends containing silicon-based polymers (**Table 13.2**). Different methods were used for the acquisition of the required experimental data, such as pressure, volume and temperature measurements (to obtain the specific volume and to calculate the reduction parameters (pressure, specific volume and temperature)) [12, 13], differential scanning calorimetry (DSC), optical and electron microscopy (phase transition behaviour, morphology) [12, 13], spinning- and sessile-drop measurements (surface tension) [38, 39], neutron scattering (observation of miscibility limit in symmetric isotopic blends of PDMS [40]).

Table 13.2 Thermodynamically characterised blend systems containing silicon-based polymers

Blend no.	Silicon-based polymer			Organic or silicon polymer			Blend characteristics	Ref.
	Code	M_w (kg/mol)	T_g (°C)	Code	M_w (kg/mol)	T_g (°C)		
1	PDMSiM	721	−88	PIB	1.5−85	−81, −64	- Completely miscible for $M_{PIB} \leq 1.5$ - UCST and LCST to 'hour glass' behaviour with increasing molecular weight	[12]
2	PDMS	11.3 and 19.6	-	PIB	1.5−5	-	UCST, closed miscibility gap behaviour	[38]
3a	PSiαMS	188	22	PS	3−180	80−110	UCST behaviour	[13]
3b	PSiαMS			PDMPO	30.7	213	Completely immiscible	[13]
3c	PSiαMS			PVME	78	-	Completely immiscible	[13]
4	ODMS	0.46	-	OS	0.6	-	UCST behaviour	[39]
5	PDMS	15, 25 and 75	−123	d-PDMS	15, 25, 75	−123	UCST behaviour (UCST increases with increasing molecular weight)	[40]
6	PDMS	2.5−625		PVC		−123	Completely immiscible	[41]
7	PDMS	1.11 (linear)	−123	PPhMS	0.7 (cyclics) 1.9 (linear)	−30, −48	UCST behaviour	[42−45]

ODMS = oligodimethylsiloxane; PDMPO = poly(2,6-dimethylphenylene oxide); PVME = polyvinyl methyl ether; OS = oligostyrene; d-PDMS = deuterated polydimethylsiloxane

Comparing the blend systems based on polyisobutene (PIB) or polystyrene (PS) and their carbosilane analogues (obtained by the replacement in the backbone of a carbon atom with a silicon atom) to those containing PDMS (blends 1, 2, 3a and 4 in **Table 13.2**), the following general features can be deduced:

Polycarbosilane-based blends [12, 13] present an extended miscibility when compared to the blends containing siloxane polymers [38, 39], due to the closer similarity of the chemical structure of vinyl polymers to that of their carbosilane analogues. As the χ_{tot} (Equation 13.3), **Figure 13.1**) is positive and therefore unfavourable to mixing, the observed miscibility for blends containing low molecular weight vinyl polymers [12] was explained by the gain in combinatorial entropy of mixing.

The phase behaviour of the polymer blends studied is a result of a balance of interactional and free-volume contributions. The nature of the backbone rather than the nature of the constituents controls the repulsive enthalpic interactions between the vinyl polymers and their carbosilane analogues as χ_{int} values are similar in the systems studied (blends 1 and 3a, **Table 13.2**). Also, in PDMSiM/PIB blends χ_{int} and χ_{fv} are of the same order of

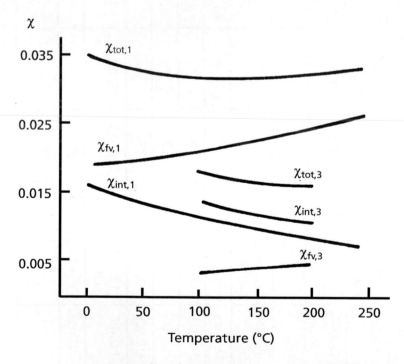

Figure 13.1 $\chi_{x,n}$ (x = tot, int or fv) of PDMS/PIB (n = 1) and PSiXαMS/PS (n = 3) systems as a function of temperature (redrawn from data of references [12, 13])

magnitude; the simultaneous occurrence of upper critical solution temperature (UCST) and lower critical solution temperature (LCST) (UCST < LCST) behaviour can be envisaged and 'hour glass' phase diagrams can be calculated [12], whereas PDMS/PIB system shows a closed miscibility gap [38]. The driving force for LCST behaviour of PDMSiM/PIB blends is a large difference between the thermal expansion coefficients characterising the blend components. For PSiαMS/PS system (blend 3a, **Table 13.2**) χ_{int} is one order of magnitude larger than χ_{fv} resulting in an exclusive occurrence of UCST behaviour [13], similar to ODMS/OS blend (blend 4, **Table 13.2**) [39].

A computational method based on the Flory-Huggins lattice model, the extended random copolymer theory and the molecular modelling method were used to show a non-negligible effect of end groups on the χ value and temperature of phase separation in blends of PDMS and polymethylphenylsiloxane of low molecular weight [46].

13.2.1.2 Surface/Interfacial Tension

The interfacial tension, σ, between (incompatible) polymers is of great theoretical and practical interest. Even if a theoretical background was established some years ago [47-49], there is a considerable lack of understanding linked to the fact that the qualitative relations between σ and the thermodynamic interaction parameter still can not be drawn into reliable and experimentally verified theoretical equations. The discussion of the influence of the molecular weight and temperature on σ is difficult since the existing relations correlate σ and χ for infinitely large polymers, the χ values cannot normally be measured (another set of theoretical relationships are required for their calculation) and few of the existing theories try to consider the dependence of χ on composition.

The variation of interfacial tension with temperature and molecular weight in polymer blends is strongly influenced by the miscibility of its components, measured by the segregation strength [50]. Three regimes of segregation are characterised by the differences between χ and its value at the critical point, χ_c: weak segregation, $\chi \geq \chi_c$; strong segregation, $1 > \chi \gg \chi_c$; very strong segregation $\chi > 1$.

The study of σ of partially miscible polymer blends as a function of segregation strength implies the overcoming of two major experimental difficulties, i.e., obtaining macroscopic phases coexisting in equilibrium (polymer blends are high viscosity systems) and the very high critical temperatures. PDMS, with its very low T_g value and relatively low viscosity is a good candidate for the preparation of model blends suitable for interfacial tension studies. Several PDMS containing polymer blends have been investigated recently (**Table 13.3**).

Table 13.3 Siloxane blends considered for interfacial tension investigation

Blend no.	M_{PDMS} (kg/mol)	Partner polymer		Verified theories	Experimental methods	Discussion	Ref.
		Type	Molar mass (kg/mol)				
1	0.46	PS	0.6	Square gradient relation [51]	Sessile-drop measurements	- UCST = 101.92 °C - critical composition, w_{PDMS} = 0.5 - in critical mixtures temperature dependence of σ is weak as compared to pure PS	[39, 52, 53]
2	11.3 or 19.6	PIB	1.5 or 3.3 or 5.0	'Harmonic mean' relation [54]	Density measurements Spinning-drop and sessile-drop measurements	UCST > 200 °C (increases with increasing M_{PIB}); for $PDMS_{11.3}/PIB_5$, UCST = 262 °C	[38]
3	3.0	PHxMS	11.3	Mean field theory Ising-3D model		σ increases with the length of the corresponding tie line	[55]
4	2.5 or 150 or 625	CR			XPS, SEM		[56]
5	2.5 or 150 or 625	PVC	120 or 190 or 260		XPS, SEM	The degree of surface segregation decreases with increasing M_{PDMS}	[41]
6	4-177	PEO	41		Sessile-drop measurements		[57, 58]

PHxMS = polyhexylmethylsiloxane; CR = polychloroprene; PVC = polyvinyl chloride; PEO = polyethylene oxide; XPS = X-ray photoelectron spectroscopy; SEM = scanning electron microscopy; M_{PDMS}: molecular weight of PDMS

The values and temperature dependence of interfacial tension between PDMS and organic polymers are strongly dependent on the nature and on the molecular weight of the organic partner (**Figure 13.2**). For mixtures of dimethylsiloxane and styrene or hexylmethylsiloxane oligomers [39, 55] it was possible to measure the interfacial tension near the critical point. The ODMS/OS blend system is characterised by a value of critical temperature slightly higher than 100 °C.

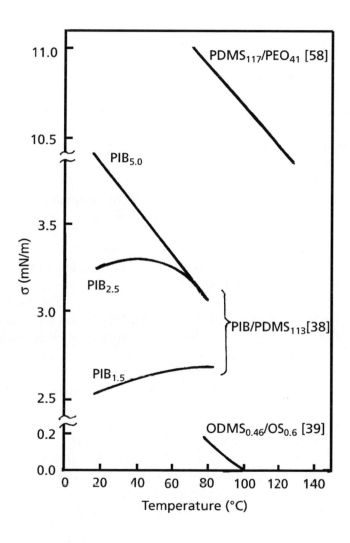

Figure 13.2 Temperature dependence of interfacial tension of PDMS containing blend systems (redrawn from data of references [38, 39, 58])

In critical mixtures the temperature dependence of σ is weak when compared to that of pure oligostyrene and a dramatic decrease of σ with the increase of the siloxane weight fraction between 0 and 0.2 is observed [53]. PIB containing blends [38] showed a linear decrease of σ with temperature only for the highest studied homologue, $PIB_{5.0}$ (a critical temperature, $T_c = 262$ °C, was obtained by extrapolating the straight line), whereas the blends obtained with lower oligomers presented $\sigma(T)$ curves characterised by maxima values indicating the existence of a closed miscibility gap. The total polar and dispersion components of surface tension (γ^p and γ^d) for each blend partner were determined (γ_{PDMS} = 20.9 mN/m, γ^p_{PDMS} = 0.8 mN/m; γ_{PIB} = 33.4 mN/m, γ^p_{PIB} = 0.4 mN/m). The mixtures of the hydrophobic PDMS with a more polar polymer (PEO) are characterised by relatively high values of σ and no reasonable value of the critical temperature of this mixture is to be envisaged [57, 58].

The very low solubility parameter and surface energy of polysiloxanes and especially PDMS, determine their immiscibility with almost all organic polymers. In polymer blends they migrate at the air contact surface of the material. Since the blend system tends to decrease its energy by enriching the surface with the component of the lower surface tension and/or by diminishing the number of energetically disadvantageous contacts of different components in its bulk by squeezing out one component to the surface [59], the surface composition of siloxane blends is strongly different from the bulk composition.

XPS associated with SEM proved to be appropriate methods for the investigation of surface profile of siloxane blends. For both PDMS/CR [56] and PDMS/PVC mixtures [41] the solvent cast film blends contain PDMS enriched surface layers. PDMS surface enrichment dramatically decreases with increasing PDMS molecular weight (from 100% for $PDMS_{2.5}$ - an oligomer - to less than 50% for $PDMS_{625}$) and no evidence of a clear influence of the molar mass of organic polymer upon surface segregation was observed. The explanation of this strange dependence of surface segregation on the molecular weight was thought to be the higher mobility of low molecular weight molecules during solvent evaporation (the samples were prepared by solution casting method).

The migration of the siloxane sequences to the air/blend surface was also observed in siloxane containing block or graft copolymer/PS systems [60]. A deep insight into the surface activity of siloxane copolymers was performed by Teyssie and co-workers on the poly(styrene-block-dimethyl siloxane (poly(S-b-DMS)/PS) system [61]. Both surface tension (measured by wettability and contact angles) and surface composition (determined by X-ray photoelectron spectroscopy and secondary ion mass spectrometry) investigations showed the accumulation of PDMS blocks at the air/polymer interface when small amounts of the block copolymer (0.2%-2.0%) were added to the PS matrix and practically complete covering of the surface with PDMS blocks. A rather surprising observation was the block copolymer enrichment of the interface between polymer and the high energy substrate (glass, stainless steel or aluminium), a specific behaviour for copolymers containing a polar sequence, for non equilibrium spinning-cast films. However, after annealing, the copolymer migrated to the polymer/air interface.

Because of the difficulties in measuring σ, Wolf and co-workers proposed semiempirical relationships for a rough prediction of σ, useful especially in technical applications [62] and verified them on model systems containing polysiloxanes as one of the components. One of the relationships introduces the reduced lump energy ε (area determined by the curve DG/RT - composition and the double tangent to that curve; ΔG = the reduced segment molar Gibbs energy of mixing):

$$\sigma = E\varepsilon^F$$

$$\varepsilon = A\tau^B$$

F being close to 0.5 for all systems; B = 2.4; A and E are variable parameters that are not universal; τ is the relative distance from the critical point.

Another proposed relationship offering a simple means to grasp the effect of polydispersity correlated σ with the length of the corresponding tie line, Δφ (φ = volume fraction):

$$\sigma = A\ M^{0.5}\Delta\phi^B$$

13.2.2 Influence of Additives on σ

The interfacial tension between two polymer components of a blend can be modified by the addition in their mixture of low or high molecular weight additives such as solvents or copolymers with various molecular structures.

The theory of Broseta and Leibler [48] was verified by the addition of a good solvent (toluene) for both components of the PDMS/PIB blends [38]. A sharp decrease of σ was observed with increasing the weight fraction of the solvent. Methyl ethyl ketone (MEK), a good solvent for PDMS and a poor one for PIB, produces a $\sigma(w_{MEK})$ curve characterised by a minimum value for the weight fraction $w_{MEK} \approx 0.4$. A rather inverse effect was obtained by adding a small amount of an incompatible polymer (PDMS or PS-PDMS block copolymer) to a phase separated polymer solution of PS in cyclohexane, when an important increase of σ was observed [63].

The modification of σ values of PDMS/PEO [57, 58] and PDMS/PPhMS [64] blends in the presence of copolymers was also investigated. Different chemical structures of polymeric additives were determined, i.e., di-, triblock and random copolymers having structural units of the same nature as those of the blend components or copolymers with chemical structures of the blocks different from those of the blend components. The addition of relatively small amounts of all these copolymers induced a substantial decrease of σ. For PDMS/PEO blends σ decreases from 10 mN/m to 1-4 mN/m by adding 0.2% by weight of di- or triblock PDMS-PEO copolymers with different architectures [57, 58]. Contrary to theoretical expectations, it was established that the total number of Si-O units in additives is more decisive for their efficiency than copolymer architecture. PS-polymethyl methacrylate (PMMA) diblock copolymers proved to be less effective additives for the same PDMS/PEO blends (the addition of 4% by weight of PS-PMMA copolymer determines a decrease of only three units for σ). The efficiency of PDMS-methylphenylsiloxane random copolymers was demonstrated by the decreasing of σ value of PDMS/PPhMS blend from 2.2 to 1.6 mN/m [64]. Two different kinetic processes were observed during the equilibration, (a) interdiffusion of the components in the vicinity of the phase boundary and (b) hydrodynamic relaxation of the droplet.

Very recently Nose and co-workers [65] reported on the influence of oligodimethylsiloxane additives upon the thermodynamic and interfacial tension behaviour of PDMS/polytetramethyldisiloxanylethylene blend characterised by weak segregation. The addition of ODMS produces a decrease of the critical temperature and shifts the σ - T curves to lower temperatures as the concentration of the additive increases, suggesting that the adsorption effects of additive on interfacial tension are weak. Both square gradient theory and dynamic mean field calculations compare well with the experimental behaviour.

The surface activity of PDMS containing block and graft copolymers in bicomponent polymer/copolymer systems at air/polymer interfaces was also determined in relatively older reports [7, 66-73]. The surfactant behaviour was observed to be dependent on copolymer architecture and on the nature of the polymeric partner of PDMS in the copolymer structure. The complete surface coverage by PDMS occurs at copolymer concentrations as low as 0.1%-2% by weight.

Recently, it was established that multiblock copolymers possess a general tendency to form transparent microheterogeneous systems by blending with a chemically dissimilar homopolymer known to be immiscible with each of the polymer sequences of the multiblock additive. This behaviour was also reported for mixtures of PVC and polydimethylsiloxane - poly(bisphenol A carbonate) segmented copolymers [74] and for blends of PVC, PS or PMMA with polycarbonate (or polysulfone) - PDMS multiblock copolymers [75, 76]. The blend transparency is not the consequence of the thermodynamic miscibility of the components, but of the microphase separation with a dispersion of block copolymer in the form of small-sized particles (up to 100 nm in diameter). The 'transparency window' is dependent on the nature of blend components, on their molecular characteristics, on composition and the solvent used for blend preparation.

13.2.3 Miscibility – Compatible Blends

Very few mixtures composed of low molecular weight silicon-based oligomers were found to be miscible with other polymers in narrow intervals of compositions. Because of the increased similarity between the chemical structures of carbosilane polymers and organic analogues, a higher miscibility (larger composition intervals at reasonable temperatures) was observed for their blend systems (see **Table 13.2**). Moreover, blend systems of PDMS of different molecular weights [51] or PDMS isotopic blends [40] are characterised by small but positive values of interaction parameter, UCST behaviour and phase separation at room temperature for a symmetric composition. As for blend systems based on polysilane components, a negative value of χ parameter and a poor degree of miscibility was found for polymethylpropylsilane/polydi(*n*-hexyl)silane system, while polydi(*n*-hexyl)silane and polymethyloctadecylsilane yield immiscible blends [77]. Existing methods to improve blend miscibility using specific molecular interactions such as hydrophobic interaction or hydrogen bonding were not often used for blends containing silicon-based polymers. However, an increased miscibility window was reported for blends of 4-vinylphenyldimethylsilanol homopolymer or its styrene copolymers (hydrogen donors) with polybutyl methacrylate [24, 25] or poly(*N*-vinylpyrrolidone) [78] (hydrogen acceptors based on ester or amide carbonyl groups). Highly miscible blends were also obtained by solvent-casting of polyhydroxymethylsiloxane and poly(*N*-vinylpyrrolidone) or PEO mixtures [6].

An increased miscibility window based on hydrogen bond formation was also reported for other organic polymer/siloxane copolymer mixtures. For example, miscible blends of poly(siloxane-imide) segmented copolymers and polybenzimidazole were prepared and evaluated as potential high performance aerospace materials [27].

Few compatible blends containing silicon-based polymers were reported in recent publications. Peculiar structures like PTMPS or polydimethylsilphenylene (PDMSiPh), partially crystalline polymers, were blended with PS of different molecular weights and poly(2,6-dimethyl-1,4-phenylene oxide) (PDMPO) [8]. Except for the low molecular weight PS/PDMSiPh and PDMPO/PDMSiPh (80:20) systems that are miscible on the entire interval of compositions, all other blend systems were proved to be phase separated, with morphologies more or less similar to those of pure PDSiPh or PTMPS. Co-occurrence of liquid-liquid phase separation and crystallisation phenomena could explain the observed dependence of blend morphology on the nature of organic partner and blend composition.

Rigid-rod polyimides grafted with PDMS were synthesised and proved to possess a good compatibility with linear PDMS elastomers for PDMS side chain in the grafted copolymer longer than 40 siloxane units [79]. Electron microscopic imaging techniques were used to study the microphase separated mixtures of poly(styrene-phenylmethylsiloxane) block copolymer with both styrene and siloxane homopolymers [80].

A simple method to provide compatible blends is to mix appropriate PDMS copolymer structures. PDMS-b-polymethacrylic acid (PDMS-PMAA) and PDMS-b-poly(*N*-vinylpyrrolidone) (PDMS-PNVP) segmented copolymers were solution blended [81]. Starting from phase separated prepolymers (two T_g values; curves 1 and 2 in **Figure 13.3**), a phase separated material with T_g values characterising PDMS and polyvinyl rich domains were obtained (curves 3 and 4 in **Figure 13.3**).

As a result of the hydrogen bonding between the vinyl sequences, for an almost 1:1 MAA/NVP molar ratio, a single T_g for a complexed specimen at a value with 40 °C higher than the T_g of PMAA blocks was observed. Copolymer complexes characterised by MAA/NVP ratios far away from the equivalence present DSC curves with three different T_g values in the positive temperature interval. They correspond to uncomplexed polyvinyl sequences (PMAA and PNVP uncomplexed sequences) and to PMAA-PNVP complexed domains, respectively.

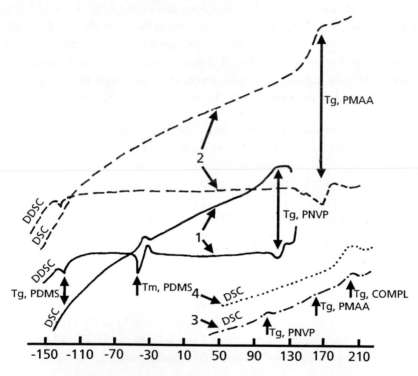

Figure 13.3 Typical DSC curves for (1) PDMS-poly(*N*-vinyl pyrrolid) (PNVP) copolymer; (2) PDMS-polymethylacrylic acid (PMAA) copolymer; (3) mixture of (1) and (2), MAA/*N*-vinyl pyrrolidone (NVP) = 1/1 molar ratio; (4) mixture of (1) and (2), MAA/NVP = 0.6/1 molar ratio (T_g = glass transition temperature; COMPL = PNVP-PMAA complexed specimen)

Binary PDMS-b-PMAA or ternary PDMS-b-(PS-random-PMAA) copolymers of low molecular weight were proved to provide compatibility in silicone greases based on PDMS oil and fatty acid lithium salts [81].

13.2.4 Rheology

Rheology of polymer blends is a subject of great interest, particularly for their processing behaviour in the two-phase state. Recent publications provide correlations of rheology parameters and interfacial tension [82-84], descriptions of structure development during spinodal decomposition [85] or as a result of flow stress [86-97] and the influence of the phase state on the rheology in the vicinity of phase separation or far from the phase separation [94-103]. Near the phase separation temperature and far from the T_g the linear viscoelastic behaviour of a polymer blend is controlled by critical concentration fluctuations and the rheological determination of both the binodal and spinodal diagrams is possible by an appropriate quantitative account of these fluctuations.

PS/PPhMS blends were studied as model binary blends of unentangled polymers characterised by UCST behaviour [104]. The rheology of this system was proved to be sensitive to phase separation very much like blend systems producing LCST type phase diagrams. In the homogeneous state, the viscosity - composition dependence follows a mixing rule (Equation 13.4) accounting for the surface fractions instead of volume fractions:

$$\ln\eta = \theta_{PS}\ln\eta_{PS} + \theta_{PPhMS}\ln\eta_{PPhMS} + \ln(\eta_{PS}/\eta_{PPhMS})\,[g\theta_{PS}\,\theta_{PPhMS}/(1 + g\theta_{PS})] \qquad (13.4)$$

where η = viscosity; θ = surface fraction; g = geometric factor.

In the phase separated state the blend system consists of isolated droplets of the less viscous PPhMS-rich phase dispersed in the matrix of the more viscous PS-rich coexisting phase and follows the scaling relation established by Onuki [105] (Equation 13.5):

$$N_1\text{-}N_2 \sim (\eta_{PS})^{3/2}\,(\eta_{PPhMS})^{-1/2}\,w_{PPhMS}\,\dot{\gamma} \qquad (13.5)$$

where N = normal stress; η = viscosity; w = weight fraction; $\dot{\gamma}$ = shear rate.

The rheology of the PDMS/PIB blend systems is also discussed [106-109].

13.2.5 Properties and Applications of Blends Containing Silicon-based Polymers

Blending of two or more polymers/copolymers is largely used to modify the physical properties of a polymeric material in a desired direction. General theoretical rules to predict the properties of polymer blends as verified on systems containing silicon-based polymers were discussed in previous sections. This section deals with peculiar properties and end-use properties of polymer blends containing silicon-based macromolecular compounds.

13.2.5.1 Thermal Behaviour

Based on their thermal stability, silicones were earlier tested as non-halogenated fire retardants in blends with polyolefins but their efficiency was proved to be poor in uncrosslinked polyolefins [110, 111].

Generally, the thermal stability of polymer blends is a result of the behaviour and relative proportion of blend components. However, some peculiarities not directly linked to their partner's thermal stability could arise after blending. A deep insight into the effect of blending of different vinyl polymers (low density polyethylene, polyethyl acrylate and ethylene-ethyl acrylate copolymer) with vinyl-terminated PDMS was found by McNeill and Mohammed [112] by using different experimental techniques (thermogravimetry, volatilisation analysis, DSC). Both polymer components were stabilised by mixing and the degradation of the blend systems was observed to occur at temperatures higher than predicted, accounting for the interaction between blend partners before starting the decomposition processes. Mass spectrometry data on the resulting volatile products led to a stabilisation mechanism based on the trapping of the formed radicals by the siloxane chains through their methyl or vinyl groups or on the reaction of ester/carboxylic groups with siloxane units (**Scheme 13.2**).

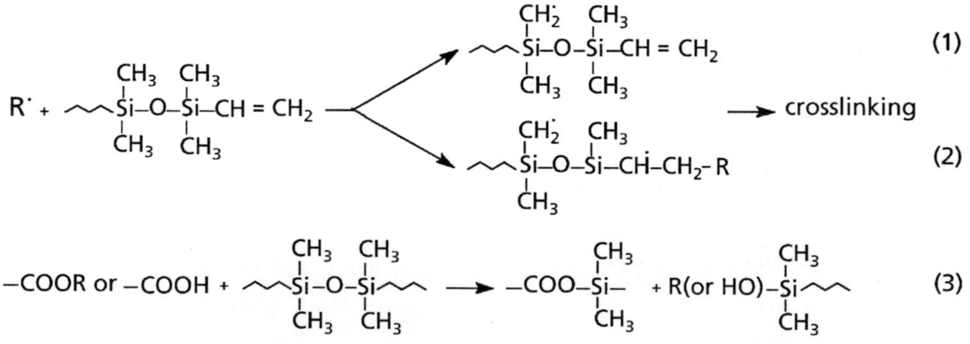

where $R^{\cdot}$ is a radical species formed during the decomposition of vinyl polymer

Scheme 13.2 Reactions involved in the stabilisation of vinyl polymers by blending with vinyl-terminated PDMS [112]

Siloxane polymers (PDMS [113], polydiphenylsiloxane and PDMS-diphenylsiloxane copolymers [114]) were also used as stabilising agents for PVC. For low concentrations of siloxane polymer, a destabilisation of the blend was observed due to a delay in the release of chain carriers (HCl or Cl radical) determined by the presence of siloxane. Blends containing higher proportions of siloxane polymers (more than 50%) showed an augmentation of the thermal stability when compared to pure PVC and their stability increases with increasing siloxane concentration. The stability of siloxane/PVC blends was explained by the accumulation of HCl or Cl radicals at the boundary of the blend heterogeneous system, consequently decreasing their catalysing effect on the dehydrochlorination of PVC. Moreover, PDMS-diphenylsiloxane copolymer, able to participate in crosslinking reactions (hydrogen abstraction from the CH_3-Si groups by the radical species formed during PVC decomposition and migrating across the phase boundaries followed by the recombination of newly formed macroradicals), showed a greater stabilisation efficiency as compared to polydiphenylsiloxane. The same stabilising effect of siloxane polymers/copolymers was observed in blends of polyvinyl acetate [115].

13.2.5.2 Mechanical Properties

Contradictory results were obtained from the investigation of mechanical properties of blends containing organic and siloxane polymers. Yamada and co-workers [116] reported an improvement of 10% of thermoplastic polyurethanes (PU) properties by blending with small amounts of PDMS, whereas other authors observed significant decrease of their performance after blending [117-120]. Recently, a study of the properties of PU/ PDMS blends [121, 122] showed a large improvement (25%) in mechanical properties for PDMS concentrations lower than 2%, while for higher proportions of PDMS (more than 3%) the physical properties were proved to be adversely affected. A packing model that explains the effect of PDMS upon mechanical properties of PU blends by an antiplasticisation of PDMS in low concentration and a subsequent plasticisation of the PU soft segment for higher PDMS concentration was proposed.

The addition of small amounts (~1%) of polysilastyrene (PSiS) in polypropylene or PS was proved to enhance their hardness, in a greater extent for PS, due to a higher compatibility between blend components [9]. As a consequence of the protective effect of PSiS against light degradation, a marked improvement of mechanical properties of the blend systems (tensile strength, tensile modulus, elongation at break) after UV irradiation as compared to unblended polymer was observed.

Highly oriented fibres were prepared by gel drawing of mixtures of ultrahigh molecular weight polyethylene and poly(di-*n*-pentylsilane) known to form columnar liquid crystalline phases [10, 11]. The mechanical deformation of the irradiated blends showed the improvement of the fibre strength [11].

13.2.5.3 Electrical Properties

The introduction of flexible siloxane sequences into the main chain of polymer electrolytes was proved to provide an increase of the ionic conductivity [123-126]. An improvement of mechanical properties of rigid electroconducting polymers without the degradation of the electrical characteristics was also reported [127]. The same beneficial influence was observed by blending siloxane polymers (PDMS-phenylmethylsiloxane copolymers (PDMPhMS) of different molecular weights) or siloxane-organic copolymers (polydimethylsiloxane-g-alkylene oxide, PDMS-AO) with poly(ether–urethane) (PEU) [128]. An increase of the ionic conductivity from 10^{-7} up to 10^{-5} S/cm at room temperature was achieved. Blend morphology analysis (PEU/PDMPhMS blends are phase separated, while PEU/PDMS-AO blends are compatible), dielectric constant and ion mobility measurements suggested that the improvement of the conductivity was determined by the increase in ion mobility that compensates for the decrease of ion density in incompatible blends and by the augmentation of both characteristics in compatible electrolytes.

The surface resistivity of polypropylene and PS was reduced from 10^{17} to 10^{16} and 10^{13} ohm·cm, respectively, by their blending with low proportions of polydimethylsilylene-*co*-diphenylsilylene [9].

Mixtures of imine-based side chain liquid crystalline polysiloxanes or ladder-like polysilsesquioxanes with corresponding low molecular weight liquid crystalline analogues were prepared and their performances as materials for electro-optical devices were evaluated [129].

13.2.5.4 Biomedical Applications

The exceptional biostability and high oxygen permeability of polysiloxanes, as pure polymers/copolymers or blend systems, promoted their use in a wide range of medical devices, such as oxygenators, contact lenses, finger joints, catheters, blood pumps, tubing, plastic surgical materials, breast implants, ophthalmologic implants, adhesives and heart valves. Siloxane polymers and copolymers were also used as additives to improve blood compatibility and surface properties of organic polymers [130-137]. To evaluate potential applications such as implantable biomaterials, the anti-inflammatory properties of blend systems of PVC and polycaprolactone-PDMS-polycaprolactone triblock copolymers were recently investigated [138]. It was found that at least 2.4% by weight of siloxane copolymer in PVC gives sufficient modification of the surface properties to influence tissue responses to the substrate material. Bausch & Lomb Inc., announced recently the approval of Food and Drug Administration, USA for selling PureVision, the first silicone-hydrogel contact lenses [139]. Review articles on this subject are also available [140, 141].

13.3 Copolymer Networks and Interpenetrating Networks

Silicon-based polymers, especially PDMS, undergo phase separation phenomena in multicomponent and/or multiphase polymer systems. When mixed with most organic polymers, they tend to migrate towards the surface of the material and are slowly eliminated from the bulk. However, such phenomena are used in high performance materials. To maintain the siloxane chains in polymeric alloys and/or blends, two practical approaches were applied (a) their incorporation through chemical links in block and graft copolymers or copolymer networks (CPN) and (b) their permanent entanglement in IPN. The first approach not only yields new polymeric materials combining the properties of individual components (sometimes highly opposite, i.e., hydrophobic-hydrophilic, flexible-rigid), but also provides additives for the compatibilisation of incompatible mixtures. As for the IPN, in most cases the interpenetration of different polymer chains does not reach the molecular level. IPN behave as dual phase continuity systems with finely divided phases of hundreds of Angstroms in size. If IPN represent a combination of two or more polymers in a network form whose individual chains are maintained together through entanglements, CPN are single networks having two components. Such CPN exhibit features of both polymer alloys and polymer networks/gels.

An extensive literature, including review papers [7, 142], deals with the synthesis, morphology, properties and application of block and graft copolymers containing silicon-based sequences. Section 13.3 is concerned only with crosslinked structures such as copolymer networks and interpenetrating networks.

13.3.1 Copolymer Networks (CPN)

Different polymers were considered as partners for such complex structures. Thus, Shiomi and co-workers [143] reported the synthesis and swelling behaviour of a phase separated two-component polymer network of PS and PDMS as copolymer components starting from telechelic PDMS (pyrrolidinium end groups) and PS (containing a low proportion of main chain acrylic acid groups) and Hill and co-workers [144] described the dynamic mechanical properties of crosslinked siloxane-divinylbenzene copolymers.

The combination of siloxane polymers and modified epoxy resins focused the interest of fundamental and applied research in this area. Epoxy resins are widely used in electronic applications, because of their properties, excellent heat, solvent, moisture and chemical resistance, good adhesion to many substrates, superior mechanical and electrical characteristics. To encapsulate microelectronic devices through crosslinked protective layers, epoxy resins seem to be one of the preferred materials. To flexibilise such rigid structures, the incorporation of siloxane moieties is an alternative. In this respect, the literature reports the development of naphthalene-type epoxy resins modified with amino-terminated PDMS to reduce internal stress while conserving a high T_g [145, 146]. The crosslinked material was evaluated for semiconductor encapsulation. The same effect was observed for three functional epoxy resins modified by hydrosilation with side chain methylhydrogensiloxanes [147], the copolymer network being proposed for electronic applications. Zheng and co-workers used an epoxy-terminated PDMS as modifier for bisphenol A-type epoxy resins [148]. The crosslinking was performed with aromatic amine curing agents. It was observed that the toughness of the epoxy resin was enhanced without loss of the thermal resistance and of substantial modulus. The crosslinked epoxy-rich phase (matrix) presented a higher T_g when compared to pure epoxy resins cured with aromatic amines.

Poly(ester–siloxane)urethane copolymer networks possessing Si-O-C linkages were found to exhibit elasto-plastic behaviour in static conditions [149]. The structure-morphology-properties (thermal stability, dynamic mechanical properties) relationships are dependent upon the chemical nature of both PS soft segment and PU hard sequence, as well as on the crosslinking density [150-155].

Hydrophobic-hydrophilic PDMS-polymethacrylic acid CPN were also prepared by radical polymerisation of MAA in the presence of chain glycidoxypropyl functionalised PDMS when a simultaneous ring opening of epoxy groups through the reaction with carboxylic units takes place [156].

13.3.2 Interpenetrating Networks (IPN)

Polysiloxanes possessing different organic substituents to the silicon atoms were widely used as components of IPN being combined with many organic polymers such as vinyl or diene polymers (PS, polymethyl methacrylate, polymethacrylic acid, polyethylene, polypropylene, melt-processable fluorocarbons, polybutadiene), PU, PEO, polyoxazolines, polyesters, polycarbonates, polyamides. All types of IPN were obtained, full- and semi-IPN, by both sequential and simultaneous preparation, as well as thermoplastic, latex and gradient IPN [157].

As a consequence of their strong incompatibility with almost all organic polymers, the formation of siloxane-based IPN presents peculiar aspects. Special care has to be taken to ensure the appropriate dispersion of the components all along the network formation process, i.e., to avoid gross phase separation before the network formation. Owing the propensity of siloxane chains to migrate to the air surface of the polymeric materials, to stabilise the bulk structure of IPN, the siloxane component is introduced as crosslinked polymer. The crosslinked polysiloxane is formed independently of the matrix resin thus imparting their desirable characteristics (release, wear, thermal or physiological properties) to the final material. Crosslinking of the siloxane component was achieved by the known chemical processes specific to siloxane chemistry, like hydrosilation or silanol condensation with silyl ethers or chlorosilanes and by reacting organofunctional groups linked to the siloxane chains with appropriate organic reagents (hydroxy- or amino-alkyl with isocyanate or epoxy) (**Scheme 13.3**).

The hydrosilation reaction presents the advantages of a high rate, a high final conversion and the absence of low molecular weight by-products. The condensation of silanol groups is accompanied by the evolution of low molecular weight alcohols or hydrochloric acid. The simultaneous presence of radicals and oxygen (implied in the crosslinking of organic polymer) was proved to diminish the efficiency of stannous octoate catalyst [158].

The morphology and properties of IPN are depending on IPN type and processing technique, on polymer compatibility and interfacial tension, on crosslinking density and relative proportion of components. The compatibility between components controls the degree of phase separation and of molecular interpenetration. The modification of the chemical nature of the substituents to the silicon atoms was used to achieve an appropriate degree of mixing of IPN components, in other words, to overcome their extreme incompatibility which results in a gross phase separation or the full solubility leading to alloys with poor properties. The degree of phase separation is also controlled by crosslinking density. Usually, the increase of this parameter determines the decrease of domain size and the augmentation of molecular interpenetration. The crosslinking density of siloxane component is determined by the molecular weight of linear polymer precursor and/or by its functionality.

Every interpenetrating system has to be treated in a quite specific manner. While for some vinyl monomers (styrene) the swelling of a preformed siloxane network into the monomer and the formation of the IPN by the subsequent polymerisation of the vinyl

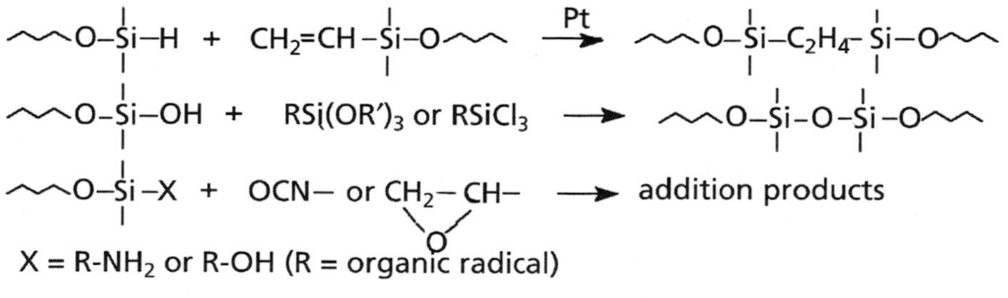

Scheme 13.3 Crosslinking of polysiloxane component

monomer is quite evident, for polar monomers one has to use a common solvent to avoid phase separation. For IPN made from mixtures of prepolymers, the compatibilisation of the system is achieved by linking different organofunctional groups to the siloxane chains. Despite the method of preparation, the final idea is to obtain materials with a high degree of the entanglement of different macromolecular chains in order to improve the mechanical or surface properties of the final material.

Table 13.4 summarises recent advances in IPN containing silicon-based polymers.

13.4 Conclusion

It was the intention of this review presentation to underline the role and importance of silicon-based macromolecular structures in building up new and performant polymeric materials such as blend systems and copolymer/interpenetrating networks. All these materials take advantage of the quite unique properties of the Si-O, Si-Si and Si-C skeleton and are highly diversified through the attachment of a large variety of organic radicals to the silicon atoms. While for polymer blends the applications are limited by the strong incompatibility of mixture components, in copolymer and interpenetrating networks the crosslinking and/or permanent entanglement determine a more stable macrostructure and long-term end use properties.

Table 13.4 Interpenetrating polymer networks

No.	Organic polymer	Siloxane polymer	Properties	Applications	Ref.
POLYSILOXANE/POLYSILOXANE IPN					
1	-	Polysiloxane 1 Polysiloxane 2	Elastomeric properties	Model IPN (static moduli - network structure relationship)	[158-164]
2	-	PDMS *In situ* prepared SiO₂		Electronic device Encapsulants	[165]
VINYL or DIENE POLYMERS/POLYSILOXANE IPN					
3	PMMA	PDMS	Increased toughness, lower modulus and tensile strength *versus* PMMA; high gas permeability	Optical devices	[157, 166-169]
4		PPhMS			[169, 170]
5		PDPhS			[169]
6		PDMS-PU	Good thermal resistance		[171-173]
7	PMMA, PVyAc, PNVP, PAAM	*In situ* prepared SiO₂	Improved mechanical properties and thermal resistance		[174]
8	P(MMA-*co*-BMA)	PDMS	Higher toughness, lower modulus and tensile strength *versus* PMMA		[175-177]
9	PMAA	PDMS	Flexible, transparent, high oxygen permeability, increased hydrophilicity	Hydrophilic contact lenses	[178, 179]
10	P(MA-*co*-AA) P(MA-*co*-HEA)	*In situ* prepared SiO₂			[180, 181]
11	PGPTA	PDMS	Reinforced PDMS network with small amounts of PGPTA		[182]

Table 13.4 Continued

No.	Organic polymer	Siloxane polymer	Properties	Applications	Ref.
12	PTFE	PDMS	Breathable and waterproof membranes	Biomaterials in filtration or skin-like devices	[183, 184]
13	PS	PDMS	Improved behaviour in ultimate tensile strength, elongation and modulus		[182, 185-187]
14	PB	PDMS	Phase separated systems with minimal interpenetration		[182]
15	P((B-co-S)-g-AN)	PDMS			[188]
16	P(S-B-S)	P(DMS-DPhS-VyMS)		Unique sterilisable thermoplastic systems	[189]
17	PP	PDMS	Limited interpenetration; high impact strength facilitating injection molding		[189, 190]
18	PSi-PU	PDMS	Modified impact and surface properties as compared to S-AN copolymer; gas permselectivity	Gas separation	[191]
OTHER IPN					
19	PU	PDMS or PPhMS	Lower phase separation for IPN containing higher soluble PPhMS; lower wear and friction, increased elastic recovery; high-temperature resistance; improved dielectric properties; reduced blood-polymer interaction; increased oxygen permeability	Imaging and biomedical applications Synthetic fabrics Composite foams Adhesive layers	[171, 172, 189, 191, 192]

Table 13.4 Continued

No.	Organic polymer	Siloxane polymer	Properties	Applications	Ref.
20	Polyamides	Differently substituted polysiloxanes	Low shrinkage rate, self-lubricating properties	Bearing applications	[189]
21	Polycarbonates	PHMS	Biocompatibility	Drug release	[193, 194]
22	Epoxy resins	Organofunctional polysiloxanes or siloxane copolymers	Improved toughness and thermal stability as compared to epoxy resins	Non-linear optical materials coatings	[195-199]
23		*In situ* prepared SiO$_2$	Strong phase separated; high thermal resistance	Biomaterials coatings	[200]
24	PDMPO	PDMS	Increased elongation at break, decreased tensile strength as compared to PDMPO		[201, 202]
25	PEO	PDMS	IPN filled with lithium salts exibit high ionic conductivity	Solid electrolytes	[203]
26	PS	PHMS crosslinked with allyl functionalised cyclodextrins	Complexing ability	Stationary phase in liquid chromatography	[204]
27	Polydiacetylene	PDMS	Iodine-doped IPN show an independent electrical hopping conduction mechanism		[205]

PVyAc = polyvinyl acetate; PNVP = poly(N-vinylpyrrolidone); PAAM = polyacrylamide; P(MMA-co-BMA) = poly(methyl methacrylate-co-butyl methacrylate); PMAA = polymethacrylic acid; P(MA-co-AA) = poly(methyl acrylate-co-acrylic acid); P(MA-co-HEA) = poly(methyl acrylate-co-hexyl acrylate); PGPTA = polyglycerylpropoxy triacrylate; PTFE = polytetrafluorethylene; PB = polybutadiene; PSi-PU = Poly(siloxane-urethane); P((B-co-S)-g-AN) = poly((butadiene-co-styrene)-g-AN); P(S-B-S) = poly(styrene-butadiene-styrene); PP = polypropylene; PDMPO = poly(2,6-dimethylphenylene oxide); PDMS-PU = poly(dimethylsiloxane-urethane); P(DMS-DPhS-VyMS) = poly(dimethyl-diphenyl-vinylmethyl)siloxane; PHMS = polyhydromethylsiloxane

References

1. E.G. Rochow, in *Silicon-Based Polymer Science - A Comprehensive Resource*, Eds., J.H. Zeigler and F.W.G. Fearon, Advances in Chemistry Series, No. **224**, 1990, XVII.

2. W. Noll, *Chemistry and Technology of Silicones*, Academic Press, New York, NY, USA, 1968.

3. *Polymer Handbook*, Eds., J. Brandrup, E.H. Immergut and E.A. Grulke, John Wiley & Sons, Inc., New York, NY, USA, 1999.

4. I. Yilgor and J.E. McGrath in *Reactive Oligomers*, Eds., F.W. Harris and H.J. Spinelli, *ACS Symposium Series*, 1985, **282**, Washington, DC, USA, Chapter 14.

5. K.E. Polmanteer, *Journal of Elastoplastics*, 1970, **2**, 165.

6. S. Lu, M.M. Melo, J. Zhao, E.M. Pearce and T.K. Kwei, *Macromolecules*, 1995, **28**, 4908.

7. I. Yilgor and J.E. McGrath, *Advances in Polymer Science*, 1988, **86**, 1.

8. M.M. Werlang, M.A. De Aranjo, S.P. Nunez and I.V.P. Yoshida, *Journal of Polymer Science, Part B: Polymer Physics*, 1997, **35**, 2609.

9. T. Asuke, J. Chien-Hua and R. West, *Macromolecules*, 1994, **27**, 3023.

10. S. Sheiko, H. Frey and M. Möller, *Colloid and Polymer Science*, 1992, **270**, 440.

11. M. Möller, H. Frey and S. Sheiko, *Colloid and Polymer Science*, 1993, **271**, 554.

12. R.D. Maier, M. Kopf, D. Maeder, F. Koopmann, H. Frey and J. Kressler, *Acta Polymerica*, 1998, **49**, 356.

13. R.D. Maier, J. Kressler, B. Rudolf, P. Reichert, F. Koopmann, H. Frey and R. Mulhaupt, *Macromolecules*, 1996, **29**, 1490.

14. S.H. Jacobson, D.J. Gordon, G.V. Nelson and A. Balazs, *Advanced Materials*, 1992, **4**, 198.

15. O. Olabisi, L.M. Robenson and M.T. Shaw, *Polymer - Polymer Miscibility*, Academic Press, New York, NY, USA, 1979.

16. *Polymer Blends*, Eds., D.R. Paul and S. Newman, Academic Press, New York, 1978.

17. K. Binder, K. Kremer, I. Carmesin and A. Sariban in *Chemistry and Physics of Macromolecules*, Eds., E.W. Fischer, R.C. Schulz and H. Sillescu, VCH, Weinheim, Germany, 1991, 499.

18. R.G. Kirste and B.R. Lehnen, *Die Makromolekulare Chemie*, 1976, **177**, 1137.

19. P. Bajaj, D.C. Gupta and S.K. Varshney, *Polymer Engineering and Science*, 1983, **23**, 820.

20. G. Grigorescu, S. Ioan, V. Harabagiu and B.C. Simionescu, *Revue Roumaine de Chimie*, 1997, **42**, 701.

21. G. Grigorescu, S. Ioan, V. Harabagiu and B.C. Simionescu, *European Polymer Journal*, 1998, **34**, 827.

22. G. Grigorescu, S. Ioan, M. Pinteala and B.C. Simionescu, *Journal of the Serbian Chemical Society*, 1998, **63**, 961.

23. M. Benmouna, M. Duval, C. Strazielle, F.I. Hakem and E.W. Fischer, *Macromolecular Theory and Simulations*, 1995, **4**, 53.

24. S. Lu, E.M. Pearce and T.K. Kwei, *Journal of Macromolecular Science A*, 1994, **31**, 1535.

25. S. Lu, E.M. Pearce and T.K. Kwei, *Journal of Polymer Science, Part A: Polymer Chemistry*, 1994, **32**, 2607.

26. S. Petitjean, G. Ghitti, P. Teyssie, J. Merien, J. Riga and J. Verbist, *Macromolecules*, 1994, **27**, 4127.

27. C.A. Arnold, D.H. Chen, Y.P. Chen, R.O. Waldbauer, M.E. Rogers and J.E. McGrath, *High Performance Polymers*, 1990, **2**, 83.

28. M.L. Huggins, *Annals of the New York Academy of Science*, 1942, **43**, 1.

29. P.J. Flory, *Journal of Chemical Physics*, 1944, **12**, 425.

30. I.C. Sanchez in *Polymer Blends*, Eds., D.R. Paul and S. Newman, Academic Press, New York, 1978.

31. I. Prigogine, *The Molecular Theory of Solutions*, North-Holland Publishers, Amsterdam, The Netherlands, 1959.

32. P.J. Flory, R.A. Orwoll and A. Vrij, *Journal of the American Chemical Society*, 1964, **86**, 3507.

33. P.J. Flory, R.A. Orwoll and A. Vrij, *Journal of the American Chemical Society*, 1964, **86**, 3515.

34. P.J. Flory, *Journal of the American Chemical Society*, 1965, **87**, 1833.

35. D. Patterson, *Journal of Polymer Science C*, 1968, **16**, 3379.

36. D. Patterson and A. Robard, *Macromolecules*, 1978, **11**, 690.

37. G.T. Dee and D.J. Walsh, *Macromolecules*, 1988, **21**, 811 and 815.

38. M. Wagner and B.A. Wolf, *Macromolecules*, 1988, **26**, 6498.

39. T. Nose, *Polymer*, 1995, **36**, 2243.

40. G. Beaucage, S. Sukumaran, S.J. Clarson, M.S. Kent and D.W. Schaefer, *Macromolecules*, 1996, **29**, 8349.

41. I.O. Volkov, A.I. Pertsin, L.V. Filimonova, M.M. Gorelova and E.M. Belavtseva, *Polymer Science, Series B*, 1997, **39**, 310.

42. C.M. Kuo and S.J. Clarson, *Macromolecules*, 1992, **25**, 2192.

43. C.M. Kuo and S.J. Clarson, *European Polymer Journal*, 1993, **29**, 661.

44. C.M. Kuo and S.J. Clarson, *Polymer*, 1994, **35**, 4623.

45. C.M. Kuo and S.J. Clarson, *Polymer*, 2000, **41**, 5993.

46. C. Qian, S. Grigoras and L. Kennan, *Macromolecules*, 1996, **29**, 1260.

47. E. Helfand, *Polymer Compatibility and Incompatibility, Principles and Practices*, Ed., K. Solc, MMI Press, Khur, 1987.

48. D. Broseta and L. Leibler, *Journal of Chemical Physics*, 1987, **87**, 7248.

49. A. Vrij, *Journal of Polymer Science, Part A: Polymer Chemistry*, 1968, **6**, 1919.

50. G.H. Fredrickson, *Physics of Polymer Surfaces and Interfaces*, Ed., I.C. Sanchez, Butterworth - Heinemann, London, UK, 1992.

51. G.T. Dee and B.B. Sauer, *Macromolecules*, 1993, **26**, 2771.

52. T. Nose, *Macromolecules*, 1995, **28**, 3702.

53. T. Nose and N. Kasemura, *Polymer*, 1998, **39**, 6137.

54. S. Wu, *Journal of Macromolecular Science, Reviews in Macromolecular Chemistry C*, 1974, **10**, 1.

55. A. Stammer, *Phase Behaviour and Interfacial Tension of Polysiloxane Mixture*, University of Mainz, 1997. [Ph. D. Thesis]

56. M.M. Gorelova, A.J. Pertsin, I.O. Volkov, L.V. Filimonova and E.S. Obolonkova, *Journal of Applied Polymer Science*, 1996, **60**, 363.

57. M. Wagner and B.A. Wolf, *Polymer*, 1993, **34**, 1461.

58. U. Jorzik and B.A. Wolf, *Macromolecules*, 1997, **30**, 4713.

59. F. Schmid, *Journal of Chemical Physics*, 1996, **104**, 2689.

60. Y. Cai, D. Gardener and G.T. Caneba, *Journal of Adhesion Science and Technology*, 1999, **13**, 1017.

61. S. Petitjean, G. Ghitti, G. Fayt, R. Jerome and P. Teyssie, *Bulletin des Societes Chimique Belges*, 1990, **99**, 997.

62. B.A. Wolf, *Macromolecular Symposia*, 1999, **139**, 87.

63. A. Schneider and B.A. Wolf, *Macromolecular Rapid Communications*, 1997, **18**, 561.

64. A. Stammer and B.A. Wolf, *Macromolecular Rapid Communications*, 1998, **19**, 123.

65. Y. Sakane, K. Inomata, H. Morita, T. Kawakatsu, M. Doi and T. Nose, *Polymer*, 2001, **42**, 3883.

66. Y. Kawakami, R.A. Murthy and Y. Yamashita, *Polymer Bulletin*, 1983, **10**, 368.

67. Y. Kawakami, R.A. Murthy and Y. Yamashita, *Die Makromolekulare Chemie*, 1984, **185**, 9.

68. M.J. Owen and T.C. Kendrick, *Macromolecules*, 1970, **3**, 458.

69. Y. Tezuka, A. Fukushima, S. Matsui and K. Imai, *Journal of Colloid and Interface Science*, 1986, **114**, 16.

70. N.M. Patel, D.W. Dwight, J.L. Hedrick, D.C. Webster and J.E. McGrath, *Macromolecules*, 1988, **21**, 2689.

71. W. Litt and B. Huang, *Polymer Bulletin*, 1988, **20**, 531.

72. I. Yilgor, W.P. Steckle Jr., E. Yilgor, R.G. Freelin and J.S. Riffle, *Journal of Polymer Science, Polymer Chemistry Edition*, 1989, **27**, 3673.

73. S.D. Smith, J.M. DeSimone, H. Huang, G. York, D.W. Dwight, G.L. Wilkes and J.E. McGrath, *Macromolecules*, 1992, **25**, 2575.

74. M.M. Gorelova, A.J. Pertsin, V.Y. Levin, L.I. Makarova and L.V. Filimonova, *Journal of Applied Polymer Science*, 1992, **45**, 2075.

75. V.S. Popkov, G.G. Nikiforova, I.M. Raygorodsky and I.P. Storozhuk, *Polymer Science*, 1995, **37B**, 428.

76. V.S. Popkov, G.G. Nikiforova, V.G. Nikol'sky, I.A. Krasotkina and E.S. Obolonkova, *Polymer*, 1998, **39**, 631.

77. J. Radhakrishnan, N. Tanigachi and A. Kaito, *Polymer*, 1999, **40**, 1381.

78. S. Lu, E.M. Pearce and T.K. Kwei, *Polymer Engineering and Science*, 1995, **35**, 1113.

79. M. Itoh and I. Mita, *Journal of Polymer Science, Part A: Polymer Chemistry*, 1994, **32**, 1581.

80. A. Du Chesne, G. Lieser and G. Wegner, *Colloid and Polymer Science*, 1994, **272**, 1329.

81. V. Harabagiu, *New Initiators for the Synthesis of Polymers and Copolymers*, 'Gh. Asachi' Technical University, Iasi, Romania, 1992. [Ph. D. Thesis]

82. M. Kapnistos, A. Hinrichs, D. Vlassopoulos, S.H. Anastasiadis, A. Stammer and B.A. Wolf, *Macromolecules*, 1996, **29**, 7155.

83. H. Gramespacher and J. Meissner, *Journal of Rheology*, 1992, **36**, 1127.

84. D. Graebling and R. Muller, *Journal of Rheology*, 1990, **34**, 193.

85. J. Lauger, R. Lay, S. Maas and W. Gronski, *Macromolecules*, 1995, **28**, 7010.

86. I. Vinckier, P. Moldenaers and J. Mewis, *Journal of Rheology*, 1996, **40**, 613.

87. J.A. Zawada, G.G. Fuller, R.H. Colby, L.J. Fetters and J.E.L. Roovers, *Macromolecules*, 1994, **27**, 6851.

88. J.A. Zawada, G.G. Fuller, R.H. Colby, L.J. Fetters and J.E.L. Roovers, *Macromolecules*, 1994, **27**, 6861.

89. R.M. Kannan and J.A. Kornfield, *Journal of Rheology*, 1994, **38**, 1127.

90. B.H. Arendt, R.M. Kannan, M. Zewail, J.A. Kornfield and S.D. Smith, *Rheologica Acta*, 1994, **33**, 322.

91. R.G. Larson, *Rheologica Acta*, 1992, **31**, 497.

92. S. Mani, M.F. Malone and H.H. Winter, *Macromolecules*, 1992, **25**, 5671.

93. R. Wu, M.T. Shaw and R.A. Weiss, *Journal of Rheology*, 1992, **36**, 1605.

94. R. Horst and B.A. Wolf, *Macromolecules*, 1993, **26**, 5676.

95. R. Horst and B.A. Wolf, *Rheologica Acta*, 1994, **33**, 99.

96. E.K. Hobbie, A.I. Nakatari and C.C. Han, *Modern Physics Letters B*, 1994, **8**, 1143.

97. M.L. Fernandez, J.S. Higgins and S.M. Richardson, *Polymer*, 1995, **36**, 931.

98. A. Ajji and L. Choplin, *Macromolecules*, 1991, **24**, 5221.

99. Y. Takahashi, H. Suzuki, Y. Nakagawa and I. Noda, *Macromolecules*, 1994, **27**, 6476.

100. Y. Takahashi, H. Suzuki, Y. Nakagawa and I. Noda, *Polymer International*, 1994, **34**, 327.

101. M. Kapnistos, D. Vlassopoulos and S.H. Anastasiadis, *Europhysics Letters*, 1996, **34**, 513.

102. A.R. Nesarikar, *Macromolecules*, 1995, **28**, 7202.

103. S.K. Kumar, R.H. Colby, S.H. Anastasiadis and G. Fytas, *Journal of Chemical Physics*, 1996, **105**, 3777.

104. D. Vlassopoulos, A. Koumoutsakos, S.H. Anastasiadis, S.G. Hadzikiriakos and P. Englezos, *Journal of Rheology*, 1997, **41**, 739.

105. A. Onuki, *Europhysics Letters*, 1994, **28**, 175.

106. V.T. Tsakalos, P. Navard and E. Peuvrel-Disdier, *Journal of Rheology*, 1998, **42**, 1403.

107. M. Astruc and P. Navard, *Macromolecular Symposia*, 2000, **149**, 81.

108. M. Astruc and P. Navard, *Journal of Rheology*, 2000, **44**, 693.

109. J. Mewis, I. Vinckier and P. Moldenaers, *Macromolecular Symposia*, 2000, **158**, 29.

110. J.L. Falender, S.E. Lindsey and J.C. Saam, *Polymer Engineering and Science*, 1976, **16**, 54.

111. A.L. Schroll and M.R. MacLaury, *Journal of Applied Polymer Science*, 1984, **29**, 3883.

112. I.C. McNeill and M.H. Mohammed, *Polymer Degradation and Stability*, 1995, **50**, 285.

113. I.C. McNeill and S. Bassan, *Polymer Degradation and Stability*, 1993, **39**, 139.

114. S. Zulfigar and S. Ahmad, *Polymer Degradation and Stability*, 1999, **65**, 243.

115. S. Zulfigar and S. Ahmad, *Polymer Degradation and Stability*, 2001, **71**, 299.

116. M. Yamada, T. Makumoto, T. Kurosaki, Y. Iwasaki, A. Sejimo and H. Okada, *Meiji Rubber and Chemical Company, Technical Report*, 1991, **14**, 27.

117. S. Sakurai, S. Nokuwa, M. Morimoto, M. Shibayama and S. Nomura, *Polymer*, 1994, **35**, 532.

118. M. Shibayama, M. Suetsugu, S. Sakurai, T. Yamamoto and S. Nomura, *Macromolecules*, 1991, **24**, 6254.

119. X. Chen, J.A. Gardella, T. Ho and K.J. Wynne, *Macromolecules*, 1995, **28**, 1635.

120. R. Benrashid and G.L. Nelson, *Journal of Polymer Science, Part A: Polymer Chemistry*, 1994, **32**, 1847.

121. D.J.T. Hill, M.I. Killeen, J.H. O'Donnell, P. Pomery, D. St. John and A.K. Whittaker, *Journal of Applied Polymer Science*, 1996, **61**, 1757.

122. T. Bremner, D.J.T. Hill, M.I. Killeen, J.H. O'Donnell, P.J. Pomery and D. St. John, *Journal of Applied Polymer Science*, 1997, **65**, 939.

123. P.G. Hall, G.R. Davies, J.E. McIntyre, I.M. Ward, D.J. Bannister and K.M.F. LeBrocq, *Polymer Communications*, 1986, **27**, 98.

124. D. Fish, I.M. Khan, E. Wu and J. Smid, *British Polymer Journal*, 1988, **20**, 281.

125. K. Nagaoka, H. Naruse, I. Shinohara and M.J. Watanabe, *Journal of Polymer Science, Polymer Letters Edition*, 1984, **22**, 659.

126. Z. Ogumi, Y. Uchimoto and Z. Takehara, *Solid State Ionics*, 1989, **35**, 417.

127. E. Kalaycioglu, L. Toppare, Y. Yagci, V. Harabagiu, M. Pinteala, R. Ardelean and B.C. Simionescu, *Synthetic Metals*, 1998, **97**, 7.

128. M. Shibata, T. Kobayashi, R. Yosomiya and M. Seki, *European Polymer Journal*, 2000, **36**, 485.

129. P. Xie, L. Sun, D. Dai, D. Liu, Z. Li and R. Zhang, *Molecular Crystals and Liquid Crystals*, 1995, **269**, 75.

130. *Polyurethanes in Biomedical Engineering II*, Eds., H. Planck, I. Syre, M. Dauner and G. Egbers, Elsevier, Amsterdam, The Netherlands, 1987, p. 213.

131. R.S. Ward, P. Litwack and R. Rodvien, *Transactions of the Society of Biomaterials*, 1982, 5, 46.

132. K.J. Quinn and J.M. Courtney, *British Polymer Journal*, 1988, **20**, 25.

133. B.D. Ratner in *Treatise on Clean Surface Technology*, Volume 1, Ed., K.L. Mittal, Plenum Press, New York, NY, USA, 1987, p.247.

134. T. Kumaki, M. Sisido and Y. Imanishi, *Journal of Biomedical Materials Research*, 1985, **19**, 785.

135. M.D. Lelah and S.L. Cooper, *Polyurethanes in Medicine*, CRC Press, Boca Raton, FL, USA, 1986.

136. Y. Kawakami and Y. Yamashita in *Ring-opening Polymerisation*, Ed., J.E. McGrath, ACS, Washington, DC, USA, *ACS Symposium Series*, 1985, No. **286**, Chapter 19.

137. B. Arkles, *Chemtech*, 1983, **13**, 542.

138. L. Tang, M-S. Shen, T. Chu and Y.H. Huang, *Biomaterials*, 1999, **20**, 1365.

139. R.L. Powers, *Rubber and Plastics News*, 1999, **28**, 21, 5.

140. G.D. Friends, J.F. Kunzler and R.M. Ozark, *Macromolecular Symposia*, 1995, **98**, 619.

141. J.F. Kunzler, *TRIP*, 1996, 4, 52.

142. B.C. Simionescu, V. Harabagiu and C.I. Simionescu in *The Polymeric Materials Encyclopedia: Synthesis, Properties and Applications*, Ed., J.C. Salamone, CRC Press, Boca Raton, FL, USA, 1996, **10**, 7751.

143. T. Shiomi, Y. Tezuka, K. Okada and K. Imai, *Macromolecular Symposia*, 1994, **84**, 325.

144. D.J.T. Hill, M.C.S. Perera, P.J. Pomery and H.K. Toh, *Polymer*, 2000, **41**, 9131.

145. C.S. Wang and M-C. Lee, *Journal of Applied Polymer Science*, 1998, 70, 1907.

146. J-Y. Shieh, T-H. Ho and C-S. Wang, *Die Angewandte Makromolekulare Chemie*, 1997, **245**, 125.

147. L-L. Lin, T-H. Ho and C-S. Wang, *Polymer*, 1997, **38**, 1997.

148. S. Zheng, H. Wang, Q. Dai, X. Luo, D. Ma and K. Wang, *Macromolecular Chemistry and Physics*, 1995, **196**, 269.

149. A. Stanciu, A. Airinei and S. Oprea, *Polymer*, 2001, **42**, 6081.

150. S. Ioan, G. Grigorescu and A. Stanciu, *Polymer*, 2001, **42**, 3633.

151. A. Stanciu, A. Airinei, D. Timpu, A. Ioanid, C. Ioan and V. Bulacovschi, *European Polymer Journal*, 1999, **38**, 1959.

152. V. Bulacovschi, A. Stanciu, I. Rusu, A. Cailean and F. Ungureanu, *Polymer Degradation and Stability*, 1998, **60**, 487.

153. S. Ioan, G. Grigorescu and A. Stanciu, *Polymer Plastics Technology and Engineering*, 2000, **39**, 807.

154. A. Stanciu, V. Bulacovschi, M. Lungu, S. Vlad, S. Balint and S. Oprea, *European Polymer Journal*, 1999, **38**, 2039.

155. A. Stanciu, V. Bulacovschi, V. Condratov, C. Fadei, A. Stoleriu and S. Balint, *Polymer Degradation and Stability*, 1999, **64**, 259.

156. C.I. Simionescu, V. Harabagiu, D. Giurgiu, V. Hamciuc and B.C. Simionescu, *Bulletin des Societes Chimiques Belges*, 1990, **99**, 991.

157. R. Bischoff and S.E. Cray, *Progress in Polymer Science*, 1999, **24**, 185.

158. X.W. He, J.M. Widmaier, J.E. Herz and G.C. Meyer, *Polymer*, 1989, 20, 364.

159. S. Wang and J.E. Mark, *Journal of Polymer Science, Part B: Polymer Physics*, 1992, 30, 801.

160. P. Xu and J.E. Mark, *Polymer*, 1992, 33, 1843.

161. W. Shuhong, X. Ping and J.E. Mark, *Macromolecules*, 1991, 24, 6037.

162. E.E. Hamurcu and B.M. Baysal, *Polymer*, 1993, 24, 5163.

163. T. Iwahara and A.J. Kotani, *Chemistry Letters*, 1995, 6, 425.

164. G.J. Gibbons and D. Holland, *Journal of Sol Gel Science and Technology*, 1997, 8, 599.

165. D.W. McCarthy, J.E. Mark and D.W. Schaefer, *Journal of Polymer Science, Part B: Polymer Physics*, 1998, 36, 1167.

166. X.W. He, J.M. Widmayer, J.E. Hertz and G.C. Meyer, *Polymer*, 1992, 33, 866.

167. X.W. He, J.M. Widmayer, J.E. Hertz and G.C. Meyer in *Advanced Interpenetrating Polymer Networks*, Eds., D. Klempner and K.C. Frisch, Technomic Publishing Company, Lancaster, PA, 1994, 4, 321.

168. P. Laurienzo, M. Malinconico, E. Martuscelli, G. Ragosta and M-G. Volpe, *Journal of Applied Polymer Science*, 1992, 44, 1883.

169. H.L. Frisch, L. Wang, W. Huang, Y.H. Hua, H.X. Xiao and K.C. Frisch, *Journal of Applied Polymer Science*, 1991, 43, 475.

170. T.C. Gilmer, P.K. Hall, H. Ehrenfeld, K. Wilson, T. Bivens, D. Clay and C. Endreszl, *Journal of Polymer Science, Part A: Polymer Chemistry*, 1996, 34, 1025.

171. J-R. Caille, inventor; Dow Corning Ltd., assignee; FR 2 270 289, 1997.

172. P. Zhou, Q. Xu and H L. Frisch, *Macromolecules*, 1994, 27, 938.

173. P. Zhou, H-L. Frisch, L. Rogovina, L. Makarova, A. Zhdanov and N. Sergeienko, *Journal of Polymer Science, Part A: Polymer Chemistry*, 1993, 31, 2481.

174. C.J.T. Landry, B.K. Coltrain, J.A. Wesson, N. Zumbulyadis and J.L. Lippert, *Polymer*, 1992, 33, 1496.

175. F.O. Ecshbach and S.J. Huang, *Polymer Preprints*, 1992, **33**, 2, 526.

176. D. Knauss, T. Yamamoto and J-E. McGrath, *Polymer Preprints*, 1992, **33**, 2, 988.

177. J.P. Kennedy and G.C. Richard, *Macromolecules*, 1993, **26**, 567.

178. B. McGarey, A.D.W. McLenaghan and R.W. Richards, *British Polymer Journal*, 1989, **21**, 227.

179. C. Robert, C. Bunel and J-P. Vairon, Inventors; Essilor International, assignee; E.P. 643 083, 1997.

180. W. Yin, J. Li, J. Wu and T. Gu, *Journal of Applied Polymer Science*, 1997, **64**, 903.

181. W. Yin, T. Gu, H. Liu and B. Yin, *Polymer*, 1997, 38, 5173.

182. E.E. Hamurcu and B-M. Baysal, *Macromolecular Chemistry and Physics*, 1995, **196**, 1261.

183. M.E. Dillon and D.K. Lange, *Polymer Materials Science and Engineering*, 1990, **62**, 814.

184. M.E. Dillon and D.K. Lange, *Polymer Materials Science and Engineering*, 1991, **65**, 84.

185. E. Pearce, T.K. Kwei and S. Lu, *Polymer Preprints*, 1997, **38**, 2, 392.

186. T. Miiata, J.I. Higuchi, H. Okuno and T. Uragami, *Journal of Applied Polymer Science*, 1996, **61**, 1315.

187. D.S. Lee, W.T. Kim, H.J. An, S.C. Kim and J. Kim, *Pollimo*, 1994, **18**, 815.

188. M. Okaniwa, *Polymer*, 2000, **41**, 453.

189. B. Arkles and J. Crosby, *Advanced Chemistry Series*, 1990, **224**, 181.

190. M. Zolotnitsky, Proceedings of ANTEC '95, Boston, MA, USA, 1995, Volume 3, 3576.

191. S.A. Ali, D.J. Hourston, K. Manzour and D.F. Williams, *Journal of Applied Polymer Science*, 1995, 55, 733.

192. H. Xiao, Z.H. Ping, J.W. Xie and T.Y. Yu, *Journal of Polymer Science, Part A: Polymer Chemistry,* 1990, **28**, 585.

193. S. Boileau, L. Bouteiller, R.B. Khalifa, Y. Liang and D. Teyssie, *Polymer Preprints,* 1998, **39**, 1, 457.

194. S. Boileau, L. Bouteiller, R.B. Khalifa, Y. Liang and D. Teyssie in *Silicones and Silicone-Modified Materials,* Eds., S.J. Clarson, J.J. Fitzgerald, M.J. Owen and S.D. Smith, Washington, DC, USA, *ACS Symposium Series,* 2000, **729**, 383.

195. U. Buchholz and R. Mülhaupt, *Polymer Materials Science and Engineering,* 1990, **74**, 339.

196. L. Price, R. Ryntz, K. Frisch, H. Xiao, V. Gunn, R. van den Heuvel, K. Baars and H. van den Reijen, *Journal of Coatings Technology,* 1996, **68**, 65.

197. P-H. Sung and C-Y. Lin, *European Polymer Journal,* 1997, **33**, 231.

198. S. Marturunkakul, J. Kumar and S.K. Tripathy, *Polymer Materials Science and Engineering,* 1993, **69**, 436.

199. I. Gorodisher and M.C. Palazzotto, inventors; Minnesota Mining and Manufacturing, assignee; US 5494 981, 1996.

200. C.J.T. Landry, B.K. Coltrain, J.A. Wesson, N. Zumbulyadis and J.L. Lippert, *Polymer,* 1992, **33**, 1496.

201. W. Huang and H.L. Frisch, *Die Makromolekulare Chemie - Supplement,* 1989, **15**, 137.

202. H.L. Frisch and W. Huang, *Journal of Polymer Science, Part A: Polymer Chemistry,* 1991, **29**, 131.

203. M. Grosz, S. Boileau, P. Guegan, H. Cheradame and A. Deshayes, *Polymer Preprints,* 1997, **38**, 1, 612.

204. A. Du Chesne, K. Wenke, G. Lieser and G. Wenz, *Acta Polymerica,* 1997, **48**, 142.

205. H. L. Frisch and X. Chen, *Journal of Polymer Science, Part A: Polymer Chemistry,* 1993, **31**, 3307.

14 Lignin-Based Blends

Georgeta Cazacu and Valentin I. Popa

14.1 Introduction

The rapid development of petrochemistry in the late five decades was dominated by the research and production of the synthetic polymers to the detriment of compounds arising from renewable resources. Today, several reasons justify the necessity to pay an increasing attention to the **biomass** as a provider of micro- and macromolecular products, namely:

i. the availability of a vast array of chemical structures and molecular architectures in all size and shapes;

ii. the renewable character of these sources independently on geographical and geophysical situations;

iii. the huge quantities of raw materials and

iv. the fact that solar energy ensures continuously their regeneration. However, research and production efforts in the direction of the biomass complex upgrading are still modest in comparison with its enormous potential, both quantitatively and qualitatively, for the three major compounds, cellulose, lignin and hemicelluloses.

Lignin is the main aromatic component of the vegetable tissues; it represents about 20-30 wt% of mass of higher plant tissues and it exists into cell walls and intercellular spaces. The lignin notion does not reflect a defined structure substance, it refers to a family of heterogenous biopolymers resulting from the oxidative polymerisation of: *trans*-coniferyl, *trans*-sinapyl and *trans*-coumaryl alcohols, that contain limited number of branches and/or crosslinks [1, 2]. About 10,000-12,000 scientific papers, patents, and books have been published concerning this important natural polymer.

From wood, lignin may be produced in many ways using different procedures such as: sulfite [lignosulfonate (SSL)] and alkaline (soda or Kraft) pulping and by processes based on the involvement of mineral acid (acid hydrolysis lignin, AHL), by water and steam treatments at various temperatures and pressures (autohydrolysis and steam explosion lignin, SEL), organic solvent mixtures (organosolv lignin, OSL) and by mechanical wood milling (MWL). Production of lignin products is concentrated in a few companies and is dominated by lignosulfonate derived by spent sulfite liquor. In **Table 14.1** are listed the main lignin producers.

The recent interest concerning renewable resources, and also, the concern about reducing the pollution of our environment have encouraged many efforts into researching the potential applications of lignins as a phenolic raw material and as a structural material.

Efforts devoted to incorporate lignin preparations into useful polymeric materials have started in 1944, but intensive work has been done in this field, especially in the last twenty years.

Thus, for the first time a laminated plastic product named Tomlinite that contains a lignin adhesive was developed in 1944 by Tomlinson [3]. In the following period, studies on the lignin had continued to increase the production of lignin-based materials.

During 1975-1980, many more studies have been dedicated to the examination of those lignin properties such as polyfunctionality, brittleness and rigidity (glassy nature) that appear to be useful for the development of thermosetting and thermoplastic materials [4-9]. In 1990-2000, significant advances have been made in the study of compatibility and structure-properties relationship of lignin-based materials [10-18] (**Scheme 14.1**).

14.2 Lignin/Epoxy Resin Blends

In commercial applications epoxy resins are rarely used without the incorporation of some other materials. Thus, a way to enhance their performance is by providing additional mechanical properties or modifying the physical characteristics of the blends. Several approaches have been chosen to incorporate non-modified or chemically modified lignin (see **Table 14.2**) into epoxy resins (ER), aiming at partially replacing oil-based materials.

In 1988, Feldman and co-workers [19] obtained an adhesive system by simply blending of the kraft lignin powder (up to 20% lignin) with the liquid epoxy prepolymer and hardener.

DSC, DMA for 5%-20% lignin content in epoxy resin-lignin blends shown a single glass transition temperature (T_g), which is characteristic for monophasic systems [20] but two T_gs for blends having higher content of lignin [21, 22]. The significant improvement in the adhesion of the lignin-epoxy resin blends (LER) to aluminium substrate could be explained by an interaction on the one hand between lignin-polyamine hardener and on the other hand between lignin-epoxy prepolymer (LE) [19, 22, 23].

Table 14.1 Major lignin producers in 1998 [23, 24]				
Type of lignin	Commercial name	Producer	Country	Annual capacity (solid t/year)
Kraft lignin		Borregard LignoTech	Norway Sweden	160,000
Lignosulfonate		LignoTech, Sweden	Sweden	60,000
Lignosulfonate		Borregard, Germany	Germany	50,000
Kraft lignin	EUCALIN	LignoTech, Iberica	Spain	30,000
Lignosulfonate		LignoTech, Finland	Finland	20,000
Lignosulfonate		LignoTech, USA Rotschill	USA	60,000
Lignosulfonate	LIGNOSITE	Georgia, Pacific	USA	220,000
Kraft lignin	INDULIN A	Westvaco	USA	35,000
Ca-lignosulfonate		Flambeau Paper	USA	60,000
NH_4, Na-lignosulfonate Steam explosion lignin Kraft lignin	TOMLINITE	Tembec Forintek Canada Corp. Domtar Corp.	Canada	75,000
Lignosulfonate		Avebene	France	40,000
Lignosulfonate		Tomezzo	Italy	30,000
Lignosulfonate Steam explosion lignin		Sanyo Kokusaka Nippon Paper Ind.	Japan	50,000
Lignosulfonate		LiniTech, South Africa	South Africa	200,000
Ethanol lignin	ALCELL	Repap Company	USA	Analytical grade samples
Organosolv lignin Autohydrolysis lignin Steam explosion lignin		Aldrich Chemical Company	USA	Analytical grade samples
Others				150,000

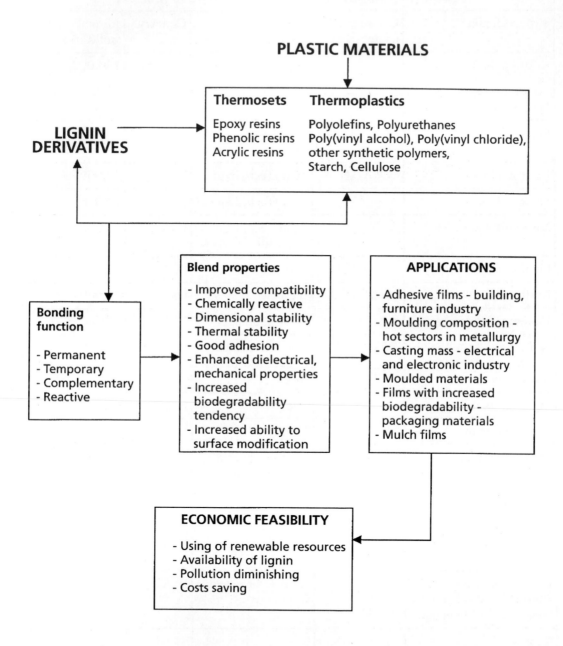

Scheme 14.1 Binders (polymers) evaluated as blend matrix with lignin

Table 14.2 Epoxy resin (ER) and lignin or epoxy-modified lignin-based blends

System	Investigation methods	General remarks	Source
L/additive/ER Lignin: kraft lignin; SEL; Additives: crystalline or amorphous silica, aluminum	- DSC, DMA; - solid-state PC-MS NMR; - adhesive properties and weatherability tests	- ER/L – up to 40% L; - miscible system – thermally cured for 20% L content - superior mechanical properties; - improved adhesivity	[21, 22, 25]
LE/ER Lignin: hydroxyalkylated organosolv; Hardening agents: aromatic diamine)	- DSC - mechanical tests	- LE content 57 wt%; - miscible system; - good strength	[26, 27]
L-Ph-E/ER Lignin: kraft lignin: Phenol: bisphenol A	- DSC, DMA	- improved glueability	[28]
L-Ph-ER /ER Lignin: Fe-LS - 2.24% Fe; Phenol; 2-hydroxy-naphthalene; 2,2 bis(4-hydroxyphenylpropane); phenolformaldehyde resin; Hardening agents: phthalic anhydride	- IR and Mössbauer spectroscopy; - viscometry; - optical and electron microscopy; - thermogravimetry; - mechanical and dielectric properties measurements - biodegradability tests	- epoxy phenolated-lignin resin content up to 25 wt%; - moulding mass with good dielectric and mechanical properties; - biodegradability/biodisintegration	[29]
L-Ph-ER or LER/ER Lignin: NH$_4$-LS;Ca-LS; Phenol; 2,2 bis(4-hydroxyphenyl)propane); phenolformaldehyce resin; Hardening agents: anhydride, amine, imide	- viscometry; - thermo-optical analysis; - thermogravimetry; - optical and electronic microscopy; - mechanical and dielectric properties measurements; - biodegradability tests	- epoxy lignin resin content up to 50 wt%; - good compatibility after crosslinking at high temperatures; - homogeneous structure; - moulding mass and adhesive films; - good electric and mechanical properties; - high biodisintegration rate of binary mixtures cured in the mild conditions	[30, 31, 32]

Table 14.2 Continued

System	Investigation methods	General remarks	Source
LER/additives/ER Lignin: NH₄-LS; Additives: plasticiser; stabiliser; pigment; filler; Hardening agents: anhydride, diamine	- viscometry; - thermo-optical analysis; - thermogravimetry; - optical and electronic microscopy; - mechanical and dielectric properties measurements; - biodegradability tests	- LER content up to 66 wt%; - optimal compatibility compositions are 25 and 50 wt%; - homogeneous structures; - adhesive lignin composites – shear strength values=18.2-26.5 kgf/cm²; - moulding mass; - improved Charpy impact strength;	[21, 30, 33, 34]

DSC: *differential scanning calorimetry*
L-Ph-E: *phenolated lignin epoxy resin*
L: *lignin*
LE: *lignin epoxy*
DMA: *Dynamic mechanical analysis*
NMR: *nuclear magnetic resonance*
PC-MS: *permeation chromatography - mass spectroscopy*
IR: *infra red spectroscopy*
LER: *lignin epoxy resin*
Fe-LS: *iron lignosulfate*

The study on the epoxy resin-lignin blends was extended using different lignin types, epoxy prepolymers, hardeners and other additives. The results obtained show that epoxy resin-lignin blends containing hardwood lignin (Tomlinite (TO), Eucalin (EU), SEL) separated by the Kraft process or isolated by steam explosion, impart a better adhesion than those having softwood lignin (Indulin AT (AT)) [30]. The improvement can be correlated with the structure and the molecular weight of lignin fraction. A significant improvement of the adhesive strength and of the weatherability of epoxy was observed by the introduction of organosilanes (2 wt% epoxyalkylsilane) into a lignin/epoxy blend (10% lignin) [35].

Chemical modification of lignins represents an alternative way to combine the higher characteristics of uniform reactivity, and compatibility with other epoxy resins. Functionalisation of lignin hydroxyalkylation or phenolation, followed by reaction with epichlorohydrin allowed to obtain of epoxy-functional lignins, which may be cured with diamines or anhydrides [26, 27, 36-40].

Tomita and co-workers [39] described a new L-EP adhesive in which lignin was modified by ozonisation. Nieh and Glasser [31] reported the synthesis of an EP resin based on hydroxypropyl lignin. Hofmann and Glasser [26, 27, 31] prepared the lignin-epoxy resins from hydroxyalkyl lignin derivatives with varying degrees of alkoxylation. It was shown that these lignin-based epoxy resins can be crosslinked with aromatic diamine (*m*-phenylene diamine, *m*-PDA) to form strong thermosets whose properties depend on lignin content and polyether chain length, and have T_g ranging from 0 °C to 100 °C. The results of the tensile stress-strain tests reveal that the value of breaking stress and tensile modulus are higher than those for the epoxy resin control. A trend to stiffer and stronger materials can be observed with increasing lignin content, while the ultimate extension simultaneously decreases [26].

Ito and Shiraishi [40] treated Kraft lignin with bisphenol A and then with epichlorohydrin in the presence of a catalyst which gives a waterproof adhesive with the improved glueability. The increase in degree of lignin phenolation led to increase of the T_g and the storage modulus in the rubbery plateau region of the cured films developing a better three-dimensional structure.

Tai and co-workers [41] synthesised lignin-based epoxy resin from Kraft lignin, Kraft lignin derivatives (including bisguiacyl lignin) and phenolated Kraft lignin. The epoxy content was found to be approximately 16 eq/100 g. The solubility of lignin epoxides in organic solvents was reported to be in the order:

Phenolated lignin > lignin epoxides > bisguiacyl lignin epoxide

These lignin epoxides were tested as adhesives on aluminium and beech wood using anhydride and diamine as curing agents, and the results were good [41].

Simionescu and co-workers [29, 33, 38, 42, 43] reported the synthesis of the lignin epoxy resins by coupling reactions with simple (phenol, β-naphthol, bisphenol) and macromolecular (novolak) phenols of the lignosulfonate (ammonium, calcium, iron and chromium lignosulfonate) followed by reaction with epichlorohydrin in alkaline medium. Two fractions of lignin-epoxy resin resulted: a liquid and a solid one. The physico-chemical properties [29, 33] of the liquid fractions permit their utilisation in **binary mixtures containing epoxy resin,** as new materials.

Incorporation of some epoxy and phenolic hydroxyl groups by successive reactions with phenol and epichlorohydrin in lignosulfonate macromolecule led to improvement of their compatibility with epoxy resin. This is reflected in the obtaining of some uniform and homogeneous structures formed from ER/LER cured with phthalic anhydride, which have a T_g at about 50 °C and good dielectric and mechanical properties (**Table 14.3**). Further phenolation had a positive effect on the dielectric properties. It was noted that the volume resistivity increase was associated with the decrease of dielectric constant and loss dielectric tangent angle [29].

Compatibility between the lignin-epoxy resin and epoxy resin is very poor at room temperature, but at higher temperatures (>75 °C) it is much improved (**Figure 14.1**) [30].

The resin mixtures containing 25-50% LER present a high uniformity, while the blends with <20 wt% and >75% wt% LER have a heterogeneous structure. These results are in good agreement with those obtained by thermo-optical analysis (TOA) (**Figure 14.2**).

The blends with various LER/ER ratios (1:3, 1:2 and 1:1) show only one T_g (**Figure 14.2**), indicating the development of a single phase during the crosslinking reaction and post-treatment, as a result of irreversible chemical linking. Incorporation of the phenolated lignin-epoxy resin (Ph-LER) in epoxy resin matrix gave also a uniform and homogeneous polymer network with a high glass transition temperature (52 °C), indicating that the phenolation reaction has a positive effect on degree of crosslinking [30].

Crosslinking system agents, phthalic anhydride (PhA); 4,4′-diaminodiphenylmethane (DDM); bismaleimide (BM); *N,N′*-diamethylamine, (DMA) and curing conditions are very important for the quality of materials. [30, 33]. Thus, the utilisation of a longer curing time, followed by a post-treatment at high temperatures assures a complete crosslinking. The ending of the tridimensional structure of the epoxy-modified lignin/epoxy blends has been demonstrated by the improvement of the mechanical properties, i.e., Charpy impact strength, and dielectric properties, such as resistivity, rigidity, dielectric constant and dielectric losses, and increasing the Martens thermal stability [33].

Table 14.3 Dielectric and mechanical properties of phenol-lignosulfonate-epoxy resin/epoxy resin blends (Ph-LER/ER; 1:3) cured with phthalic anhydride

Sample	Volume resistivity (ohm cm)	Surface resistivity (ohm)	Dielectric constant 50 Hz	Loss in dielectric tangent angle 10^3 Hz	Dielectric rigidity (kV/mm)	Charpy impact strength (kJ/mm^2)	Bonding strength (N/m^2)
ER	1.0×10^{10}	1.0×10^{10}	10.00	0.2000	10.00	3.00	55.00
Fe-LER/ER	1.7×10^{14}	7.0×10^{14}	5.60	0.0100	8.30	2.80	45.00
Ph-Fe-LER/ER	5.3×10^{14}	3.2×10^{14}	8.18	0.0080	9.86	2.50	37.00
Fe-Cr-LER/ER	3.4×10^{14}	4.0×10^{14}	5.96	0.0030	6.5	2.43	35.00
Ph-Fe-Cr-LER/ER	8.2×10^{13}	3.0×10^{12}	8.63	0.0100	9.3	2.50	37.00
Ca-LER/ER	2.3×10^{14}	1.6×10^{12}	6.10	0.0200	13.3	2.43	39.00
Ph-Ca-LER/ER	3.1×10^{14}	8.0×10^{12}	5.76	0.0060	10.2	2.40	35.00
NH$_4$-LER/ER	2.8×10^{15}	1.4×10^{12}	7.30	0.0200	8.00	2.42	48.00
Ph-NH$_4$-LER/ER	6.2×10^{14}	1.8×10^{14}	8.31	0.0010	9.54	2.20	40.00

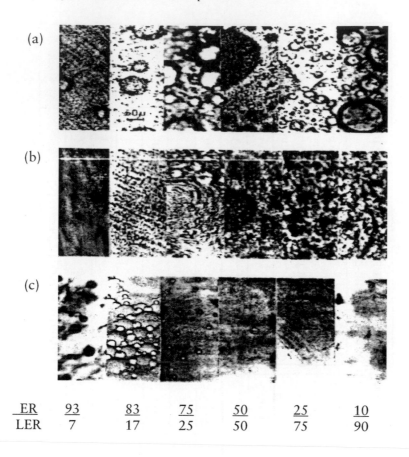

(a)

(b)

(c)

$\underline{ER}$	93	83	75	50	25	10
LER	7	17	25	50	75	90

Figure 14.1 Microstructures of LER/ER binary mixtures *versus* composition: (a) initial uncured blends;(b) uncured blends after heating for 1 h at 100 °C; (c) electron micrographs of LER/ER casting crosslinked with phthalic anhydride, 20 h at 80 °C [30]

In the severe conditions of the post-treatments (temperatures ranging from between 80 °C and 160 °C and curing time from 3 to 23 h), Charpy impact strength values were between 8 and 25 kJ/m² for the samples cured with a mixture of two hardeners.

In mild conditions of curing the increase in Charpy impact strength is significant, but the samples had a poor weathering resistance being degraded in about six months of soil burial compared with the samples cured in severe conditions [34].

The action of environmental factors, such as: soil type, microorganisms, humidity, temperature, type of plant culture, is more pronounced on the samples containing epoxy-modified lignin, which has a greater ability for water absorption and favours the

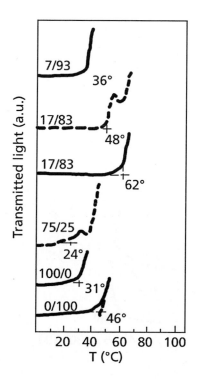

Figure 14.2 Thermo-optical curves for LER/ER blends for the indicated mixing ratios: —— film; - - - cured mixtures 20 h, 80 °C

microorganism's growth and, therefore, degradation-disintegration of the samples crosslinked in the mild conditions. The study of the mutual influence resins-environment led to the following conclusions:

a. the plant's growth has been less influenced by the presence of samples of LER/ER blends;

b. environmental conditions determined important modifications in the chemical structure and properties of resins.

The samples from the LER/ER blend crosslinked under severe conditions (25%-40% of crosslinking agent, curing time of 45-240 minutes, 150 °C) followed by post-treatment (120-160 °C and reaction time of 18-24 h), presented a better resistance than those samples crosslinked in the mild conditions (without post-treatment), tested using the soil burial test [34].

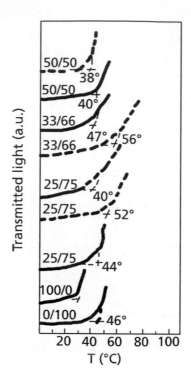

Figure 14.3 The variation of transition temperatures for different LER/ER blends for the indicated mixing ratios: —— film; - - - cured mixtures 20 h, 80 °C

Including some fillers (lead soap, alum earth, aluminium trioxide, talc, chalk, sand, silica, glass fibres), plasticiser (polyester C_6, dibutylphthalate), pigments (iron oxide, titanium oxide) in lignin/epoxy blends in a proportion of 10-40 wt% gives epoxy-modified lignin/epoxy composite materials without affecting crosslinking reactions and properties [10, 30, 44].

The shear strengths of the samples glued with adhesive from LER/ER (1:2) blend cured with phthalic anhydride had values ranging from 18.2 to 26.5 kgf/cm², in many cases exceeding the corresponding value for epoxy resin adhesive, 21 kgf/cm².

14.3 Lignin/Phenolic Resin Blends

An important application of lignin is its use as a substitute of phenol in the phenol-formaldehyde resin synthesis (PFR), or as a partial replacement of phenol-formaldehyde resin in adhesive formulations with good glueability [45-50]

Three kinds of lignin phenolysis reaction are reported, namely:

i. lignin and phenol are pre-reacted and the lignin-phenol adduct is reacted with formaldehyde [51, 52];

ii. lignin is reacted with phenolformaldehyde and then the resulting product is reacted with phenol or a phenol-formaldehyde resin previously prepared by standard procedures [8, 53];

iii. reaction of lignin with phenol-formaldehyde [54].

The phenolation of lignin may take place in the presence both of catalysts (HCl, H_2SO_4 or BF_3) [28] and alkaline hydroxide [51], at elevated temperatures, obtaining the soluble and fusible phenol-modified lignin resins, which by treatment with formaldehyde are transformed in thermoreactive resins. It is known that a condensation reaction occurred between the *o*- or *p*- position of phenol and α- position substituted by OH-, O-R_1, =O, or =C-R_2 (R_1 and R_2 lignin residues) of the side-chain of the phenylpropane units of lignin [55].

Kraft lignin, acid hydrolysis, OSL, SEL and lignosulfonates (LS) as lignin component, and phenol, cresol, resorcinol or bisphenol as phenol have been used for manufacturing lignin-phenolic resins obtaining adhesives with satisfactory strength properties. Lignin-polyphenolic resins are predominantly used for production of weather-resistant wood products, such as plywood, particleboard, fibreboard, flake board and strandboards. Generally, these board products are manufactured using lignocellulosic materials (wood veneer, random geometric configuration wood chips or wood fibres) and an adhesive. These blends are cured under the heat and pressure conditions and predetermined time, see **Table 14.4**.

Cook and Sellers [71] reported the adhesive preparation by replacement of phenol-formaldehyde (PF) resin with 35-40% OSL. Results indicated that the replacement up to 35%-40% of PF resin with crude OSL strength values are comparable with those obtained with commercial PF, resin while using of pure OSL determines a decrease of strength values.

Ono and Sudo [72] prepared an adhesive resin by treatment of SEL with phenol in the first stage, and in the second stage with formaldehyde. Adhesive formulations consisting of lignin-modified resins, extender (wheat flour), additive and water were used to glue of panels and tested by TBA. Results revealed that phenolated steam explosion lignin-based resin (LP resin) presents an intrinsic retardation in curing. The tensile strength measurements shown that phenolated SEL adhesive provides excellent strength after repeated boils indicating a post-cure process. The cure rate is improved by the increase of pH of the resin. Muller and co-workers [8] have investigated the effect of phenolated

Table 14.4 Phenolic resins and lignin blends

System	Investigation methods	Properties and Applications	Source
L/PF resin blend Lignin: -crude and purified OSL -kraft lignin -Na, K or NH_4 lignosulfonate	Adhesive properties: - maple block screening method (ASTM D905, [56]); - flake board screening method: internal bond (IB); modulus of rupture (MOR); accelerated ageing MOR (American Plywood Association performed 6 cycle test); - dimensional stability tests	- 30-45% PF resin replaced with lignin; - OSL purity influences the bond strength values; - Possibility of using of lignin adhesive systems for structural wood panels; - Adhesive characteristics comparable with those of commercial PF resin - Satisfactory results obtained by flake board and block lap-shear tests	[46, 53, 57]
Lignin/phenol-formaldehyde resin Lignin: steam explosion lignin, SEL; Ca-lignosulfonate Phenolation/treatment with formaldehyde of lignin	Phenolated-SEL formaldehyde resin (LPF resin): ^{13}C NMR spectroscopy; gel permeation chromatography; Adhesive formulation: - Torsional braid analysis (TBA); - Tensile shear strength measurement (JIS K 6851, [58])	- Functionality of phenolated lignin $\cong 2.9$; - 0.95-1.31 mol phenol/lignin C_q unit - The cure rate depends on pH resin; increase of pH of the LPF resin improves curing; - Glueability is influenced by the phenolation conditions and purity of lignin; - Lignin-resol resin adhesive satisfied the JIS requirements	[8, 28, 48]
LPF resin/PF resin L: - hydrolysis lignin; -lignosulfonate Treatment with HCHO/ phenolation of lignin;	- LPF resin: viscometry, gel filtration chromatography; free HCHO, free phenol; alkalinity; - LPF resin adhesives: mechanical properties tests: IB; MOR; modulus of elasticity (MOE) (ASTM D1037-80, [59]); dimensional stability	- Good compatibility of L in LPF resin synthesis at the 35% phenol substitution; - LPF resin can be used in oriented strandboard (OSB) and panel manufacture; - Physical strength and dimensional stability properties of strandboard bonded with LPF were similarly with those of PF-bonded	[60, 61]

Table 14.4 Continued

System	Investigation methods	Properties and Applications	Source
LPF resol resin Lignin: Ca, Na, NH$_4$-lignosulfonate (LIGNOSITE) Pre-condensation of HCHO with phenol/addition of lignin	- Lignin–modified PF resol resin: viscosity; free formaldehyde content; non-volatile percent; gel time; adhesive formulations: internal bond strength) (ASTM D1037-80) [59]; dimensional stability	- 15-30% of phenol replaced by the lignosulfonate; viscosity of the LPF precursor resin – 50-100 cSt; M$_n$ – 180-250; M$_w$ 3,600-14,000; 15-30 min gel time of LPF resin; - Dispersion-type adhesive formulations used to bond wood chips, veneers and sheets of plywood; - Mechanical properties similarly with PF resin; cost saving	[54, 62]
Lignin/novolak resin Lignin: alkali or kraft lignin Novolak – Durez type – molecular weigh	- L-novolak resin (blend or synthetic): softening point; - Molecular weight; - Moulding index test and cup flow test (ASTM D731, [64]); Physical properties (ASTM D256 [65], D570 [66], D648 [67], D790 [68], D955 [69]); - Resistance to boiling solvents (ASTM D543 [70])	- L-novolak blend formulation: 10-40% lignin content; - L-novolak resins: 10-40% of the phenol replaced with lignin - The moulding test – satisfy minimum industry standard – particularly excellent surface gloss - Properties equivalent and in some cases superior those compounds from commercial novolak resin: heat deflection temperature - Good resistance to boiling solvents - High crosslinking in the presence of curing agent of lignin-novolak resins	[48, 63]

lignin/phenol-formaldehyde resin on the adhesion quality and they found that the strengths have the following order in relation to lignin type:

phenolated-steam explosion lignin < kraft lignin < neat phenolic resin.

This behaviour is explained by the presence of considerable amount of syringyl propane units in the SEL which present a great reactivity to phenol.

Klasnya [46] prepared an LPF resin adhesive by polycondensation of kraft lignin with phenol and formaldehyde at 100 °C in alkaline medium. The strength properties of plywood bonded with this adhesive depend on temperature and the time of pressing. Values obtained for shear strengths for LPF resin (30%-50% lignin replacement of phenol) are 3.31-3.59 N/mm^2 in comparison with 2.83-3.21 N/mm^2 in the case of control-PF resin. If a high lignin content (>50%) is used, the shear strengths decreased (1.81-2.05 N/mm^2).

Oh [60] and Sellers [61] reported the substitution of phenol with lignin in the synthesis of lignin-modified PF resin. The resin synthesis includes a first stage of reaction for a maximum condensation of formaldehyde with lignin to introduce of hydroxymethyl groups, and a second stage of reaction designed for copolymerisation between the hydroxymethylated lignin and phenolic resin molecules. Oh and Sellers used the lignins extracted with methanol, ethanol or dioxane, from various solid residues of the acid hydrolysis process of newsprint from municipal solid wastes (MSW). The lignins presented a satisfactory compatibility at a 30%-35% phenol substitution level in PF resins used as wood binders. The characteristics and molecular weight of the LPF resins were comparable with those of control PF resins, see **Table 14.5**.

The strandboards glued with LPF resin using four resin solids, level of concentrations (3.5, 4, 4.5 and 5%) and two pressing times (4 and 5 minutes) were tested for physical strength and dimensional stability properties The physical property tests included tension shear perpendicular to the surface (IB), and static bending (MOR strength retention and MOE) (**Table 14.6**).

Senyo [73] found that by replacing 20% of PF resin with methylolated-low molecular weight OSL, the formaldehyde emissions decrease during the pressing process of the chipboards. MOR and IB properties of the wood chipboards bonded with 20% methylolated-lignin/80% PF resin are comparable to those glued with 100% PF resin (MOR 17.8 and 19.9, respectively; IB 13.3% and 14.0%, respectively).

In the past, spent sulfite liquor has been used as adhesive, but because it is soluble in water, its use is rather limited. Roffael [74, 75] and Allan [76] reported preparation of some adhesives with good mechanical properties and dimensional stability by mixing PF resin with the spent sulfite liquor.

Property		Control PF resin	LPF resin	LPF$_1$ resin	LPF$_2$ resin	LPF$_3$ resin	LPF$_4$ resin
Table 14.5 Physical properties of LPF and control-PF resins							
Non-volatile solids (%)		48.2-50	55.2	51.8	53.4	53.0	56.2
Viscosity (mPa-s)		195-325	278	298	260	270	298
pH		10.7-11.2	10.3	10.2	10.3	10.2	10.2
Specific gravity (g/cm^3)		1.22-1.23	1.22	1.23	1.23	1.23	1.23
Gel time (min)		17.6-22	27.5	22	22	22	25
Alkalinity (%)		5.6	3.4	3.6	3.8	2.6	4.4
Free HCHO (%)		0.05	0.24	0.23	0.19	0.17	0.23
Urea (%)		5.3-5.4	5.0	4.5	4.6	4.2	4.5
Molecular weight characteristic	M_w	5516-6229	3074	9431	9916	5816	648
	M_n	824-1059	336	772	885	831	980
	M_w/M_n	5.8-6.7	9.1	12.2	11.2	7.0	6.6

According to Stout and Ludwig [51] the sodium lignosulfonate isolated from spent sulfite liquor and phenol are pre-reacted under alkaline conditions, and then the lignosulfonate adduct is reacted with formaldehyde. Up to 50% of phenol was substituted with Na-LS. Coyle [53] and Döering [54] reported the preparation of LPF-based adhesive by reacting of a PF pre-condensed with methylolated-lignosulfonate or lignosulfonate to form a lignin-modified PF precursor resin. The chemical interactions of lignin with PF resin are responsible for the improvement of mechanical properties of the end-products [47].

Another adhesive for the manufacture of plywood, fibre board, particle board and similar products have been reported by Forss and Fuhrmann [77]. This adhesive contains PF resin (resoles) and a lignin derivative, such as lignosulfonates and alkali lignin in the form of alkali or earth alkali salts. Forss found that the molecular weight distribution of the lignin derivatives is a significant factor which influences the adhesive characteristics [78]. The weight ratio of the lignin derivative to the PF resin depends on their applications, for example in fibre board production the adhesive may be added as a dilute water solution with a solid content of 5%, in plywood manufacture of 30% and in particle board production of 50%. The strength properties of end-products manufactured with lignin/PF resin adhesive (bending and tensile strength) meet the requirements of the standards.

Resin solid (%)	Resin type	Panel density (kg/m³)	IB (kPa)	MOR (MPa)	MOR ret. (%)	MOE (GPa)
3.5	PF	769	579	27.5	78.1	3.7
	LPF	783	538	32.6	65.0	3.8
4	PF	692	476	28.8	60	-
	LPF_1	668	586	24.1	67	-
	LPF_2	689	324	21.2	77	-
	LPF_3	698	441	24.7	61	-
	LPF_4	678	483	24.5	62	-
4.5	PF	750	558	28.5	83.5	3.7
	LPF	774	531	25.4	87.9	3.8
5	PF	686	710	29.1	67	-
	LPF_1	670	641	25.9	72	-
	LPF_2	681	538	25.8	71	-
	LPF_3	682	565	28.0	53	-
	LPF_4	668	558	24.5	61	-

Table 14.6 Physical properties for strandboard panels bonded with a commercial PF resin and LPF resins at four resin solid level of concentration for 5 minute pressing time

Each value of the physical properties represents an average of n test specimens
All IB values are normalised to a 673 kg/m³ density
MOR ret is the retention percentage of MOR versus MOR after an accelerated ageing exposure test

Shen, Fung and Calvé [79] have shown that the use of low molecular weight lignosulfonate fractions offer the best conditions to glue the particle boards with a thickness of 6 and 13 mm, at low pressing time obtaining the higher mechanical properties (MOR=19.4-23.2 MPa; MOE = 3.5-4 GPa; IB = 320-389 kPa) in comparison with the standard requirements (MOR = 14 MPa; MOE = 2.7 GPa; IB = 280 kPa).

Sodium, calcium or sodium-ammonium lignosulfonate types were used for the 15% phenol substitution in the resol-type liquid phenol-formaldehyde resin, involving a methylolated lignin condensation in the first step and then treatment with phenol-formaldehyde resin in the second step [62]. Shear test results (wood failure and shear strength) of plywood bonded with lignin-modified PF resins shown an influence of the lignosulfonate nature. Values are in the following order:

Na-lignosulfonate (Borresperse) < Na-lignosulfonate (Reebax) < kraft lignin <

< Na-NH$_4$-lignosulfonate (Tembind), Ca-lignosulfonate (Lignosite 100).

Replacement of phenol with 15% lignin-based materials in PF resin adhesive for plywood production gives cost savings of up to 30% [62].

Another use of lignin-modified PF resins reported in the literature is in moulding compounds. Gobran [63] reported the production of ligno-novolak resins, either in the form of physical blends of lignin and novolak resins or as synthetically derived from the reaction of lignin, phenol and formaldehyde in the presence of an acidic catalysts. These resins are curable by thermoset, completely crosslinked resins being obtained by the action of a curing agent such as hexamethylentetramine. Lignin/novolak resins can be used to produce compression moulding compounds possessing acceptable water absorption and mechanical strength, higher heat deflection temperatures and higher electrical properties compared to those of commercial novolak moulding products [48, 63, 78, 80].

Ysbrandy and co-workers [81] reported obtaining a synthetic mixture with interesting properties formed from phenolated pitch/lignin resin to moulding products using certain catalyst system (anhydrous ZnCl$_2$, phenolsulfonic acid or a combination of (Na$_2$S + NaOH)). The tensile properties of mouldings were the best for 25% resin content. At a higher resin level the phenolated pitch/lignin systems present a poor compatibility due to the substituted nature of the phenolic pitch. The dimensional stability of the mouldings decreased with decrease in resin content.

14.4 Lignin/Polyolefin Blends

The lignin role in polyolefin (PO) blends depends on the lignin nature, separation procedure and content. Due to its phenolpropanoic structure similar to that of hindered phenols, lignin can act as stabiliser [11] or initiator of PO degradation [82-84]. Incorporation of lignin in PO material influences the useful properties of the polymer and its biodegradation characteristics, see **Table 14.7**.

By incorporation of lignin, an increase in thermal resistance of thermoplastics has been found by Bono [100] and Bubnova and co-workers [101]. The biodisintegrable nature of lignin-filled polyethylene has been claimed in another patent by Bono and Lambert [102].

583

Table 14.7 Lignin/polyolefin blends with useful properties and/or biodegradation characteristics

System	Investigation methods	Properties	Source
L/PO/additive PO: LDEP; HDPE; PP L: kraft lignin Additive: ethylene acrylic acid copolymer (EAA); titanate; conventional additives	- mechanical properties determination - dynamic viscosity - spectrophotometry - DSC - electrical and thermal properties	- reduced tensile strength - coupling agent incorporation improved mechanical properties - melt viscosity increase with increasing lignin content - good electrical resistance	[85, 86]
L/PP-PP L: kraft lignin	- tensile testing (ASTM D638 [87])	- tensile strength values depend on the lignin content and molecular weight of PP - lignin acts as a filler - poor adhesion between lignin and PP	[88]
L/PE LDPE L: ammonium lignosulfonate	- mechanical properties (ASTM D638 [87]) - thermal behaviour - biodegradation tests	- reduction of the physico-mechanical properties with the increase of lignin - water absorption increase - thermal stability reduced - increased susceptibility to the environmental degradation	[89-92]
PE/L-g-PE HDPE L: sodium lignosulfonate	- DSC - tensile strength (ASTM D638 [87]) - impact properties - SEM	- L-g-PE up to 50% - improved mechanical properties - better lignin PE matrix adhesion	[92]
L/PO/EVA PO: LDPE; HDPE, PP L: prehydrolysis kraft lignin	- SEM - mechanical properties	- LDPE/L system presents phase separation - EVA incorporation in the LDPE improves compatibility, tensile strength and elongation at break	[93, 94]

Table 14.7 Continued

System	Investigation methods	Properties	Source
L/PE LDPE L: LS-NH$_4$; LS-NH$_4$E; LS-NH$_4$O	- mechanical properties - TG and DTG - DSC - optical and electron microscopy - biodegradation tests	- improved mechanical indices - environmental degradation susceptibility increased	[10, 95-97]
LS /PP/additive iPP L: LS-NH$_4$E Additive: PP-g-GMA	- physico-mechanical properties - DSC - TG and DTG - SEM - X-ray diffraction - Contact angle measurements - biodegradation test	- improved compatibility by to use PP-g-GMA - contact angle of iPP/LER blends decreases with the increase lignin content - thermodynamic work of adhesion increase - improved miscibility and adhesion of the components - components interact during processing - biocompatibility - physical treatments (UV-irradiation, plasma, electron beam) modified surface properties and increaseed environmental degradation susceptibility	[10, 11, 98, 99]

LDPE: *low density polyethylene*
DTG: *derivative thermogravimetry*
SEM: *scanning electron microscopy*
TG: *thermogravimetry*
PP: *polypropylene*
HDPE: *high density PE*
iPP: *isotactic PP*

Kharade and Kale [85] have incorporated up to 30 wt% dry lignin powder in low-density polyethylene (LDPE), high-density polyethylene (HDPE) and polypropylene (PP). Table 14.8 shows that the tensile strength and elongation at break values decreased with lignin content for all three polymer blends indicating a poor compatibility or interaction between lignin and the matrix polymer.

The use of ethylene acrylic acid copolymer (EAA) and a titanate coupling agent lead to the improvement of tensile properties.

Gonzalez-Sánchez and Alvarez [88] obtained a PP/L blend in a thermokinetic mixer by melt blending of PP with different molecular weights and kraft lignin powder (10 to 55 wt%). Results reveal an increase of the tensile modulus (E_t) with lignin content from 0.73-0.80 GPa for PP to 1.31-1.34 GPa for PP/L (55 wt% lignin). Maximum tensile strength decreases indicating the poor adhesion between non-polar structure of PP and more polar structure of lignin and a poor dispersion into the matrix. Elongation at break decreased as lignin content increased.

Table 14.8 Mechanical properties of lignin-polyolefin blends [85]					
Polymer matrix	Lignin (%)	Tensile strength (kg/cm²)	Elongation at break (%)	Tensile strength of blend/Tensile strength of virgin polymer ratio	Izod impact (J/cm)
PP	0	337	22	1.0	0.20
	5	271	9	0.80	0.18
	15	242	6	0.72	0.20
	30	137	5	0.41	0.16
	30*	195	9	0.58	0.18
LDPE	0	110	255	1.0	2.20
	5	96	47	0.87	1.0
	15	84	35	0.76	0.65
	30	67	15	0.61	0.55
	30*	83	19	0.75	0.75
HDPE	0	323	601	1.0	1.90
	5	211	283	0.65	0.70
	15	184	143	0.57	0.60
	30	153	53	0.47	0.60
	30*	189	65	0.59	0.70
* with 5 phr ethylene acrylic acid copolymer (EAA) and 0.5 phr titanate coupling agent					

Rusu and Tudorachi [89, 90] reported results regarding the mechanical properties and thermal stability of mixtures based on the various amounts of lignin (5-30 wt%) from wood sulfite pulping (lignin) and LDPE.

Incorporation of lignin (up to 10 wt%) in PE leads to the decrease of melt flow index (MFI), but addition of larger quantities of lignin didn't change the MFI values more. The tensile strength is reduced by 25%-28%, from initial value for an increased content up to 5 wt% lignin, but for higher content (up to 30%) is decreased by a percentage of 10-12 wt% only. The elongation at break decreases drastically up to 5% lignin, after which the decreases becomes insignificant. Regarding the modulus of elasticity, a characteristic providing information on the rigidity of the polymeric materials, one can see that an increase in the lignin content of the system induces an increasing in the values corresponding of this property. The introduction of lignin and its increased content induced a decrease in the Izod impact strength values. The water absorption which serves as an indication of the ability of PE/L blends to biodegrade in the environment, increases with increasing of the L content [90, 91].

Interpretation of the TG and DTG diagrams of the PE/L shows that in the temperature domains of 0-220 °C and 220-340 °C, the weight loss was around of 3%-4% and 7%-9%, respectively. This can be explained by the fact that lignin acts as stabilisation agent for LDPE degradation.

The biodegradability tests achieved both '*in vitro*' (in the presence of *Aspergillius niger* and *Trichoderma viride* fungi) and '*in vivo*' (in the presence of a compost sludge sewage mixture) showed that the susceptibility to biodegradation increases with increasing the lignin content in blends [91]. The burial soil test (after 24 months) shows that the LDPE/ 20 wt% L blends presents a decrease in tensile strength of about 30% and in elongation at break of 56%.

Compatibility between PO and lignin can be improved by introduction of reactive functional groups in each or in only one blending compounds. Thus, Casenave and co-workers [92] developed a new technique. In their method, ethylene or propylene monomers are first grafted by a catalytic reaction on the lignin surface and then it is mixed with the polyethylene matrix. This procedure offers the possibility of improving the interfacial adhesion by establishing the chemical bonds between the matrix and the additive and to ensure a better dispersion and adhesion to the synthetic matrix [92, 93]. Properties of PE/L-*g*-PE are always higher than those of PE/L. Incorporation of grafted lignin (L-*g*-PE) or L produces the decrease of mechanical properties with increasing lignin content. Young's modulus for PE/L-*g*-PE remains constant and closed to that of pure PE up to 64% lignin content after it decreases.

SEM has revealed the different morphologies of the broken sections of PE/L-*g*-PE blends where it can see either the particles that were extracted from other half of the samples, but with enhanced roughness of the surface of spheres, or broken particles indicating a better lignin-grafted/matrix adhesion [92].

Another way to improve of the compatibility between lignin and polyolefins is utilisation of a compatibilising agent such as ethylene-vinyl acetate copolymer (EVA) [93]. Alexy and co-workers [94] showed by SEM that a homogeneous structure is obtained when EVA (28% vinyl acetate units content) is used in an LDPE/L blend. The concentration of lignin was varied from 15 wt% to 35 wt% and for EVA from 2 wt% to 25 wt%. Results show that tensile strength and elongation at break are significantly improved by EVA copolymer addition. Ciemniecki and Glasser [103] have observed that the increasing amount of vinyl acetate groups incorporated into the backbone of PE produces the increase of interactions between lignin (hydroxypropyl lignin) and PE matrix. This leads to improvement of compatibility of blend components. In another work Alexy [86] observed that the addition of lignin at 10% level and conventional additives (amide of erucic acid; phenolic antioxidant; phosphitic antioxidant) has a similar effect on the processing stability of LDPE and PP. A different degradation behaviour between PE/lignin and PP/lignin blends has been observed mainly at higher concentration of lignin. In the case of PE blends, lignin acts as an initiator of the degradation process after short UV light exposure while at long-term heat stress as a stabiliser. For PP blends having high lignin content it has been observed that lignin acts as an initiator of the degradation process at long UV light exposure. The mechanical properties of PE and PP decrease with increasing lignin content.

Vasile and co-workers [95-99, 104] showed that the incorporation of low amounts of lignosulfonates (3%-8%) or epoxy-modified lignosulfonates in PO leads to an important increase of the physico-mechanical indices, i.e., the tensile strength is 2 times higher and elongation at break 6 times higher to those of LDPE or iPP (**Table 14.9**). The percentage of lignin can be raised up to 30 wt% by using EVA or C_5 (petroleum resin as compatibilising or dispersion agents). Also, to achieve an improved compatibility and increased mechanical resistance, a partially cure by the addition of crosslinking agents, such as DDM or PhA is recommended.

Increasing of the elongation at break, by the modified lignosulfonates indicates their plasticising or impact modifier effect, as already shown for the increase of the impact strength of other mixtures. Although the amounts of lignosulfonates used in mixtures with polyolefins are very small (< 5%) the thermal characteristics are modified. The T_g of the mixtures are higher than to those of LDPE, while the melting temperatures are a little lower [98] (see **Table 14.10**).

Table 14.9 Physico-mechanical indices of PO/LS blends [95]		
Sample	Tensile strength (MPa)	Elongation at break (%)
LDPE	13.0	120.5
LDPE/5% LS-NH$_4$	19.9	173
LDPE/5% LS-NH$_4$/1.5% C$_5$	18.1	134
LDPE/5% LS-NH$_4$O*	11.8	171
LDPE/5% LS-NH$_4$O/1.5% C$_5$	13.0	124
LDPE/3% LS-NH$_4$E**	26.4	631
LDPE/3% LS-NH$_4$E/1.5% C$_5$	22.8	728
LDPE/3% LS-NH$_4$E/1.5% DDM	24.1	599
LDPE/20% LS-NH$_4$E/10% EVA	4.4	33
iPP	15.3	85
iPP/2% LS-NH$_4$E/1.5% PhA	19.1	98
iPP/4% LS-NH$_4$E/1.5% PhA	20.7	636

LS-NH$_4$O: oxyammonolised lignosulfonate at T = 162 °C, p = 0.6 MPa, t = 3.3 h [96]
**LS-NH$_4$E: epoxy-modified ammonium lignosulfonate/epichlorohydrin-1/10 ratio, 35% NaOH, 75 °C, 5 h [38]*
C$_5$: petroleum resin modified with maleic anhydride

Decomposition of LDPE under dynamic conditions of heating is significantly changed in the presence of even 3% LS; DTG curves are shifted to higher temperatures and the weight loss rate is slower, so it can be concluded that the mixtures have a higher thermal stability, lignin functioning as a stabiliser [95, 96].

Dispersion of the components in the epoxidised lignosulfonates-containing mixtures is more advanced, the compatibility being improved either by addition of the C$_5$ dispersing agent or by crosslinking with DDM or PhA for very low LS content [104].

In other experiments, Vasile and co-workers [10, 95, 96] have improved the compatibility of PO with lignin by mixing epoxy-modified lignin with glycidylmethacrylate-grafted-polypropylene (PP-g-GMA) as compatibilising agent and partially crosslinking. The data obtained by DSC, SEM, contact angle measurements, X-ray diffraction have confirmed the improvement of compatibility due to the chemical reaction between components and also due to their good dispersion in the blends [104].

Table 14.10 Thermal characteristics of several LDPE/LS blends [98]				
Sample	T_g (°C)	Δc_p (J/kg)	T_m (°C)	ΔH_m (J/g)
LDPE	-15	0.054	140.2	231.51
LDPE/3% LS-NH$_4$E	-6......-7.6	0.039	138.9	172.60
LDPE/3% LS-NH$_4$E/1.5% C$_s$	-4......-10.6	0.293	141.7	209.46
LDPE/3% LS-NH$_4$E/1.5% DDM	-4......-9	0.015	135.0	199.40
T_g: glass transition temperature; Δc_p: difference of heat capacities in the T_g interval; T_m and ΔH_m: melting temperature and heat of melting				

Pascu and co-workers [105, 106] showed that by a combination of the chemical procedures of component modification with physical ones of surface treatment, e.g., photooxidation, plasma and electron beam treatment, bio/environmental disintegrable polyolefin materials can be obtained. UV-irradiation of the PO/L blends induced the change of the surface properties [105]. The results obtained by IR spectroscopy, DSC and X-ray diffraction have shown that plasma and electron beam treatments on the iPP/LER/PP-*g*-GMA films are very efficient imparting a high polarity and an increased hydrophylicity of blends [105].

After degradation tests by soil burial or in the presence of microorganisms (*Paecilomyces varioti* and *Chaetomium globosum Kunze Fries*), all values of thermal characteristics decreased, probably due to the morphological modification, phase separation and crystallinity changes due to the formation of the degradation products [10, 95-97, 99]. Behaviour after soil burial degradation depends on the mixture composition, namely most of the LDPE containing samples are resistant to 12 months soil burial, while the iPP containing ones are stable for only 6 months a quantity of only 4% LS-NH$_4$E being enough for significant degradation [95, 96]. The LDPE-containing blends show an increase of the sample's weight due to the water absorption, a modification only being observed at the beginning of degradation, while iPP-containing ones present important weight losses (2.35-4.21 wt%) which increase with duration of degradation.

The surface of iPP is modified due to both the initiation of lignin degradation and superficial modification of blends. Superficial and deep cracks appear, starting with LER-particles and the changes in all components of the free surface energy are characteristic for the unstable surfaces for photooxidation [11, 107]. The weight loss increased and elongation at break decreased.

In PP/OSL or oxidised lignin with up to 2-3 wt% lignin an increase in photostability was shown (radical scavenger) [82] while at higher lignin content it initiates radical reactions [84]. Other effects of lignin incorporation in PO are: increase in conductivity, hydrophilicity, improved printability imparts biodegradability characteristics [108, 109]

14.5 Lignin/Polyurethane Blends

Use of lignin or lignin derivatives as an additive or as an active components into polyurethanes has been one of the most intensively investigated application. Lignin incorporation modifies its cure rate by increasing the concentration of aromatic groups within the polymeric network and may contribute to increasing in the degree of crosslinking density of polyurethane (PU) (**Table 14.11**).

Kelley [110] has presented the thermal and mechanical properties of various PU which contain only lignin as the polyols in relation to the polyurethanes which contain both lignin and poly(ethylene glycol) (PEG). The T_g, MOE and tensile strength increase with increasing lignin content (25%-40% lignin), while the ultimate strain decreased. Properties of lignin-based PU depend on the preparation method, lignin content and molecular weight of PEG and the requirements of end product application. For the network with high ultimate strain is recommended the preparation of chain-extended hydroxypropyl lignin. Thus, PU derived from a toluene diisocyanate (TDI)-PEG-kraft lignin exhibited very high values of the ultimate strain [110].

From TG curves of solvolysis lignin-PEG-diphenylmethane diisocyanate system has been observed that lignin retarded the thermal degradation of PU in air in comparison with degradation in nitrogen [111]. This is explained by the oxidative condensation reactions of lignin. An activation energy value (121 kJ/mol) for PU with a lignin content of 20 wt% is higher than of the control PU. It can be concluded that the PU is thermal stabilised due to the presence of lignin.

The effects of crosslinking density, the molecular weight of lignin or lignin derivatives on the PU were investigated by Yoshida using a kraft lignin-polyether triol-poly(methylene diphenyldiisocyanate) system [112, 113]. It was found that at low and intermediate NCO/OH ratios, the tensile strengths and Young's modulus increase with lignin content up to 25%-30%. For high NCO/OH ratios, the tensile strengths and Young's modulus attained values of 45 MPa and 1.2 GPa, respectively, when kraft lignin content is between 5% and 10%. The ultimate strains of the PU prepared from higher molecular weight fractions decrease monotonically with kraft lignin content, while those of the PU produced from lower molecular weight fractions attained a maximum, then they decreased rapidly with the increase of lignin content. It can be concluded that behaviour of mechanical properties

Table 14.11 Lignin-polyurethane (L-PU) blends

System	Investigation methods	General remarks	Source
L/PU with/without PEG; L: kraft lignin; organosolv lignin; steam explosion lignin; solvolysis lignin; hydroxypropyl kraft lignin; PU: polyols and diisocyanate (hexamethylene diisocyanate, aromatic diisocyanate) PEG: polyethylene glycol	- DSC - Mechanical properties: modulus of elasticity (MOE); ultimate strength; ultimate strain. - Thermogravimetry; - Swelling tests	Polyurethane networks properties vary directly with lignin content; T_g, MOE and tensile strength increase with lignin content; Chain-extended hydroxypropyl lignin – high ultimate strain; Presence of lignin increases thermal stability of PU; Molecular weight of PEG influences tensile strength of PU.	[110-113]
L/PU PU: isocyanate prepolymer; branched polyol prepolymers L: kraft lignin (Tomlinite, Indulin AT, Eucalin)	- tensile testing (ASTM C719-79 [114]) - weathering tests: artificial (AW) (temperature shock, UV radiation or/and natural (NW) weathering - SEM - DSC - swelling tests - hardness testing (ASTM D2240 [115]) - mechanical properties (ASTM D412 [116], CGSB CAN2-190-M77) - compression testing (ASTM D395 [117])	- Addition of lignin modifies the curing mode of PU - Tensile strength of PU-L sealant blend depend on lignin content, adhered substrate and ageing conditions - SEM micrographs reveal the uniform distribution of lignin particles but and the different morphologies of the components - DSC confirms immiscibility between L and PU matrix - Lignin not interacts with PU - Lignin acts as a filler	[35, 118-121]
PU/MA-g-L PU formulation: polyisocyanate prepolymers; polyether MA-g-L	- IR spectroscopy - GPC - DSC - swelling test - mechanical properties (ASTM D412 [116])	- MA-g-L no interacts with PU elastomer but acts as a good reinforcing agent - MA-g-L increases swelling degree	[122]

EW: equivalent weight; MA-g-L: maleic anhydride grafted lignin; GPC: gel permeation chromatography of the PU did not present any uniformly systematic relationship with respect to the molecular weight of the lignin fractions [113].

Feldman [118] investigated the viability of blending kraft lignin with PU-based sealants. Tensile strength values of a L-PU sealant blends vary according to the amount of lignin present in the matrix, to the type of substrate to which it is adhered and exposure conditions. Toughness and modulus values of blended sealants increase with the addition of lignin

The effects of NW are more severe than that of the AW because moisture, pollutants and other environmental factors act combined on the PU-L sealant blends. Loss of modulus is 20% in AW specimens and a 30% for NW samples in comparison to control PU sealant. The addition of small quantities of lignin reduces the loss of modulus in L-PU blends [118].

The results of tensile tests showed that lignin acts a reinforcing agent in the two-phase polymeric particulate system. SEM micrographs show both the uniform distribution of lignin particles, and the different morphologies of the constituent phases [119]. The effect of AW (thermal shock and UV radiation) on the surface of L-PU specimens was illustrated by the presence of lignin particles entrapped into the polymeric matrix.

The L/PU blends have been characterised by several thermal transitions occurring at approximately $-52\ °C$ (attributed to the T_g of the soft segments), at $9\ °C$ (attributed to the micro-crystalline melting of the soft segments), at $90\ °C$ (range ordered hard segments) and at $125\ °C$ (T_g of lignin) [119].

Lignin as a filler was found to restrict the degree of swelling of PU less than other inorganic fillers, i.e., siliceous clay/TiO_2 mixture. Data based on the swelling method shows that lignin does not interact with elastomer matrix to a great extent [120, 121]. The extent of the restraint varies with type of lignin and the interaction potential of different lignins decreases in the following order: Tomlinite > Indulin AT > Eucalin [35].

The addition of mineral fillers or different kinds of lignins to the elastomer formulation increases the hardness at a given time contributing to the increase in the Young's modulus [120]. The modulus ratio (modulus of the modified to the nonmodified elastomer E/E_0) is dependent on the filler loading. The addition of lignins decreases both the stress and strain at break of this elastomer in relation to the unfilled blend. The results obtained for L-PU blends are 2.5 times greater than that unfilled elastomer (2.7%). The interactions between the PU-based elastomer and various kraft lignin fillers were investigated by the surface analysis establishing the contributions of the filler to the work of adhesion. The modulus E values appears to be a function of the work adhesion at different volumetric loadings indicating that the modulus values are proportional with the work adhesion [123], see **Figure 14.4.**

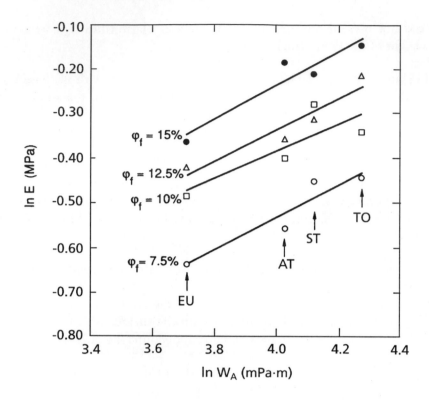

Figure 14.4 Dependency of the work adhesion, W_A, on the tensile modulus at different volumetric filler loadings [123]

Lignin modification by grafting with maleic anhydride was used to improve its compatibility with PU elastomer [122]. Swelling studies indicated that MA-*g*-L produces a reduction in the effective number of crosslinking and consequently an increase in swelling, with a corresponding increase in MA-*g*-L content in PU blends. PU/MA-*g*-L blends present an improvement in mechanical properties (ultimate strain, E_B, and ultimate stress, T_B) compared to the PU control. The best results were found for PU/MA-*g*-L (15% or 20%) blends. Tensile properties are in function of MA-*g*-L content. Ultimate strain E_B shows a maximum at 15% MA-*g*-L, which could be explained by the fact that MA-*g*-L produces a reduction of the intermolecular bonds per unit volume of PU.

14.6 Lignin/Polyester Blends

The preparation of the poly(ethylene terephthalate) (PET)/LER and PET-co-isophthalate (COP)/LER blends with a little higher melting temperatures than those of base polymer has been reported by Agafitei and co-workers [124].

The microscopic aspects show a non-uniform globular distribution of LER which becomes much more uniform by mixing of components in a Brabender plastograph at 200 °C and by partially crosslinking with DDM. The different behaviour of the PET-containing blends with respect to COP-containing ones have been shown by DSC (see **Table 14.12**).

In the LER-containing blends, two separated regions (T_g at 74 °C and 55 °C) are present indicating incompatibility of the components. The DDM crosslinking reaction is much more efficient in COP-containing blends, when a single T_g interval appears, the compatibility of components being improved. The 'cold' crystallisation temperatures of the blends shift towards low values, while the melting temperatures increase especially after crosslinking. The crystallisation heats increase with LER content, while those of melting decrease. These modifications indicate the morphological change by LER incorporation.

Table 14.12 DSC results for PET or COP/LER blends [124]					
Sample	T_g (°C)	$T_{\text{'cold' cryst.}}$ (°C)	T_{melting} (°C)	$\Delta H_{\text{cryst.}}$ (kJ/kg)	$\Delta H_{\text{melting}}$ (kJ/kg)
Mechanical mixing					
PET	75	160	256	24.43	44.84
PET/7% LER	40	132	-	35.38	-
PET/7% LER/DDM	55	132	-	46.06	-
	74				
COP	75	165	210	13.70	23.23
COP/6% LER	57	140	215	39.95	26.23
	72				
COP/6% LER/DDM	70	140	219	38.87	16.81
Göttfert mixer					
COP/3% LER/DDM	73	150	213	20.06	14.07
COP/4.5% LER/DDM	70.5	145	210	23.31	20.35
COP/7% LER/DDM	68	137	212	29.61	19.29
COP/10% LER/DDM	68	141	210	25.07	20.95

DTG data show a slight increase of peak temperature in the blends that means a thermal stabilisation of the polyesteric component by LER incorporation.

Presence of lignin in polyester matrix leads to the improvement of superficial properties and of the dielectric properties (**Table 14.13**) [124].

In another paper, Cazacu and co-workers [125] studied behaviour and UV- and non-exposure of the COP/LER blends to the sewage sludge test, observing the important changes in their bulk and surfaces properties after only two months of ageing. The mass of the blend samples increase with about 0.5-1 wt%/month, while the thickness of the sheets also increases with about 10%-12% due to the water absorption from medium. The water absorption favours the degradability starting with the hydrophilic component (LER) from blends. After ageing, the T_g becomes less evident and shifted to higher temperatures. The 'cold' crystallisation temperatures and 'cold' crystallisation heats of the degraded blends shifted towards lower values, while the melting temperatures increase with about 20 °C especially in the case of the UV irradiated and degraded COP/LER samples [126].

After degradation, the surface properties of the blends were modified. Results indicated that more susceptibility of surface to the attack of microorganisms are those with high LER content and UV-irradiated.

Table 14.13 Dielectric properties of COP/LER blends obtained on Göttfert mixer [124]					
Sample	Resistivity		Dielectric constant ε (50 Hz)	$t_g\,\delta$	E_{bte} (kV/mm)
	Volume ρ (ohm.cm)	Surface (ohm)			
COP	2.3 x 10^13	-		0.006	20.5
COP/LER 3%	3.1 x 10^16	6.1 x 10^13	4.38	0.01	19.6
COP/LER 4.5%	2.1 x 10^16	6.2 x 10^14	4.42	0.029	24
COP/LER 7%	3.1 x 10^15	1.7 x 10^14	4.25	0.028	20.7
COP/LER 10%	1.2 x 10^12	8.3 x 10^9	8.81	0.160	18.3
E_{bte}: breakdown voltage $t_g\,\delta$: loss in dielectric angle					

14.7 Lignin/Poly(Vinyl Chloride) Blends

Poly(vinyl chloride) (PVC) is known for its ability to form miscible systems with other low or high molecular weight compounds, acting as plasticisers. The selection of lignin for obtaining the particular blends was justified by the criteria of miscibility of two polymers, such as their close solubility parameters, the difference between their molecular weight and their functionality [127].

Feldman reported preparation of the L/PVC blends from PVC compound formulated with several amounts of TiO_2 (10%) and a kraft lignin and an OSL [127]. Although, the two polymers are either not miscible or partially miscible, the kraft lignin presence in a proportion up to 7.5 phr does not affect thermal stability and processability of PVC. In the case blending of the OSL with PVC SEM micrographs indicate their non-miscibility.

The DSC thermograms recorded for the L/PVC blends after 19 minutes of annealing time present only one T_g located at around 85 °C. This could be correlated with an increased miscibility by thermal treatments. Mechanical properties of unweathered blends depend on the lignin load. The maximum tensile strength is about 9% higher and the elongation at break is reduced about 6% than the respective values of the PVC control [128].

The results of DMA analyses, (i.e., a gradual decrease of T_g and a gradual increase of tan δ of the blends as function of lignin loading), suggest that the lignin presence led to a certain degree of breaking of the intermolecular bonds existing within PVC structures. Higher relaxation energy data of the blends in comparison with PVC control indicated a reduction of the free volume, probably due to the polar-polar interaction between carbonyl or hydroxyl groups of lignin and α-hydrogen or the chlorine of PVC.

Feldman [129] extended this study using different lignin grades. The influence of various lignin types on the processability of the L/PVC blends is reflected in the fusion characteristic and equilibrium torque. The equilibrium torques are slightly lower and reflects a lower viscosity of the melt for the blend containing partial water-soluble lignin (PWSL), OSL and kraft lignin. The strength at yield and break and IR spectra demonstrated an interaction occurring between OH groups of lignin and a-hydrogen of PVC. This interactions decreases in the following order:

Kraft lignin > OSL > PWSL > lignosulfonate

Low weathering stability of L/PVC is determined by slight decomposition of lignin during processing at high temperatures.

Feldman [130] reported preparation of a new adhesive sealant based on PVC/poly(dimethyl siloxane) (PDMS) and kraft lignin. By adding small amounts of lignin (up to 5%), the modulus of the blends is improved what is reflected by the increases of the stress-strain values of the joints. Moreover, the application of commercial primers on the surface of wood substrates increases the adhesion properties of these blends.

Compatibility of PVC with lignin can be improved by modification of lignin. Oliveira and Glasser [131] reported an achievement of a blend based on PVC with lignin-caprolactone copolymer (LCL) synthesised by grafting poly(ε-caprolactone) (PCL) onto the lignin backbone.

The thermal analysis data for the LCL/PVC blends reveals a single T_g whose value is intermediary between T_g of the individual blend component, being an indicator for their compatibility. For concentration up to 50% LCL copolymer a plasticisation effect of PVC is observed.

The stress-strain properties showed a decrease in modulus from 250 MPa to 10 MPa and the increase of elongation from 200% to 600% with the rise of LCL load up to 50%. For higher LCL content than 50% an increase of modulus and a reduction in elongation, due to partial crystallisation of PCL component, was observed.

For LCL copolymer/PVC blends aged for up to 6 months at room temperature, the modulus data changed with ageing time and blend composition. SEM micrographs reveal a single-phase morphology, where the fine lignin-copolymer particles were uniformly dispersed in PVC matrix, forming a homogeneous blend from the two miscible polymers. Transmission electron microscopy (TEM) micrographs of thin films of LCL/PVC blends illustrated at two-phases at microstructure level with domain sizes ranging from 10-30 nm.

14.8 Lignin/Other Synthetic Polymer Blends

Results have been reported in the literature of some attempts focused on the ability of lignin or lignin derivative to interact with another macromolecule having a linear backbone, such as poly(vinyl alcohol) (PVA), poly(methyl methacrylate) (PMMA) and polystyrene (PS).

In last years, preparation of *lignin/PVA blends* by mixing of polymers either in the form of solution [103] or solid [132] have been reported. SEM results suggested a partially miscibility for hydroxypropyl lignin (HPL)/PVA blends explained by a secondary association between hydroxyl groups of HPL and partially hydrolysed PVA [103]. The presence of polymer-polymer interaction between the two types of macromolecules is shown by reduction in melt endothermic area of blends.

Corradini [132] in his study emphasised the importance of the hydrolysis degree of PVA and the blend preparation manner (with or without grinding). Determination of heat of fusion and of apparent activation energy of the thermal decomposition reaction showed dependence on the chemical structure of the L/PVA blend.

Another polymeric system composed from HPL and PMMA has been studied by Ciemniecki and Glasser [103]. Observations obtained by SEM revealed that HPL/PMMA blends are characterised by a two-phase morphology regardless of solvent, weight of HPL fraction, molecular weight of PMMA or method of preparation [133]. DSC and dynamic mechanical thermal analysis (DMTA) thermograms revealed the presence of two endothermic transitions in the initial run and a single, broad transition in the second run. Low molecular weight HPL fractions are a plasticisation effect on the material while high-molecular weight HPL fractions have anti-plasticisation effect on the molecular composite structure [133].

Introduction of acrylate functionality in lignin macromolecule has been reported by Glasser and Wang [134]. Isocyanatoethyl methacrylate, methacrylate chloride and methacrylic anhydride were used to introduce methacrylate groups into lignin [134, 135]. Acrylated lignin derivatives were copolymerised with vinyl monomers to produce network materials with gel structures and copolymerisation with methylmethacrylate to highly crosslinked, hard and glassy acrylic copolymers.

Feldman studied the L-acrylic copolymers by DSC and he observed the incompatibility of those polymers; however the presence of lignin produce an improving of the mechanical properties and durability of an acrylic sealant [136].

The grafting of synthetic polymers to lignin represents another way to obtain of engineering products with controlled structure and properties. Thus, Narayan and co-workers [137] reported synthesis of lignin-polystyrene graft copolymers (L-*g*-PS) by anionic polymerisation. The L-*g*-PS graft copolymers can function as compatibilisers/interfacial agents in preparing blends of kraft lignins with PS to obtaining new materials.

Due to native macromolecular structure, the specific intermolecular associative interactions between lignin components can favour the formation of cohesive domains in the lignin-based materials [138, 139]. Starting with this supposition Li and co-workers [140, 141] proved that it is possible to produce of thermoplastics with very high lignin content. The blends containing 85% industrial kraft lignin, PVA and two plasticisers (diethyleneglycol dibenzoate and indene) by solvent-casting (in aqueous 82% pyrrolidine) or injection moulding have been prepared [142].

The strength properties of very high lignin content plastics depend on the degree of intermolecular association between kraft lignin components. The ultimate strain extended to over 65% in the case of plastics containing the dissociated kraft lignin preparation in comparison with the materials based on the associated lignin preparation. The tensile strength and Young's modulus increased linearly with the average molecular weight of kraft lignin preparations incorporated at 85% (w/w) level in polymeric materials [142]. These results show the birth of the new generation of polymeric materials that are based directly on the molecular structure of lignin or lignin derivatives.

14.9 Lignin/Natural Polymer Blends

14.9.1 Lignin/Starch Blends

The use of lignin (hydrophobic materials) as filler in starch films was studied to improve the hydrophilic-hydrophobic balance and also to obtain a new lignin carbohydrate system [143].

Baumberger [144, 145] obtained thin films by solvent casting or by thermal moulding of 100/0 to 60/40 wt/wt starch/lignin mixtures using various lignin preparations (lignosulfonates, kraft lignin, OSL, and Granit lignin).

The macroscopic investigation of starch/lignin blends revealed that lignin had either a plasticising or reinforcing effect on the mechanical resistance of starch matrix. The use of lignosulfonates lead to increase of starch plasticity without improving its water resistance, but OSL and kraft lignins plasticised starch films reducing at the same time the water affinity of films [146, 147]. Light microscopy showed an incompatibility between starch and the high molecular weight fraction of kraft lignin. In the case of the use of the low molecular weight fraction kraft lignin an improvement of compatibility and to impart both plasticity and hydrophobicity [147]. Maximum stress and % elongation at break increase with the increase of amylose content [144].

Dynamic thermomechanical properties of the starch/lignin/glycerol films [145] revealed the fact that lignin induces a T_g decrease of 27 °C to 33 °C relative to the reference (starch/glycerol) films suggesting a plasticising effect of starch matrix and a partial miscibility of system components.

Another starch-based blend which containing lignin or lignin derivatives was reported by Ghosh [142]. Ghosh blended in various ratios lignin or its ester (lignin butyrate, LB) with a starch-caprolactone copolymer (SCL) by solvent casting or melt-processing procedure.

Presence of lignin or LB had a significant effect on crystallinity and melting temperatures of the PCL component, revealing polymer-polymer interaction between SCL and lignin. Unmodified lignin causes an increase of the crystallinity degree of PCL during melt-process, while LB has a more pronounced effect during solvent-casting processing. These results suggest that high T_g of nonmodified lignin play a role as a nucleating agent for PCL, enhancing crystallisation and melting, and low-T_g of lignin esters (LB) shows a significant compatibility with PCL reducing crystallisation and fusion [142]. Addition of lignin or LB up to a concentration of 10%-20% by weight improves the mechanical properties of SCL.

14.9.2 Lignin/Cellulose Blends

A series of blends composed from cellulose or hydroxypropyl cellulose (HPC) and an unmodified OSL or ethylated lignin (ETL), have been prepared by solvent-casting or by injection moulding procedure [148-150]. The DSC thermograms present a single T_g which increases with lignin content (up to 55%). These results show that there are strong intermolecular interaction between the two biopolymers [149]. DMTA reveals a similar rising in temperature of damping transitions with lignin content [148]. Lignin causes a significant reduction of temperature range where tan δ transition occurs, suggesting an association between L with liquid crystalline mesophase, by increasing degree of molecular interaction [149].

Rials [150] observed that the modification of hydroxy functionality of organosolv lignin by ethylation or acetylation influences the state of miscibility and phase morphology of blends prepared with HPC, obtaining a maximum interaction at a degree of substitution of 23%-40% of the OH group. The higher degree of substitution (DS > 40% diminishes the HPC/lignin interaction giving a completely incompatible system at a DS=90%.

An examination of the melting point of the HPC component in comparison to lignin revealed [149] that T_g migrates to the volume fraction of lignin. The difference is dependent on the lignin volume fraction and the polymer-polymer interaction. At the lowest degree of interaction between the lignin component and HPC phase separation occurs, leading to significant enhancements in both modulus and tensile strength of the HPC/lignin blends [150].

In the case of a blend formed from ethyl cellulose (EC) and lignin, addition of lignin (up to 40% lignin) lead to the appearance of two separate phases with creation of a liquid crystal mesophase [151]

In order to increase the phase compatibility between lignin and cellulose (derivatives) block copolymers were synthesised containing covalent-bonds between lignin and cellulose ester segments [12, 17]. Lignin-thermoplastic cellulose derivatives copolymer exhibited phase distinctions with the individual block having molecular weights as low at 10^3 dalton [17, 152].

DSC data revealed that lignin-cellulose propionate (CP) block copolymer was unable to provide an improvement in lignin/cellulose (derivatives) blend compatibility [12]. Results show that lignin favours the formation of an amorphous or liquid crystalline mesophase structure through establishing strong interactive association between lignin and the polysaccharide component.

The addition of lignin reduced the dynamic elastic modulus of cellulose ester solutions but the modulus of the cellulose solutions (in dimethylacetamide (DMAc)/LiCl) raises at all shear frequencies. This is explained by the formation of secondary interactions of lignin with cellulose in the system DMAc/LiCl solvent system [152]. However, when cellulose ester/cellulose acetate butyrate (CAB)/L mixtures were spun into continuous fibres from DMAc solution, both fibre tensile strength and modulus increased significantly. This increase was explained by a positive effect of lignin (even at small amounts of lignin, i.e., 4%), on the molecular order of the cellulose derivatives in solution and in solid state.

The addition of lignin enhances the liquid crystalline mesophase order in non-crystalline cellulose derivatives and due to this order the strength properties increase. TEM of a CAB/lignin (20%) film revealed a well-ordered cholesteric arrangement which is more pronounced in the presence of lignin [152-154].

The degree of compatibility between a commercial cellulose ester (CAB) and lignin esters having different ester groups substituents (acetate, LA; butyrate, LB; hexanoate, LH and laurate, LL) has also been investigated [151]. In the case of the use of LA and LB a compatible blends with the strongest interactions with domains sizes of 15-30 mm have been obtained. The DSC and DMTA data confirmed this observation revealing a single T_g. The T_g decreases with the increase of the lignin esters, showing a plasticising effect of lignin esters on the CAB.

The LH and LL containing blends form two distinct transitions on different levels [142]. CAB blends with LA and LB are compatible for all level compositions from 10 to 50 mm level, CAB blends with LH and LL become incompatible when lignin esters content rise to <20%.

14.9.3 Lignin/Polyhydroxyalkanoates

Polyhydroxy-alkanoates are biodegradable and biocompatible thermoplastic polymers with special medical applications, but their producing have a high cost.

Ghosh and co-workers [142] blended in melt and solution OSL and its ester (organosolv butyrate, LB) with polyhydroxy-alkanoate, such as poly-3-hydroxybutyrate (PHB) and poly-3-hydroxybutyrate-*co*-3-hydroxyvalerate (PHBV). Presence of lignin, in proportion of 10%-20%, inhibits or retards crystallisation of PHB at any cooling rate. The T_g values of the PHB phase shifted towards the glass transition temperatures of L or LB indicating a polymer-polymer interaction between blend components.

Incorporation of LB in the PHBV lead to some interaction and an important reduction of crystallinity in the samples [142]. Generally, presence of lignin improved tensile strength and elongation at break.

14.10 Concluding Remarks

Taking into account both the accessibility of the natural vegetable resources and their capacity to be environmentally friendly the possibilities to achieve materials based on natural/synthetic polymers are multiple. Lignin represents one of the main component of biomass and it could be considered as an important source in the obtaining of new materials.

The literature mentions a lot of examples concerning of the possibilities to processing of lignin or lignin derivatives with synthetic or natural polymers. Blending of these polymers is less expensive and represents an alternative way for the creation of materials with desired characteristics. Blend properties are controlled by compatibility of the components. Compatibility between lignin and polymers, matrices can be improved by modification of the initial components and/or the use of the compatibilising agents. Incorporation of lignin leads to change of physico-mechanical properties and increases of the degradation susceptibility under environmental conditions.

The use a huge quantities of lignin along with other natural and synthetic polymers could be alternative to decrease their costs by substitution of the petrochemical products and to improve the compatibility with the environment.

The last discoveries from lignin and macromolecular chemistry, specially concerning their association at intermolecular level could allow us to create some polymeric materials with very high lignin content having performed properties.

The problems which must be solved are connected with the development of effective plasticisers and compatibilising agents which could lead to the obtaining of new type of thermoplastics possessing a broad range of physico-mechanical properties.

References

1. K.V. Sarkanen and C.H. Ludwig in *Lignin, Occurrence, Formation, Structure and Reaction*, Eds., K.V. Sarkanen and C.H. Ludwig, Wiley-Interscience, New York, NY, USA, 1971, 1.

2. W.G. Glasser and H.R. Glasser, *Holzforschung*, 1974, **28**, 5.

3. G.H. Tomlinson II, *Pulp and Paper Magazine of Canada*, 1944, **45**, 817.

4. O. Hsu and W.G. Glasser, *Applied Polymer Symposia*, 1975, **28**, 297.

5. V. Saraf and W.G. Glasser, *Journal of Applied Polymer Science*, 1984, **29**, 5, 1831.

6. V. Saraf, W.G. Glasser, G. Wilker and J.M. Grath, *Journal of Applied Polymer Science*, 1985, **30**, 2207.

7. W.G. Glasser and R. Leitheiser, *Polymer Bulletin*, 1984, **12**, 1.

8. D. Muller, S. Kelley and W.G. Glasser, *Journal of Adhesion*, 1984, **17**, 3, 185.

9. J.J. Lindberg, T.H. Kuusela and K. Levon in *Lignin: Properties and Materials*, Eds., W.G. Glasser and S. Sarkanen, ACS Symposium Series No. 397, American Chemical Society, Washington, DC, USA, 1989, p.190.

10. C. Vasile, M.M. Macoveanu, V.I. Popa, M.C. Pascu and G. Cazacu, *Roumanian Chemical Quarterly Reviews*, 1998, **6**, 2, 85.

11. C. Vasile, M.M. Macoveanu, G. Cazacu and V.I. Popa, Proceedings of the 5th Mediterranean School on Science and Technology of Advanced Polymer-Based Materials, Eds., E. Martuscelli, P. Musto and G. Ragosta, 1997, Capri, Italy, p.185.

12. W.D. Oliveira and W.G. Glasser, *Polymer*, 1994, **35**, 9, 1977.

13. W.D. Oliveira and W.G. Glasser, *Journal of Wood Chemistry and Technology*, 1994, **14**, 1, 119.

14. J. Shaw, S. Buchwalter, J. Hedrick, S. Keng, L. Kosban, J. Gelorme, D. Lewis, S. Pu'uishothaman, R. Saraf and A. Vrehbeck, *Printed Circuit Fabrikation*, 1996, **19**, 11, 38.

15. G.M. Telysheva in *Ligno-Cellulosics. Science, Technology, Development and Use*, Eds., J.P. Kennedy, G.O. Phillips and P.A. Williams, Ellis Horwood Ltd., Chichester, UK, 1992, p.643.

16. W. Oliveira and W.G. Glasser in *Lignin, Properties and Materials*, Eds., W.G. Glasser and S. Sarkanen, ACS Symposium Series No.397, American Chemical Society, Washington, DC, USA, 1989, p.414.

17. W.D. Oliveira and W.G. Glasser, *Macromolecules*, 1994, **27**, 5.

18. B. Monties, B. Chabbert and S. Baumberger, Proceedings of the 215th ACS National Meeting, Cellulose Division, 1998, Dallas, TX, USA, Paper No.1.

19. D. Feldman and D. Banu, *Journal of Polymer Science: Chemistry Edition*, 1988, **26**, 973.

20. D. Feldman, D. Banu and M. Khoury, *Journal of Applied Polymer Science*, 1989, **37**, 877.

21. D. Feldman and M. Khoury, *Journal of Adhesion Science and Technology*, 1988, **2**, 2, 107.

22. D. Feldman, D. Banu, A. Natansohn and J. Wang, *Journal of Applied Polymer Science*, 1991, **42**, 1537.

23. J.L. McCarthy and A. Islam in *Lignin: Historical, Biological and Materials Perspectives*, Eds., W.G. Glasser, R.A. Northey and T.P. Schultz, ACS Symposium Series No.742, American Chemical Society, Washington, DC, USA, 2000, 2-99.

24. J.D. Gargulak and S.E. Lebo in *Lignin: Historical, Biological and Materials Perspectives*, Eds., W.G. Glasser, R.A. Northey and T.P. Schultz, ACS Symposium Series No.742, American Chemical Society, Washington, DC, USA, 2000, p.304.

25. D. Feldman, D. Banu, C. Luchian and J. Wang, *Journal of Applied Polymer Science*, 1991, **42**, 1307.

26. K. Hofmann and W.G. Glasser, *Journal of Wood Chemistry and Technology*, 1993, **13**, 1, 73.

27. K. Hofmann and W.G. Glasser, *Journal of Adhesion*, 1993, **40**, 229.

28. N. Shiraishi in *Lignin, Properties and Materials*, Eds., W.G. Glasser and S. Sarkanen, ACS Symposium Series No.397, American Chemical Society, Washington, DC, USA, 1989, p.488.

29. C.I. Simionescu, C.I. Turta, S. Bobcova, M.M. Macoveanu, G. Cazacu and C. Vasile, *Izvestia Akademii Nauk Respublik Moldova, Biologiceskie i Himiceskie Nauki*, 1994, 3, 270, 62.

30. C.I. Simionescu, V. Rusan, M.M. Macoveanu, G. Cazacu, R. Lipsa, C. Vasile, A. Stoleriu and A. Ioanid, *Composite Science and Technology*, 1993, **48**, 317.

31. W.G. Glasser, W. de Oliveira, S.S Kelley and W.L.S. Nieh, inventors; Centre for Innovative Technology, assignee; US Patent 4,918,167, 1990.

32. M.M. Macoveanu, F. Ciobanu, E. Costea, G. Cazacu and C. Vasile, *Cellulose Chemistry and Technology*, 1997, **31**, 361.

33. C.I. Simionescu, V. Rusan, M.M. Macoveanu, G. Cazacu, R. Lipsa, A. Stoleriu and A. Ioanid, *Cellulose Chemistry and Technology*, 1991, **25**, 355.

34. C.I. Simionescu, C. Vasile, M.M. Macoveanu, G. Cazacu and A. Ioanid, *Cellulose Chemistry and Technology*, 1994, **28**, 517.

35. D. Feldman, D. Banu, M. Lacasse, J. Wang and C. Luchian, *Journal of Macromolecular Science A*, 1995, **32**, 8/9, 1613.

36. D'Alelio and F. Gaetano, inventors; no assignee; US Patent 3,905,926, 1975.

37. K. Hofmann and W.G. Glasser, *Polymer Preprints*, 1990, **31**, 1, 657.

38. C.I. Simionescu, G. Cazacu and M.M. Macoveanu, *Cellulose Chemistry and Technology*, 1987, **21**, 5, 525.

39. B. Tomita, K. Kurozumi, A. Takemura and S. Hosoya in *Lignin, Properties and Materials*, Eds., W.G. Glasser, S. Sarkanen, ACS Symposium Series No.397, American Chemical Society, Washington, DC, USA, 1989, p.496.

40. H. Ito and N. Shiraishi, *Mokuzai Gakkaishi*, 1987, **33**, 6, 393.

41. S. Tai, J. Nagano and N. Migita, *Journal of Japan Wood Research Society*, 1967, **13**, 257.

42. C.I. Simionescu, V. Rusan, C.I. Turta, S.A. Bobcova, M.M. Macoveanu, G. Cazacu and A. Stoleriu, *Cellulose Chemistry and Technology*, 1993, **27**, 627.

43. C.I. Simionescu, V. Rusan, G. Cazacu, M.M. Macoveanu, R. Lipsa and G. Stoica, *Cellulose Chemistry and Technology*, 1990, **24**, 397.

44. C. Vasile, G. Cazacu, M.M. Macoveanu, E. Costea, A. Lazar, A. Stoleriu, M. Sabliovschi and L. Odochian, *Memoriile Sectiilor Stiintifice*, 1993, Series IV, **16**, 1, 117-152.

45. H.H. Nimz in *Wood Adhesives. Chemistry and Technology*, Ed., A. Pizzi, Marcel Dekker, New York, NY, USA, 1983, p.247.

46. B. Klasnya and S. Kopilovic, *Holz als Roh und Werkstoff*, 1992, 50, 282.

47. E. Roffael and B. Dix, *Holz als Roh und Werkstoff*, 1991, **49**, 199.

48. P. Benar and U. Schuchardt, Proceedings of the 5th European Workshop on Lignocellulosics and Pulp (EWLP'98), Advances in Lignocellulosics Chemistry for Ecologically Friendly Pulping and Bleaching Technologies, Aveiro, Portugal, 1998, p.45-48.

49. A. Gardziella, Proceedings of the 4th International Forum of the International Lignin Institute on Lignin from Pulping of Agricultural Plants, Lausanne, Switzerland, 1998, p.1-7.

50. N.G. Lewis and T.R. Lantzy, in *Adhesives from Renewable Resources*, Eds., R.W. Hemingway, A.H. Conner and S.J. Branham, ACS Symposium Series No. 385, American Chemical Society, Washington, DC, USA, 1989, p.13-26

51. C.H. Stout and A.W. Ludwig, inventors; Georgia-Pacific Corporation, assignee; US Patent 3,658,638, 1972.

52. Y.Z. Lai in *Chemical Modification of Lignocellulosic Materials*, Ed., D.N.S. Hon, Marcel Dekker, New York, NY, USA, 1996, p.35-95

53. R.P. Coyle, inventor; Champion International Corporation; assignee; US Patent 3,931,072, 1976.

54. G.A. Doering, inventor; Georgia-Pacific Resins, assignee; US Patent 5,202,403, 1993.

55. R.F. Christman and R.T. Oglesby in *Lignins. Occurrence, Formation, Structure and Reaction*, Eds., K.V. Sarkanen and C.H. Ludwig, Wiley-Interscience, New York, NY, USA, 1971, p.769-793.

56. ASTM D905, *Standard Test Method for Strength Properties of Adhesive Bonds in Shear by Compression Loading*, 1998

57. D.G. Naae, Proceedings of the 5th International Forum on Commercial Outlets for New Lignins & Definition of New Projects of International Lignin Institute, Bordeaux, France, 2000.

Handbook of Polymer Blends and Composites

58. JIS K 6851, *Testing Methods for Tensile Shear Strength of Wood-to-Wood Adhesive Bonds*, 1999.

59. ASTM D1037, *Standard Test Methods for Evaluating Properties of Wood-Base Fiber and Particle Panel Materials*, 1999

60. Y. Oh, T. Sellers, Jr., M.G. Kim and R.C. Strickland, *Forest Products Journal*, 1994, **44**, 2, 25.

61. T. Sellers, Jr., M.G. Kim, G.D. Miller, R.A. Haupt and R.C. Strickland, *Forest Products Journal*, 1994, **44**, 4, 63.

62. H. Olivares, H. Aceituno, G. Neiman, E. Rivera and T. Sellers, Jr., *Forest Products Journal*, 1995, **45**, 1, 63.

63. R.H. Gobran, A. Takahashi and E.P. Reilly, inventors; American Can Company, assignee; US Patent 4320036 1982.

64. ASTM D731-95, *Standard Test Method for Molding Index of Thermosetting Molding Powder*, 1999.

65. ASTM D256, *Standard Test Methods for Determining the Izod Pendulum Impact Resistance of Plastics*, 2000.

66. ASTM D570, *Standard Test Method for Water Absorption of Plastics*, 1998.

67. ASTM D648, *Standard Test Method for Deflection Temperature of Plastics Under Flexural Load in the Edgewise Position*, 2000.

68. ASTM D790, *Standard Test Methods for Flexural Properties of Unreinforced and Reinforced Plastics and Electrical Insulating Materials*, 2000.

69. ASTM D955, *Standard Test Method of Measuring Shrinkage from Mold Dimensions of Thermoplastics*, 2000.

70. ASTM D543, *Standard Practices for Evaluating the Resistance of Plastics to Chemical Reagents*, 1995.

71. P.M. Cook and T. Sellers in *Lignin. Properties and Materials*, Eds., W.G. Glasser and S. Sarkanen, ACS Symposium Series No. 397, American Chemical Society, Washington, DC, USA, 1989, 324-333.

608

72. H.K. Ono and K. Sudo in *Lignin. Properties and Materials*, Eds., W.G. Glasser and S. Sarkanen, ACS Symposium Series No. 397, American Chemical Society, Washington, DC, USA, 1989, p.334.

73. W.C. Senyo, W.A. Creamer, C.F. Wu and J.H. Lora, *Forest Products Journal*, 1996, **46**, 6, 73.

74. E. Roffael and W. Rauch, *Holzforschung*, 1971, **25**, 5, 149.

75. E. Roffael and W. Rauch, *Holzforschung*, 1972, **26**, 6, 197.

76. G.G. Allan, J.A. Dalan and N.C. Foster in *Adhesives from Renewable Resources*, Eds., R.W. Hemingway, A.H. Conner and S.J. Branham, ACS Symposium Series No. 385, American Chemical Society, Washington, DC, USA, 1989, p.55-67

77. K.G. Forss and A.G.M. Fuhrmann, inventors; Keskuslaboratorio-Centrallaboratorium AB, assignee; US Patent 4,105,606, 1978.

78. K.G. Forss and A.G.M. Fuhrmann, *Forest Products Journal*, 1979, **29**, 7, 39.

79. K-C. Shen, D.P.C. Fung and L. Calvé, inventors; Canadian Patents and Developments Ltd., assignee; US Patent 4,265,846, 1981.

80. A. Pizzi, *Advanced Wood Adhesives Technology*, Marcel Dekker Inc., New York, NY, USA, 1994, p.219.

81. R.E. Ysbrandy, G.F.R. Gerischer and R.D. Sanderson, *Cellulose Chemistry and Technology*, 1997, **31**, 213.

82. B. Kosikova, M.J. Kacurocova and V. Damianova, *Chemical Papers*, 1993, **47**, 132.

83. B. Kosikova, V. Demianova and M.J. Kacurakova, *Journal of Applied Polymer Science*, 1993, **47**, 1065.

84. B. Kosikova, K. Miklesova and V. Damianova, *European Polymer Journal*, 1993, **29**, 1495.

85. A.Y. Kharade and D.D. Kale, *Journal of Applied Polymer Science*, 1999, **72**, 1821.

86. P. Alexy, B. Kosikova and G. Podstránska, *Polymer*, 2000, **41**, 4901.

87. ASTM D638, *Standard Test Method for Tensile Properties of Plastics*, 2000.

88. C. Gonzales-Sánchez and L.A. Espósito Alvarez, *Angewandte Makromolekulare Chemie*, 1999, **272**, 65.

89. M. Rusu and N. Tudorachi, *Journal of Polymer Engineering*, 1999, **19**, 5, 355.

90. N. Tudorachi, N.C. Cascaval and M. Rusu, *Journal of Polymer Engineering*, 2000, **20**, 4, 287.

91. N. Tudorachi, M. Rusu, N.C. Cascaval, L. Constantin and V. Rugina, *Cellulose Chemistry and Technology*, 2000, **34**, 1-2, 101.

92. S. Casenave, A. Aït-Kadi, B. Brahimi and B. Riedl, Proceedings of the SPE *ANTEC '95*, Boston, MA, USA, 1995, Volume 2, 1438.

93. Q. Wang, S. Kaliaguine and A. Aït-Kadi, *Journal of Applied Polymer Science*, 1992, **44**, 1107.

94. P. Alexy, B. Kosikova, H.J. Radusch, T.L. Lupke, E. Spirk and G. Podstránska, Proceedings of the Polymerwerkstoffe 2000, Halle/Saale, Germany, 2000, p.97-100.

95. C.I. Simionescu, M.M. Macoveanu, C. Vasile, F. Ciobanu, M. Esanu, A. Ioanid, P. Vidrascu and N. Georgescu-Buruntea, *Cellulose Chemistry and Technology*, 1996, **30**, 5-6, 411.

96. C.I. Simionescu, M. Popa, P. Andriescu, V.I. Popa, A. Cernatescu-Asandei, D. Gavrilescu, V. Rusan and I. Diaconu, *Celuloza si Hartie*, 1989, **38**, 65.

97. M.M. Macoveanu, P. Vidrascu, M.C. Pascu, L. Profire, G. Cazacu, I. Mandreci, H. Darie and C. Vasile, *Buletinul Gradinii Botanice Iasi*, 1997, **6**, 2, 487.

98. C. Vasile, M. Downey, B. Wong, M.M. Macoveanu, M.C. Pascu, J.H. Choi, C. Sung and W. Baker, *Cellulose Chemistry and Technology*, 1998, **32**, 61.

99. M.C. Pascu, M.M. Macoveanu, P. Vidrascu, I. Mandreci, A. Ionescu and C. Vasile, *Buletinul Gradinii Botanice Iasi*, 1997, **6**, 2, 487.

100. P. Bono, inventor; Claude Lambert, assignee; FR 2,701,033A1, 1994.

101. T.A. Bubnova, G.S. Shifris, E.Y. Kalmykova, R.S. Salakhova, E.G. Balakhonov, V.M. Kirilets, E.Y. Plopskij, F. Tegaj and V.V. Aliulin, inventors; Nauchno-issledovatelskij i Proektno-Konstruktorskij Institut po Problemam Razvitiya Kansko-Achinskogo Ugolnogo Bassejna, assignee; SU 1,812,193, 1993.

102. P. Bono and C. Lambert, inventors; P. Bono and C. Lambert, assignees; WO9111,481A1, 1991.

103. S.L. Ciemniecki and W.G. Glasser in *Lignin, Properties and Materials*, Eds., W.G. Glasser and S. Sarkanen, ACS Symposium Series No. 397, American Chemical Society, Washington, DC, USA, 1989, p. 452-463

104. C.I. Simionescu, M.M. Macoveanu, C. Vasile, F. Ciobanu and G. Cazacu, Proceedings of the 4th International Symposium on Advanced Composites, Corfu, Greece, 1995, p.222-232.

105. M.C. Pascu, M.M. Macoveanu, C. Vasile, A. Ioanid and R.C. Oghina, *Cellulose Chemistry and Technology*, 2000, **34**, 399.

106. M. Pascu, C. Vasile, G. Popa, I. Mihaila and V. Pohoata, *International Journal of Polymeric Materials*, 2002, **51**, 181.

107. M.C. Pascu, C. Vasile, M. Gheorghiu, G. Popa, N. Georgescu-Buruntea and G. Cazacu, Proceedings of the 15th International Symposium on Plasma Chemistry, 2001, Orleans, France, Volume VI, p.2427.

108. B. Kosikova and V. Demianova, Proceedings of the 6th International Conference on Biotechnology in the Pulp and Paper Industry, Advances in Applied and Fundamental Research, Vienna, Austria, 1996, p.637.

109. B. Kosikova, V. Demianova and M. Kacurakova in *Cellulose and its Derivatives: Chemistry, Biochemistry and Applications*, Eds., J. Kennedy, G.O. Phillips, D.J. Wedlock and P.A. Williams, Ellis Horwood, New York, NY, USA, 1993, p.537.

110. S.S. Kelley, W.G. Glasser and T.C. Ward in *Lignin: Properties and Materials*, Eds., W.G. Glasser and S. Sarkanen, ACS Symposium Series No. 397, American Chemical Society, Washington, DC, USA, 1989, p.402.

111. S. Hirose, S. Yano, T. Hatakeyama and H. Hatakeyama in *Lignin. Properties and Materials*, Eds., W.G. Glasser and S. Sarkanen, ACS Symposium Series No. 397, American Chemical Society, Washington, DC, USA, 1989, p.382.

112. H. Yoshida, R. Mörck, K.P. Kringstad and H. Hatakeyama, *Journal of Applied Polymer Science*, 1987, **34**, 1187.

113. H. Yoshida, R. Mörck, K.P. Kringstad and H. Hatakeyama, *Journal of Applied Polymer Science*, 1990, **40**, 1819.

114. ASTM C719-91, *Standard Test Method for Adhesion and Cohesion of Elastomeric Joint Sealants Under Cyclic Movement (Hockman Cycle)*, 1998.

115. ASTM D2240, *Standard Test Method for Rubber Property-Durometer Hardness*, 2000.

116. ASTM D412-98a, *Standard Test Methods for Vulcanized Rubber and Thermoplastic Rubbers and Thermoplastic Elastomers-Tension*, 1998.

117. ASTM D395, *Standard Test Methods for Rubber Property-Compression Set*, 1998.

118. D. Feldman, M. Lacasse and R.St.J. Manley, *Journal of Applied Polymer Science*, 1988, 35, 247.

119. D. Feldman and N.A. Lacasse in *Electronic Packaging Materials Science IV*, Eds., K.A. Jackson, R.C. Sundahl, R. Jaccodine and E.D. Lilley, Materials Research Society Symposium Proceedings, MRS, Warrendale, PA, USA, 1989, 154, 265.

120. D. Feldman and M.A. Lacasse, *Journal of Adhesion Science and Technology*, 1994, 8, 9, 957.

121. M.A. Lacasse and D. Feldman, *Journal of Adhesion Science and Technology*, 1994, 8, 5, 473.

122. D. Feldman, C. Luchian, D. Banu and M. Lacasse, *Cellulose Chemistry and Technology*, 1991, 25, 163.

123. D. Feldman and M.A. Lacasse, *Journal of Applied Polymer Science*, 1994, 51, 701.

124. G.E. Agafitei, M.C. Pascu, G. Cazacu, A. Stoleriu, N. Popa, R. Hogea and C. Vasile, *Angewandte Makromolekulare Chemie*, 1999, 267, 44.

125. M.C. Pascu, G. Cazacu and C. Vasile, Proceedings of the Symposium on Progress in Biomaterials Science, Piatra Neamt, Romania, 2000, p.93-94.

126. G. Cazacu, M.C. Pascu, G.E. Agafitei, A. Ioanid and C. Vasile, *Macromolecular Materials and Engineering*, 2000, 283, 93.

127. D. Feldman, D. Banu and S. El-Raghi, *Journal of Macromolecular Science A*, 1994, 31, 5, 555.

128. D. Feldman, D. Banu, J. Lora and S. El-Raghi, *Journal of Applied Polymer Science*, 1996, **61**, 2119.

129. D. Feldman and D. Banu, *Journal of Applied Polymer Science*, 1997, **66**, 1731.

130. D. Feldman and A. Baskavan, *Journal of Adhesion*, 1989, **27**, 231.

131. W.D. Oliveira and W.G. Glasser, *Journal of Applied Polymer Science*, 1994, **51**, 563.

132. E. Corradini, E.A.G. Pineda and A.A.W. Hechenleitner, *Polymer Degradation and Stability*, 1999, **66**, 199.

133. S. Ciemniecki and W.G. Glasser, *Polymer*, 1988, **29**, 1021.

134. W.G. Glasser and H-X. Wang in *Lignin. Properties and Materials*, Eds., W.G. Glasser and S. Sarkanen, ACS Symposium Series No. 397, American Chemical Society, Washington, DC, USA, 1989, p.515-522.

135. W.G. Glasser in *Adhesives from Renewable Resources*, Eds., R.W. Hemingway, A.H. Conner and S.J. Branham, ACS Symposium Series No. 385, American Chemical Society, Washington, DC, USA, 1989, p.13-26 and p.40-54.

136. D. Feldman, M. Lacasse and D. Banu, *Journal of Polymeric Materials*, 1988, 5, 131.

137. R. Narayan, N. Stacy, M. Ratcliff and H.L. Chum in *Lignin: Properties and Materials*, Eds., W.G. Glasser and S. Sarkanen, ACS Symposium Series No. 397, Washington DC, USA, 1989, p.476-485

138. S-Y. Guan, J. Mlynár and S. Sarkanen, *Phytochemistry*, 1997, **45**, 911.

139. S. Dutta, T.M. Garven and S. Sarkanen in *Lignin. Properties and Materials*, Eds., W.G. Glasser and S. Sarkanen, ACS Symposium Series No. 397, American Chemical Society, Washington, DC, USA, 1989, p.155-176.

140. Y. Li and S. Sarkanen in *Lignin: Historical, Biological and Materials Perspective*, Eds., W.G. Glasser, R.A. Northey and T.P. Schultz, ACS Symposium Series No. 742, American Chemical Society, Washington, DC, USA, 2000, p.351-366.

141. Y. Li, J. Mlynár and S. Sarkanen, *Journal of Polymer Science B: Polymer Physics*, 1997, **35**, 1899.

142. I. Ghosh, R.K. Jain and W.G. Glasser in *Lignin: Historical, Biological and Materials Perspective*, Eds., W.G. Glasser, R.A. Northey and T.P. Schultz, ACS Symposium Series No.742, American Chemical Society, Washington, DC, USA, 2000, p.331-350.

143. S.J. Huang and P.G. Edelman in *Degradable Polymers: Principles and Applications*, Eds., G. Scott and D. Gilhead, Chapman & Hall, New York, NY, USA, 1995, Chapter 2, p.18-28.

144. S. Baumberger, C. Lapierre, B. Monties and P. Colonna, Proceedings of the 2nd International Electronic Conference on Synthetic Organic Chemistry (ECSOC-2), 1998, http://www.mdpi.org/ecsoc-2.htm.

145. S. Baumberger, C. Michon, G. Cuvelier and C. Lapierre, Proceedings of the 6th European Workshop on Lignocellulosics and Pulp, Advances in Lignocellulosics Chemistry Towards High Quality Processes and Products, Bordeaux, France, 2000, p.121-124.

146. S. Baumberger, C. Lapierre, B. Monties, D. Lourdin and P. Colonna, *Industrial Crops and Products*, 1997, **6**, 253.

147. S. Baumberger, C. Lapierre and B. Monties, *Journal of Agricultural Food Chemistry*, 1998, **46**, 2234.

148. W.G. Glasser, T.G. Rials, S.S. Kelly and V. Dave in *Cellulose Derivatives. Modification, Characterisation and Nanostructure*, Eds., T.J. Heinze and W.G. Glasser, ACS Symposium, Series No.688, 1998, American Chemical Society, Washington, DC, USA, p.265-282.

149. T.G. Rials and W.G. Glasser, *Journal of Applied Polymer Science*, 1989, **37**, 2399.

150. T.G. Rials and W.G. Glasser, *Polymer*, 1990, **31**, 1333.

151. I. Ghosh, R.K. Jain and W.G. Glasser, *Journal of Applied Polymer Science*, 1999, **74**, 448.

152. V. Dave and W.G. Glasser, *Polymer*, 1997, **38**, 2121.

153. P. Zugenmaier, *Journal of Polymer Science: Applied Polymer Symposia*, 1983, **37**, 223.

154. V. Dave, W.G. Glasser and G.L. Wilkes, *Polymer Bulletin*, 1992, **29**, 565.

15 Environmentally-Friendly Polymers and Blends

Cornelia Vasile and Anand K. Kulshreshtha

15.1 Introduction

Almost 100% of today's plastics are made from oil or natural gas. As concern mounts about the potential effects of petroleum-based plastics on the environment and the increased dependence on oil and gas imports, degradable polymers could become an important solution to the puzzle [1].

Environmentally degradable polymers (EDP or 'GreenPla' in Japan) and their blends are some of the materials suitable for sustainable development. The research and market in this field are developing and growing explosively, e.g., a 50 fold increase of production in the last decade in the developed countries. In the 1980s the number of references and patents was in the order of tens, in the 1990s the numbers increased to hundreds at the beginning of decade and had reached thousands by the end [2, 3].

One option may be to make plastics directly from corn starch or to develop new classes of polymers. EDP have the requirements for two main fields: biomedical applications (as absorbable sutures, internal bone fixing implants, orthopaedic repair products, polymeric drug and drug delivery systems, bioadhesives, tissue engineering, etc.), and environmental applications [4]. EDP have been developed as biorecyclable materials to reduce the plastic waste accumulation.

EDP and eco-friendly materials are fully biodegradable. Living organisms transform them into carbon dioxide and water without any harmful by-products, they are completely biodegradable within a composting cycle and are able to reduce environmental impact in terms of energy consumption and greenhouse effect, in specific applications. Their combustion does not yield any toxic gas or residue, they are recyclable, most of them are thermoplastic and can be recycled several times and they are derived from renewable resources.

The degradation mediated by microorganisms often coupled to chemical and/or physico-mechanical degradation is mainly specific to naturally occurring polymers such as:

polysaccharides (starch, cellulose, chitosan, dextran, etc.), protein (collagen, casein, albumin, etc.), polyesters (polyhydroxyalcanoates), others (lignin, natural rubber, poly-γ-glutamic acid) and several classes of synthetic polymers such as: poly(alkylene esters), polylactic acid (PLA), polycaprolactone (PCL), poly(vinylesters), poly(vinylalcohol), poly(anhydrides), polyphosphazenes and polyaspartic acid. Most of them are renewable sources based on sustainable feedstocks and are therefore environmentally favourable.

Cargill Dow Polymers is using sugars derived from corn as feedstock [5] for its commercial polymers and polymer blends (**Tables 15.1** and **15.2**) because of its low cost and abundance. But the company can also use other plant material such as rice, wheat, sugar beet and even agricultural waste. They have set up a joint venture to produce PLA at a 2700 metric tonnes per year pilot facility near Minneapolis, MN, USA. With landfill disposal costs already high and increasing almost everywhere, organic waste recycling has become an economical option for solid waste disposal.

Many large-scale commercial generators of organic wastes, such as hotels and schools, have been able to reduce their waste disposal costs by composting. The utilisation of biodegradable 'plastics' will definitely help to shift communities from landfill disposal to the recycling of organic wastes. The collection of organic wastes in biodegradable and compostable bags will eliminate the need for costly and inefficient debagging. This significant cost saving, will in some part be passed on to the customer, thereby increasing collection using biodegradable and compostable bags for organic wastes. This will be welcomed by communities considering organic waste recycling and composting as it avoids the substantial capital investment necessary for separate pickup and collection of organic and non-organic solid waste.

Bags are not the only biodegradable and compostable products on the market. There are many applications for this technology: examples include food utensils such as cutlery, plates and cups, medical items such as syringes and gloves and packaging materials. Imagine how much more of the solid waste stream could be diverted from landfills if all of the plates, cups, knives, forks and spoons used daily in the United States and Canada's fast-food restaurants were biodegradable and compostable.

Daicel is currently offering two types of products, cellulose acetate type (Celgreen PCA) and polycaprolactone type (polyhydroxybutyrate (PHB)). The former type has a high clarity, high strength and high viscosity, while the second exhibits good miscibility with other polymers, high strength and elongation. Celgreen PCA can replace polystyrene and Celgreen PH and PHB can replace PE and PP [10] in some application fields

Many different thermoplastic biopolymers have been developed and produced with industrial procedures using current levels of technology. Thermoplastic starch (TPS) [13-

Table 15.1 Commercial biodegradable plastics [6-9]			
Trade Name	Supplier	Chemical Composition	Comments
Biopol	Monsanto	Poly(β-hydroxy butyrate)	Isotactic, highly crystalline; similar to PP in properties but sinks in water.
Biopol	Monsanto	Poly(β-hydroxy butyrate-*co*-β-hydroxy valerate)	A range of properties can be obtained by varying the comonomer (HV) content.
	Cargill Dow, Minneapolis, MN	PLA	Crystalline, rigid, biocompatible and compostable.
	EcoChem, Wilmington, DE	PLA	Compostable and recyclable.
Tone Polymers	Union Carbide, USA	PCL	Crystalline, resembles medium density PE: compostable but not water soluble.
Vinex	Air Products and Chemicals, Allentown, PA	PVA	Water soluble and form blends.
	Belland, Solothurn, Switzerland	Acrylic compolymers	'Selectively soluble' polymers, dissolve in alkaline solution and can be reconstituted.
Environ Plastic	Planet Packaging Technologies (USA)	Polyethylene oxide and a compatible proprietary polymer.	As water temperature increases, solubility decreases. Recyclable.
Lacea	Mitsui Chemical, Japan	PLA	Highly transparent and readily processable.
Cell Green P-HB Cell Green PCA	Daicel Chemicals, Japan	Branched PCL Cellulose acetate	Readily processable and strong – mulching film and compost bags.
Kai Maru	Topy Green, Japan	-	Mulching film.

Table 15.1 Continued			
Trade Name	Supplier	Chemical Composition	Comments
Biopol	Monsanto, Italy	Biological polyesters	Expand the range of polymers made by plants, sold to Monsanto by ICI in 1996.
Bionelle	Showa Denko, Japan	Polybutylene succinate/ adipate biodegradable polyester – 3000 t/y	Excellent processability.
Walocomp LPN	Bayer, Germany	Polyester amine	Film extrusion and injection moulding possible.
Hydrolene	Montecatini Terme, Italy.	Polyvinylalcohol	-
Biomax	DuPont, USA	Hydro/Biodegradable polyester	-
Ecolex COPE	BASF, Germany	Copolyester	
Bio COPE and Easter Bio	Eastman Chemical, USA		Comparable.
Bioplast	BIOTEC, Germany	Thermoplastic Starch	Compostable bioplastics.
NatureWorks	Cargill Dow Polymers, Minneapolis, MN	PLA	Low cost resin for blends and expanding range of applications: fibres, thermoforming films, bottles, foams, coated paper.
Skygreen	SK Chemicals, Korea	PET and polybutylene succinate/adipate	Film extrusion, blow injection grades.
Tone	Union Carbide, USA; Zeneca Bio products, Knowsley, UK; Chemie Linz, GmbH, Austria.	Polycaprolactone P(3HB/3HV) PHB	Completely biodegradable synthetic polymer. Microbial polyesters.

Table 15.1 Continued			
Trade Name	**Supplier**	**Chemical Composition**	**Comments**
Lacty	Shimadzu, Japan	Polylactide	100 t/y.
Lunarese	Nihon Shokubai, Japan	Poly(ethylene succinate)	
Beogreen Iupec	Mitsubishi Gas, Japan	PHB – bacterial polybutylene succinate/adipate	
Eslon Green	Saehan Industries, Japan	PET and poly(butylenes succinate/adipate)	
EnPol	Ire Chemical, Korea	Aliphatic/aromatic polyesters	Injection moulding, extrusion, coatings, fibres, medical disposable goods; meet FDA application.
	EYA-GE Inc and Swen Ghu Inc. Taiwan	Photodegradable polymers	
Trisorb	Samyang Co., Korea	PET, PGA	Medical sutures.
Komagreen	Samsung Color Industries Co., Korea	Aliphatic/aromatic polyesters	Various grades, mater batch, compostability.
Ecoflex	BASF, Germany	PBS/terephthalate	
EASTER-BIO	Eastman Chemicals, USA	PBS/terephthalate	
Poval Gesenol Dorom VA	Kuraray Nihon Gosei, Japan Aicello	PVA	

PP: *polypropylene*
PE: *polyethylene*
PET: *polyethylene terephthalate*
PGA: *polyglycide*
HB: *hydroxybutyrate*

PBS: *polybutylene succinate*
PVA: *polyvinyl alcohol*
FDA: *Food & Drugs Administration*
HV: *hydroxy valerate*

Handbook of Polymer Blends and Composites

Table 15.2 Commercial starch/plastic blends [9, 11, 12]			
Trade Name	Supplier	Chemical Composition	Comments
Ecostar	St. Lawrence	Starch/LDPE or HDPE	Addition of starch improves anti-blocking and printability. Starch degrades leaving a porous structure.
Ecostar Plus	St. Lawrence Starch	Ecostar + organometallic and organic compounds.	Capable of oxidative degradation/ photodegradation.
-	USDA	Starch and ethylene acrylic acid copolymers.	With high levels of starch mechanical properties have to be compromised for biodegradability and UV stability.
Bionyl Regreen	Daesang Corp. Korea	Starch aliphatic polyesters	
Greenpol Skygreen	SK Corp., Korea	PCL/starch blend to lower cost modified starch	Modified stach foam for loose fillers, film injection grades.
	Hing Li Biotech Inc.	Modified starch	
Ecoplast	Hanwa Petrochemicals, Korea	Starch/aliphatic polyesters	
Eslon Green	Saehan Ind, Korea	Starch based	
Super Slurper	3 US companies	Starch-g-polyacrylonitrile graft copolymer	In dry solid state it can absorb several hundred times its weight of water.
Ecolan	Provair Ltd., UK	50% Ecostar and polyester-based polyurethane.	'Rapidly breathable' as it has a structure of interconnecting pores of controlled size.

620

Table 15.2 Continued			
Trade Name	Supplier	Chemical Composition	Comments
Novon	Novon Products Division, Morris Plains, NJ	Starch-based polymer	Two grades are available. Water soluble and compostable.
Mater-Bi	Novamont/Feruzzi, Italy; Nihon Gosei, Japan	60% cornstarch based thermoplastic, Starch/PCL or cellulose	Two grades are available. It has FDA acceptance.
Polyclean	Epron Industries, Oldham	Masterbatch use of packaging.	Compostable.
Starate	Tainjin Downhalo Applications, Jilin Goldeagle Industry, Nanjeaon Suchi, China Danhai Biotechnology Company	Modified starch/starch filled, photodegradable. Denaturated starch, long polymer chain	Six grades are available, compostable within 45 days.
Cornpol	Nihon Corn Starch	Chemical modified starch	
Biofil	Samyang Genex	Modified starch	
USDA: US Department of Agriculture *LDPE: low density polyethylene* *HDPE: high density polyethylene*			

15], is an intermediate product for making biopolymer blends with film forming properties. The research results for TPS bioplastics and their production processes are protected by international patent law or have patents pending [16-21].

The EDP market in Japan consists of 50% PBS, 30% PLA, 10% (starch and PCL/starch) and 10% other. It is expected that in 2003, the production of EPD will be 1% of total plastics and in 2010 will reach 60,000 tonnes.

In Korea, EDP are mainly manufactured from aliphatic/aromatic polyesters, starch-based or starch/PCL blends. PHB production is at the pilot level. The mixture PHB/ polyhydroxyalkanoate (PHA) is tested by fermentation in presence of *Escherichia coli* (Kohap Ltd, KAIST LG Chem and KIST). Other researchers are developing polylactide, chitosan-based EDP, etc.

Starch can be incorporated into the synthetic non-biodegradable plastic as an additive [22-29] or the plastic can be coextruded with starch [30] to prepare a blend that is more biodegradable than the pure thermoplastic or to increase the biodegradation rate of the synthetic polymer. Also, starch can be used together with fully biodegradable synthetic plastics [31-36], producing a biodegradable blend of low cost.

Blends of starch and polyethylene (LDPE) have been studied but their biodegradability is questionable, despite the use of additives [37-44]. Total Degradable Plastic Additive (TDPA; Environmental Products, Inc.), could suffer an oxo-biodegradation process, so their lifetime is controlled by varying the composition of the blend and the conditions of the oxidative process [45].

One way to increase compatibility in starch/PE blends is to use a compatibiliser containing groups capable of hydrogen bonding with starch hydroxyls. Ethylene-acrylic acid copolymer (EAA) is such an example [46-48] due to the hydrogen bonds that can be formed with the hydroxyl groups of starch [49-52].

15.2 Corn: A Renewable Source of Eco-Friendly Plastic

First-generation degradable polymers, which were largely commercialised in the 1980s, did not satisfy the public's view of complete degradation. Second-generation polymers which are degradable began being introduced during the last five years. During the 1980s, when so-called 'compostable' bags were introduced to the market, environmental groups and others rejoiced. Many saw the development of these bags as a magic bullet that would end problems caused by non-degradable plastics lingering in the environment. The initial euphoria faded when it was found that the polyethylene-plus-starch films from which some bags were made did not completely biodegrade. Those early bags disintegrated, but left PE residues that were invisible to the naked eye and hard to find by normal chemical analysis. More sophisticated testing methods, however, revealed that minute fragments of PE were still present in the supposedly finished compost. Soil scientists expressed concern that these fragments would accumulate in soils and hinder both microbial and plant growth. The enthusiasm for these materials declined, discrediting their advocates and, some claim, setting back the biodegradable plastics industry by a decade [2, 53]. Nonetheless research and development work on blends of biodegradable polymers, continued, spurred by the promise of substantial markets.

Cargill Dow, one of the world's largest agricultural processing companies, uses corn to make PLA for its EcoPLA biopolymer. Like some other biodegradable materials, EcoPLA can be made into film fibre and a paper coating, and can be thermoformed and injection moulded. In 1997 Cargill marketed a 115 litre compostable bag under a recent arrangement with Duro Bag, based in Kentucky. Professor Narayan at Michigan State University, and a longtime proponent of fully biodegradable polymers, is test marketing a material called Envar, made, of an aliphatic polyester called PCL that has been alloyed with a thermoplastic starch. In his opinion, regardless of whether it is made from synthetic or natural materials, what matters to users of biodegradable bags is whether they will disappear in compost without leaving a potentially harmful residue.

EnPac, a joint venture of DuPont and ConAgra, imports from Italy a biodegradable plastic film called Mater-Bi, which is made into bags in the US. It has already been successfully used for compostable bags in Europe and is now being used in the US market. Mater-Bi is based on corn starch, but it would be misleading to think that is the only or even the primary component. EnPac have been marketing Mater-Bi bags made for them by Castle Bag company. Another firm, Biocorp of California is already promoting and selling its resource-bag, also made from Mater-Bi film. Composting bags from Mater-Bi grade Z are as biodegradable and compostable as paper; mechanical properties are similar to LDPE, it is water resistant, act as an odour barrier, is transparent and has a low volume.

ICI announced in the 1990s its first commercial application for its Biopol biodegradable plastic. This is a linear polyester of 3-hydroxybutyric acid (HB) and hydroxy valeric acid (HVA). The polyhydroxybutyrate-*co*-hydroxyvalerate (PHBV) is made by fermentation with the naturally occurring bacterium, *Ralstonia eutropha*, which uses globules of PHB homopolymer as an energy storage medium which is analogous to fat in animals or starch in plants. After four weeks of development trials, Biopol has been accepted by the hair care company Wella, and it went on the market in 1990 in West Germany as a bottle for a product in Wella's Sanara shampoo range. ICI has been working on Biopol for 15 years [54]. Although the production capacity was small initially, ICI was able to bring production up to 5000-10,000 tonnes/year by the mid 1990s. Starch itself is the key to another type of biodegradable plastic that is making headway in some packaging applications. A number of companies are involved in this work, including Epron Industries of Oldham, UK. Epron's Polyclean is a masterbatch material that packaging manufacturers incorporate into conventional PE at the film blowing stage, just as they would a colour masterbatch. Polyclean is typically added at level of 15%. Epron has patents on delayed action oxidation catalysts for PE, PS, and PP, which promote the requisite breakdown of polymer chains (and reverse the action of conventional antioxidants). Although voluntary recycling schemes are being launched, they are likely to cover less than 1% of the 10 million tonnes/year of plastic packaging used.

A group of researchers [55] at the National Research Development Corporation in India have developed biopolymer packaging material containing 40% tapioca starch and 60% LDPE. This material is biodegradable under controlled conditions and has enormous potential usage in the packaging of foods, pharmaceuticals and disposable clothing. The biopolymer is degradable by a particular strain of soil bacteria which converts it to an ecofriendly powder within 90 days. The scientists have also developed a proprietary soil with the specific bacteria for disposing of the used packaging material.

Two of the biggest names in agriculture and chemicals have teamed up to commercially to produce a polymer that they hope will become the basic raw material for fibres of the future. Cargill Inc., and the Dow Chemical Company have formed a 50/50 joint venture company, Cargill Dow Polymers LLC, to manufacture the family of NatureWorks PLA polymers, the first family of polymers made out of annually renewable resources. The partnership combines Cargill's process technology and low cost manufacturing position for lactic acid and PLA resins and Dow's world-class polymer science, applications technology and global customer base. Cargill and Dow have invested more than $300 million in a new manufacturing facility in Blair, Nebraska, which came on stream at the end of 2001. It will produce 140,000 metric tons of NatureWorks PLA per year.

Cargill Dow Polymers currently operates a semi-commercial PLA polymer plant near Minneapolis, MN. That plant has a capacity to manufacture 4,000 metric tons of PLA a year and the company invested $8 million to double its capacity in 2000 to meet immediate market development needs.

In addition to Cargill Dow Polymers' plants in the USA, a plant will be built in Europe in about two years time and it is expected that other plants will be built about every 18 months elsewhere in the world. Developed by Cargill Dow Polymers, the NatureWorks process produces resin that will be used in everyday items such as clothing, cups, food containers, sweet wrappers, as well as home and office furnishings.

Mitsui Chemical has been developing markets for two biodegradable resins, PLA and a water absorbing amino acid polymer. The company considers 'value added package' uses as the most promising application for PLA, because products are not only 'highly transparent' and 'readily processable', but are also safe for both humans and the environment as they are derived from natural plant matter. Target applications include bottles for cosmetics that feature natural ingredients, and packaging film for premium vegetables grown organically. The company has received awards in recognition of the commercialisation of a PLA cutter blade attached to plastic wrapping film box in 2000. Mitsui has an annual 500 ton capacity for resin production and test fabrication in Ohmuta and is studying the possibility of resin exports.

Four US Department of Energy (DOE) laboratories have signed a $7 million agreement with Applied Carbo Chemicals, a specialty chemicals company, to manufacture chemicals feedstocks from renewable farm crops at a significant lower cost than via conventional petroleum-based chemosynthesis. The project follows the recent DOE development of a new microbe as part of a process that converts corn into the key intermediaries used to make a range of industrial and consumer products, including polymers, clothing fibres, paint, inks, food additives and automobile bumpers. Existing domestic markets for such chemicals total more than $1.3 billion a year. A rise in the number of employees is also expected over the next decade as the company builds manufacturing capacity and expands into global markets. This is a typical example of combining chemosynthesis with biosynthesis plants, an approach pursued in America to save energy.

Research being done suggests the possibility that in the near future the process might be extended to use industrial organic wastes (such as from the sugar industry) to replace corn. Metabolix (Cambridge, MA) has recently licensed the Massachusetts Institute of Technology's (MIT) patents on the insertion of the genes for the production of the key enzymes in the mechanism of production of PHB (an essential component of biodegradable polyester thermoplastics) into bacteria and transgenic corn. The transgenic bacteria and plants can also co-polyesterify β-hydroxybutyrate with β-hydroxyalcanoates up to C_{12} [56]. The potential of the production of biodegradable plastics, is a low cost, renewable production system (using corn, cassava, soybean, etc.), and this is also apparent from the spate of recent joint ventures as well as business purchases by big multinational commodity firms, like Monsanto and Cargill [57].

Biodegradable plastics, which are decomposed by microorganisms in the soil and water, have been attracting much attention worldwide, because they are incorporated into the natural ecosystem and thus cause no environmental problems.

There are numerous opportunities for acceptable environmentally degradable polymers. Opportunities include fast-food and one-way packaging and personal hygiene products which are compostable without separation, agricultural films which are mulchable and require no collection, and fishing gear which may break free in the ocean and rivers to cause danger to fish and mammalian life.

Biodegradable plastics should have the needed performance characteristic in their intended use, but after use should undergo biodegradation, in an appropriate waste management infrastructure, to environmentally compatible constituents. For example, a truly biodegradable plastic will be converted to CO_2, H_2O, and compost in a composting infrastructure, leaving no persistent or toxic residue. Degradability ensures that, after its intended use, the polymers are subject to waste management and the relationship to the carbon cycle of the ecosystem is self-evident.

15.3 The Role of Legislation

15.3.1 Japan

According to the Japanese Law on the Treatment of Industrial Wastes revised in 1997, the used mulching films were designated as industrial waste to be treated at qualified facilities and individual farmers were banned from burning their waste themselves. Collection of used film by farmer's cooperatives started, but tipping fees at the treatment contractors were generally too high even if farmers washed the debris out of the used films. Used biodegradable film does not need to be treated by such laborious and expensive methods, but can be simply buried in corners of individual farms.

Besides, conventional mulching films made of PE or other durable plastics need to be detached from plant roots before any kind of waste treatment, but such additional labour also became unnecessary. Musashino farmers who grow sweet corn with the application of Kie Maru mulch especially appreciate this labour saving result.

The product adopted was biodegradable 'Kie Maru' manufactured by Topy Green. Kie Maru was commercialised in 1997 and 1998 sales were 2,000 rolls, each 200 metres long and sold for 6,000 yen per roll. Use of Topy Green, biodegradable mulching films has begun to expand among environmentally conscious farmers in the major vegetable production centres such as Nagano and Kyushu as well as Musashino.

Since 1998, Musashino City, one of the important vegetable growing areas supplying the Tokyo metropolis, is subsidising the farmers' purchase of biodegradable mulching films from the general tax revenue (totalling 380,500 yen in 1998). This is done primarily to prevent illegal field-burning of used films that may cause harmful dioxin emissions (and draw strong complaints from surrounding residents).

The nationwide reporting in all media outlets surely strengthened the public consensus to place incineration and landfill of waste as the last option and to reduce the volume of waste by further introduction of their reuse according to the nature of the waste, including biological treatment of organic waste by composting or anaerobic digestion. Municipalities moved individually to reduce and limit the numbers of incinerators, and to accelerate tests of composting treatment.

Package recycling law dictates that the municipalities and industries must have started recycling all the waste from packaging from April of 2001. However, allowed recycling methods and share of the burden between municipalities and industries/consumers are still under political debate. Exemption of biodegradable plastic packages and containers from the current material recycling schemes is expected because the products are compostable. If it could be realised, the market growth will reach a tremendous level.

The development of the market for biodegradable resins worldwide is driven by environmental legislation [49, 53] such as the harmonisation of landfill, composting and recycling resolutions in the European Union, a law for 40% of plastics to be biodegradable in Japan, the development of degradable compost bags in the US and the expected EU Landfill Directive which will ban codisposal or the landfilling of different waste classes at the same site.

15.3.2 Germany

The German Federal cabinet and the German Parliament have agreed on an amendment to the Packaging Regulation with a rule for compostable plastic packaging. This special provision is valid for compostable plastic packaging materials that consist mainly of biodegradable materials based on renewable resources.

15.3.3 Asia

In the emerging and developing countries the legislation on EPD does not exist or is only in the initial stages. This was found from discussion during two symposia dedicated to the developments in this field. Future solutions for implementation of eco-friendly materials were discussed in the ICS-UNIDO International Workshops, *Environmentally Degradable Plastics: Industrial Development and Application*, 2000, Seoul, Korea and *Environmentally Degradable Plastics: Position of EDPs in Plastic Waste Management*, Lodz-Pabianice, Poland, 2001 The production of EPD in these countries is mainly PVA, modified PVA blended with hydrolysed collagen, and polyolefins/starch blends as blown film or fibres. A focal point is intended to be created to transfer know how and to disseminate scientific and technological information in Middle and Eastern Europe. ICS-UNIDO also supports development of environmentally friendly materials in China, Indonesia, Malaysia, etc.

This decision is an important milestone for biodegradable materials and opens the market for use of plastic packaging materials.

15.4 Biodegradability: Definitions and Standards

The obvious need for technical standards to unambiguously define what is meant by such terms as 'biodegradable' and 'compostable,' and to lay out strict testing methods,

627

prompted a flurry of work by national and international standards organisations over the past few years. By the end of 1996 a handful of groups had published their results. The effects of these standards have tremendous commercial importance, especially to multinational corporations who make and/or distribute raw materials and blended film. In the United States, the Institute for Standards Research (ISR) of the American Society for Testing and Materials (ASTM) published its Guide for Assessing the Compostability of Environmentally Degradable Plastics' in 1999 [58]. Earlier, ASTM established rules for labelling packaging materials that 'communicate environmental attributes' to consumers or users, known as ASTM D5488 [59], as well as a standard laboratory test method to measure biodegradability under composting conditions known as ASTM D5338 [60].

The following definitions, which are now part of ASTM standard definitions, apply to biodegradability [61]:

- **Degradable plastic** – A plastic designed to undergo a significant change in its chemical structure under specific environmental conditions, resulting in a loss of some properties that may vary as measured by standard test methods appropriate to the plastic and the application in a period of time that determines its classification.

- **Biodegradable plastic** – A degradable plastic in which the degradation results from the action of naturally occurring microorganisms such as algae, bacteria and fungi.

- **Photodegradable plastic** – A degradable plastic in which the degradation results from the action of natural daylight.

- **Oxidatively degradable plastic** – A degradable plastic in which the degradation results from oxidation.

- **Hydrolytically degradable plastic** – A degradable plastic in which the degradation results from hydrolysis.

The problem has been compounded by the lack of any uniformly accepted definition of the terms 'biodegradable' and 'compostable'. Waste bags made from a blend of PE and starch, of which there are several on the market in Canada and the United States today, are touted as 'biodegradable' by their manufacturers because the bags will disintegrate into minute particles under normal composting conditions. That residue, however, is PE. Thus, the compost is contaminated with PE, which is often undetectable to the naked eye. Is such a product entitled to call itself 'biodegradable?' The Federal Trade Commission (FTC) has been promulgating a definition of 'biodegradable' that would exclude such PE and starch blends. In the FTC, *Guide for the Use of Environmental Marketing Claims*

[62], a product is 'biodegradable' only if it will 'decompose into elements found in nature within a reasonably short period of time after customary disposal.'

The FTC definition presents a good base on which to build a uniform definition. Since PE is not an element 'found in nature,' a PE-starch bag that disintegrates into even invisible PE fragments would not be biodegradable.

15.4.1 Assessment of Biodegradable Polymers

Research on the synthesis of biodegradable polymer requires an analysis of biodegradability. In the ecobalance, the life-cycle assessment of plastic manufacture plays a key role. A correct analysis of the impact on the cost/effectiveness ratio should taken into account by decision markers. Standards have been developed by ISO (World) – TC 61/SC5/WG22 [63], ASTM (USA) – D2096 [64], DIN (Germany) V 54900 [65], CEN (Europe) TC/26/SC4/W22 [66], JIS (Japan) [67-69]. Some of the more important analytical methods for determining the biodegradability of organic polymers are the modified MIT method (utilising decomposition by activated sludges), the fermentation method, testing the growth of microorganisms (ASTM method), testing the biodegradability by burying the sample under the ground, and the ^{14}C tracer method. These methods produce various results such as weight reduction in the polymer samples and decline in their mechanical properties. However, the scientific criterion for biodegradability is not yet clear what extent of decomposition is necessary for a polymer to be regarded as a biodegradable.

The modified MIT method, outlined next, is one of the analytical methods to determine biodegradability of polymers, stipulated by the 'Law Concerning the Regulation of the Study and Manufacture of Chemical Substances':

Degree of biodegradability (%) = (BOD)/TOD x 100

where: BOD: biochemical oxygen demand of the sample (measured) (mg)

B: Oxygen consumption of the basic culture medium containing a standard activated sludge.

TOD: Theoretical oxygen demand when the sample is completely oxidised (mg).

15.4.1.2 Respirometry

One way to detect plastic film components that do not fully biodegrade during the composting process involves a sophisticated laboratory technology called respirometry.

Respirometry is done by a mineralisation test that looks at conversion of polymer carbon into carbon dioxide. The PE-starch (blend) will disintegrate and disappear in the compost, but the PE does not biodegrade, so it is possible to track the CO_2 that is 'given off' by it. Even though the film will disintegrate and no film pieces will be recovered, all the CO_2 that is formed should be from the starch. This study, showed that some plastic films contain starches derived from renewable plant materials such as corn, but they also contain additives that are undeniably synthetic. For example, one polymer used in biodegradable plastic bags is PCL, which is synthetic, even so, no-one questions its biodegradability. The raw material source is irrelevant, what matters is the chemical structure. Union Carbide markets PCL under the registered trade name TONE.

Environment (soil, sea water, compost, sewage), time and breakdown products need to be specified in all degradation testing because some of degradation intermediary products could affect vegetation and living organisms [69-71].

15.4.2 Biodegradability of Starch/Polymer Blends

The dominant commercial starch-based biodegradable polymers are marketed under the name Mater-Bi (**Table 15.2**). The three major Mater-Bi classes differ in composition and biodegradability (**Table 15.3**).

15.5 Biopolymer Materials for Making Blends

The first starch plastics in the marketplace were starch filled PE. They were only bio-disintegradable and not completely biodegradable in practical time frames. Data showed that only the surface starch biodegraded leaving behind a recalcitrant PE material. Products made from these resins do not meet the criteria of complete biodegradability in defined disposal systems (like composting) and within the operational time frames of the disposal system.

15.5.1 Starch Ester Technology

Modification of the starch –OH groups by esterification chemistry to form starch esters of an appropriate degree of substitution (1.5 to 3.0) imparts thermoplasticity and water resistance. Unmodified starch shows no thermal transitions except the onset of thermal degradation at around 260 °C. Starch acetate with a degree of substitution of 1.5 shows a sharp glass transition (T_g) at 155 °C and starch propionate of the same degree of substitution had a T_g of 128 °C. Plasticisers like glycerol triacetate and diethyl succinate are completely miscible with starch esters and can be used to improve processability. Water resistance of the starch esters is greatly improved over the unmodified starch. The starch ester resin reinforced with biofibres has properties comparable to general purpose PS.

Appropriately formulated starch esters with plasticisers and other additives provide resin compositions that can be used to make injection moulded products and for direct lamination onto Kraft paper. Starch acetates up to degree of substitution = 2.5 undergo complete and rapid biodegradation. In the case of starch triacetates, 70% of the carbon is converted to CO_2 at 580 °C in 45 days. The National Starch and Chemical Company offers intermediate degree of substitution starch esters for biodegradable plastics applications.

15.5.2 Microbial Polyesters

Among the various biodegradable plastics available, there is a growing interest in the group of PHA polymers made entirely by bacterial fermentation. One of the most important characteristics of these microbial polyesters is that they are thermoplastics with environmentally degradable properties. Potential useful products from these polymers have been shown to be degraded in soil, sludge and even seawater. Some microorganisms (such as bacteria and fungi) were shown to excrete extracellular depolymerases to degrade these polymers and utilised the decomposed compounds as nutrients.

Of the PHA most extensively studied, two polymers show the potential for being applied for the production of biodegradable plastic products such as agricultural mulching films, packaging film, bottles and containers. These two polymers are poly(3-hydroxybutyrate) (PHB) which have been produced from glucose by using a microorganism identified as *Ralstonia eutropha* and its copolymer, PHB-*co*-3-hydroxyvalerate), P(3HB/3HV) from a mixture of glucose and propionic acid with the use of the same microorganism. Other microorganisms have been also used both for production of various kinds of PHA and enzymic degradation [72-74]. P(3HB) has been commercialised by an Austrian company Chemie Linz GmbH while P(3HB/HVA) [75] has been commercialised by Zeneca, UK, through the use of large scale biotechnology processes in their mass production. PHB is best made from palm oil because of the low cost.

Aliphatic polyesters have a great potential to replace non-biodegradable plastics. However they have low melting point and inferior properties compared to other commercial polymers. To improve the properties of polyesters, various functional groups have been introduced and blends with other polymers have been investigated such as aliphatic or aromatic polyesters (PBS/polybutylene terephthalate) [76], or lignin [77, 78].

15.5.2.1 PLA and Polylactide (PLLA) [79, 80]

PLA with high molecular weight is a strong thermoplastic that can be injection moulded, extruded, thermoformed, or used as spun or melt-blown fibres to produce nonwoven goods. These polymers were first used commercially as medical sutures in 1970. Lactic acid polymers and copolymers are offered in semi-commercial quantities for medical applications by Hanley Chemicals (a Boehringer-Ingelheim subsidiary), by PURAC Biochemicals, Cargill Dow Polymers LLC (NatureWorks). They are glassy and transparent, similar to polycarbonates, polyesters, acrylic and general purpose PS, easily printed, readily metallised, heat sealable, heat and UV-resistant, good barriers to the transmission of flavours and odours found in food. Many microorganisms are capable of degrading PLA [81].

15.5.2.2 PHBV

PHBV is a family of biodegradable polyesters made by a fermentation process. A single bacterium species converts corn and potato feedstocks into a copolymer of PHB and HVA. By manipulating the feedstocks, the proportions of the two polymer segments can be varied to make different grades of material. All grades are moisture resistant while still being fully biodegradable. The sole world producer of PHBV is Monsanto, with its Biopol product. Monsanto purchased the Biopol business from Zeneca Bio Products (formerly ICI) in 1996 and manufactures the polymer at its Knowsley, UK, site.

15.5.2.3 PCL

PCL is a biodegradable aliphatic polyester prepared from ε-caprolactone, a seven-membered ring compound. PCL may be processed by a variety of techniques, including blown- and slot-cast film, sheet extrusion, and injection moulding.

Union Carbide in the United States, Solvay in the United Kingdom, and Daicel Chemical in Japan are the three world producers of PCL resin. The total 1996 consumption in the United States, Western Europe, and Japan is estimated to have been less than 4,500 tons. Most typically, PCL is used to make film products, either alone or as a blend with starch. Other applications are adhesive layers, orthopaedic casts, shoe components, and home crafts.

15.5.3 Property Improvements of PLLA and other Biodegradable Plastics

Researchers have succeeded in the copolymerisation of L-lactide with various 4-alkyl caprolactones and have greatly improved of PLLA's elongation at break. Thus PLLA resin reduced PLLA's brittleness, became more flexible and elastic, though its tensile strength and modulus decreased a little. Interestingly, this PLLA copolymer showed degradation by proteases that did not occur in the case of PLA homopolymer, which might show the possibility of biodegradation of lactic acid containing polymer by microorganisms.

Researchers applied samarium trialkoxide complex with methylalmoxane as the catalyst of ε-caprolactone and β-butyrolactone copolymerisation for the first time with success. Copolymer consisting of 70%-80% and 30%-20% of each of the lactones showed the highest degradation rate by lipases. Further, by using a similar catalyst system, they produced a ethylene carbonate/caprolactone copolymer and a 1,4-butanediol/adipate/dimethyl terephthalate copolymer, and found that such polymers containing about 10% of ethylene carbonate or dimethyl terephthalate were not only improved over their corresponding homopolymers in mechanical properties but also degraded by cholesterol esterase (EC 3.1.1.13) in 50 hours. Biodegradable plastics for broader applications should have an even better balance of resin properties as well as finer controllability of biodegradation.

15.6 Plan to Produce L-Lactic Acid from Kitchen and Food Waste

The Kyushu Institute of Technology, Fukuoka University, Japan, attempted to produce L-actic acid by fermentation of kitchen and food waste. They achieved about a 10% (by weight) yield of lactic acid. They are constructing a research test facility that can produce 4 kg of lactic acid per week from the cooked rice and other food waste generated at a university canteen. Their research plan includes a feasibility study of both PLA synthesis and microbial PHB synthesis from the resulting lactic acid, the small-scale manufacture of biodegradable products for in-house use, such as expanded packaging cushioning material, with the ultimate expectation of reducing the waste treatment burden of the food service activities by providing a means for 'zero emissions' from the facility (mandated by law). Cooked rice, like any other carbohydrate waste, could be a good source material for fermentation if the cost of materials could be calculated around zero, but whether sufficient amounts of good quality waste is available will remain a key question. However, it is true that at present in Japan, the recycling of kitchen and food waste through composting or anaerobic digestion is expected to generate demand for the compost produced, co-generated heat, or power generation by biogas.

15.7 Whey-Protein Films

Researchers are also developing other types of hydrophilic protein films. An active area of research is using whey proteins to make edible and biodegradable films. Made from cheese whey, whey protein is separated from lactose using ultrafiltration, after which it is evaporated and spray dried to a dry white powder. The California Dairy Foods Research Centre, University of California, Davis, is one of the research centres working on the development of these whey-protein-based films. Potential future applications include the protection of oxygen-sensitive foods (such as nuts), films to block oil migration into surrounding food components and alternatives to collagen for meat casings.

15.8 Processing of Biopolymer Blends

Many biobased resins can be processed on conventional plastic moulding equipment and, depending on the properties of the specific resin, can be converted into many types of plastic products. These include, but are not necessarily limited to single use items for example, compost bags (lawn and leaf), disposable food-service items (cutlery, plates, cups), packaging materials, for example, loose fill, films, but they also include more durable products like coatings, e.g., laminations, paper coatings, and other injection moulded and sheet extruded products (phone and other cards and sheet printed plastics). The desired properties of the product generally determine the relative amounts of additives used in the resin.

Processing of agriculturally-based materials to mimic plastics remains a largely unexplored area. Most plastic parts are produced by extrusion and then injection moulding. Injection moulding, a widely used method of polymer processing, is characterised by high production rates and accurately sized products. A thorough understanding of the moulding process is necessary to produce products of the required quality at the lowest possible cost. Furthermore, the fact that a blend is used (as opposed to a homopolymer) complicates matters significantly. During the flow process, a polymeric material undergoes simultaneous mechanical and thermal influence and changes its state. Depending on material morphology, this introduces residual stresses and shrinkage in injection-moulded products that affect the physical properties and dimensional stability of the finished products.

15.8.1 Starch-Polycaprolactone Blends

This reactive blend technology was developed for biodegradable film applications like lawn and leaf collection compost bags, agricultural mulch film, etc. The technology involves the following steps:

- Preparation of thermoplastic starch – Plasticisation of the starch in a twin-screw extruder with appropriate screw elements using glycerol as the plasticiser with little or no water in a co-rotating twin-screw extruder.

- Polymerisation (reactive extrusion) of ε-caprolactone directly in an extruder with residence times of 2-3 minutes to provide a novel, multi-arm, branched PCL polymer having number average molecular weight (M_n) values greater than 100,000 and weight average molecular weight (M_w) values of around 300 to 400,000 using aluminium trialkoxides as catalyst. Commercially, PCL is produced using a batch process with a residence time of around four hours.

- Downstream compounding of the new, branched polymer by reactive blending with thermoplastic starch during the extrusion polymerisation operation. Some grafting of the PCL chains onto the starch during this step and the *in situ* generated graft copolymer is able to compatibilise the two phases giving better properties to the resulting blend.

- Finally, a one step continuous process to prepare compatibilised PCL-thermoplastic starch blends from the caprolactone monomer and starch using two extruders in a T-configuration has been developed.

By controlling the rheology in the extruder, one can obtain a morphology in which the plastic starch is dispersed in a continuous PCL matrix phase. Good adhesion and compatibilisation is promoted between the plastic-starch phase and the modified PCL phase to obtain enhanced mechanical properties. Some of the advantages of using plasticised starch instead of granular starch are:

- A smaller domain size is possible by controlling the rheological characteristics

- Improved strength and processing characteristics

- Reduced macroscopic dimensions in certain applications, i.e., film thickness

All of the operations can be performed in the extruder. This eliminates the use of solvent, reduces the number of steps and simplifies the process. From a life-cycle perspective, the biodegradable product reduces waste and energy consumption and conserves resources.

15.8.2 Melt Rheology of Polylactide Blends

Poly(lactic acids) or polylactides (PLLA) are a family of polyesters available via fermentation from renewable resources and are the subject of considerable recent commercial attention. Cargill-Dow Polymers plan to produce 660 million kilogrammes annually of this new thermoplastic by 2003 yet little is known about the fundamental rheological properties of this material. The melt rheological properties of a family of PLA star polymers have been investigated and compared to the properties of the monomers. It is found that the entanglement molecular weight (ME) is approximately 9,000 g per mole (ME% 8,700 g/mol). Results suggest that PLLA is a semi-stiff polymer. The increase in zero shear viscosity for the branched materials is measured and quantified in terms of appropriate enhancement factors. Relaxation spectra show that the transition zone for the linear and branched materials are nearly indistinguishable while the star polymers have greater contributions to the terminal regime. The effects of chain architecture on the flow activation are found to be modest implying that small scale motion in PLA homopolymers largely control this phenomenon. Good agreement is found between the dynamic data and many aspects of the star polymers, however, a dependence of the zero shear viscosity on the number of arms is observed. The rheological properties of blends of linear and branched PLA have also been studied.

15.9 Blends of Starch with Biodegradable Polymers

The properties of blends of starch and aliphatic biodegradable polyesters were determined. The aliphatic polyesters used include: PCL, polybutylene succinate, and a butanediol adipate and terephthalate copolymer. To improve the compatibility between the starch and the synthetic polyester, a compatibiliser containing an anhydride functional group incorporated onto the polyester backbone was used. The blends were melt-compounded using a twin-screw extruder. The concentration of the starch in the blend was varied from 10% to 70% by weight. The amylopectin content of the starches varied from 30% to 99%. The addition of a small amount of compatibiliser increased the strength significantly over the uncompatibilised blend. For the compatibilised blend, the tensile strength did not vary with starch content when compared to the original polyester, while it decreased with increase in starch content for the uncompatibilised blend. Blends displayed a sharp intake of water and those containing Eastar Bio (see **Table 15.1**) had a higher water absorption than those containing the other polyester. Each blend displayed two T_g, one corresponding to the polyester and the other corresponding to starch. For compatibilised blends, the Tg of starch in the blend is lower than that observed for the uncompatibilised blend. For all polyester blends, those containing 99% amylopectin starch at a 70% level had the lowest crystallinity, which otherwise decreased with a decreased amylopectin level in the starch. Blend morphology indicates that the starch phase became finer (increased melting) as the amylopectin in the blend increased. Also, a higher starch content led to greater melting of the starch granules and at 70% starch by weight, a co-continuous phase between the starch and the synthetic polymer exists. The samples were found to be biodegradable under both aerobic and anaerobic conditions.

15.9.1 Blends of Starch with PLA [82]

Polylactic acid (PLA) is a promising synthetic biopolymer derived from starch through bioconversion and polymerisation. However, PLA is expensive due to the complex processing procedures. Another disadvantage of PLA is that it becomes too soft above its T_g (60 °C), which limits its applications. The objective of blending was to blend starch with PLA to reduce the total material costs, to improve starch's compatibility with PLA and enhancing the mechanical properties of the blend. A 45/55 (w/w) mixture of starch and PLA was blended using an intensive hot mixer in the presence of various concentrations of the compatibiliser. The tensile strength of the blend with optimum concentration of the compatibiliser was about 66 MPa, which was slightly higher than pure PLA (62 MPa). The crystallinity and microstructure of the blend was determined by using differential scanning colorimetry, dynamic mechanical analysis (DMA) and scanning electron microscopy. Results showed that the blend with the compatibiliser had higher crystallinity and significant interaction at the interfacial between starch and PLA.

However since starch and PLA are immiscible, resulting in poor blend quality, PVA containing an unhydrolytic residue group of poly (vinyl acetate) was used to enhance the interaction in another study between starch and PLA [56, 83]. With fixed starch and PLA ratio at 50:50 by weight, tensile strength of the blends increased slightly and linearly as PVA concentration increased. The smaller molecular weight of PVA gave the blend higher tensile strength and elongation than larger ones. PVA had little effect on the thermal transition behaviour of the blends, which was almost the same as that of pure PLA. DMA and water absorption are also discussed.

15.9.2 Commercial Compostable Plastic for Blending with Starch

Eastman Chemical Company has commercialised a compostable plastic Eastar Bio copolyester (see **Table 15.1**). The polymer's unique chemistry combines excellent strength and moisture-resistance with complete and safe biodegradability. Easter Bio copolyester can be blended with materials from renewable resources (starch, wood flour) or can be used as a paper board coating. The product can also be used without additives.

Properties of Eastar Bio are similar to LDPE. It can be processed in conventional thermoplastic equipment. The translucent/transparent material can be converted on blown or cast (including) oriented film equipment, and can be extrusion coated. The polymer meets compositional requirements for a variety of food-contact applications.

15.9.2.1 PHAV/EC Blend

Researchers in Vadodara, India, have developed a new biodegradable polymer blend that has potential applications in agriculture and pharmaceutical industries. The new polymer can be developed easily. It is a mixture of two chemicals, PHBV and ethyl cellulose (EC). To prepare the biodegradable blend, the MS University team in Vadodara dissolved the two chemicals in dichloromethane separately. The two solutions were mixed with constant stirring for 15 minutes. In the subsequent step, polymer films were prepared by casting the solutions onto glass with slow drying at room temperature. The group developed a number of blends with different combination of EC and PHBV and found that only one composition with 60% EC and 40% PHBV was a potential biodegradable polymer.

15.9.2.2 Starch/PHEE Blend

Starch was blended with polyhydroxy ester ether (PHEE). Better knowledge of the relationship between structure and properties in these blends is needed to design products which can successfully meet the required properties. It is also important to understand how these properties changes with the humidity and time. The blend structure was strongly dependent on the processing conditions, especially the moisture level. This led to different mechanical properties. The blends processed at high moisture levels possessed elongated starch structures (fibre and sheet-like) which improved the strength and stiffness of the blends. A model was used to predict the change in properties with the relative humidity and the ageing time. Hence, an understanding was developed which would help determine the manner in which these blends should be processed to tailor their end properties.

15.10 Applications

15.10.1 Edible Packaging Films

While a little packaging is a wonderful thing, a lot of packaging is a nightmare, particularly when landfills around the world threaten to engulf our living space. European countries have devised an incentive to reduce packaging in the form of a 'packaging tax'.

Polymers that form films can be composed of carbohydrates, protein, solid lipid/wax, or resin. Examples of carbohydrate polymers include various forms of cellulose, such as carboxymethylcellulose and hydroxypropyl cellulose (HPC), starch and dextrins, pectin, and alginates. Proteins currently used include albumen, corn derivative, soy protein isolate, collagen and whey. Waxes include beeswax and carnuba wax, while shellac is the only food-grade resin. Combinations of these materials can offer some of the benefits of each component.

Edible films are defined by two principles. First, edible implies that it must be safe to eat or that it is generally recognised as safe (GRAS) by the FDA. Second, it must be composed of a film-forming material, typically a polymer. You can dip, and you can coat, but the term edible film refers to a continuous barrier that is formed as the film adheres to the surface of the foodstuff. The key is to maintain that barrier for some period of time, in the same way that an inedible package might.

Edible films have potential in a number of different areas. They can coat food surfaces, separate different components, or act as casing, pouches or wraps. They can preserve product quality by forming oxygen, aroma, oil or moisture barriers; carrying functional ingredients, such as antioxidants or antimicrobials; and improving appearance, structure and handling. Is there such a thing as the perfect package? From an environmental point of view, it's the edible variety that fits the bill. Long-term, edible films show potential in eliminating synthetic oxygen and gas barriers now used in multi-layer packages, minimising damage to the foods and extending their shelf life. Using edible films in place of some of the current packaging might also make the packaging more recyclable. While it is improbable that edible films will ever completely replace plastics, city landfills are grateful for any fraction of the tons of plastic that they might replace today or in the future.

15.10.2 Compostable Plastic Bags

Novamont is selling polymers such as PCL/starch blend which generates 90% carbon dioxide in a controlled composting test and disintegrates in real composting conditions according to ASTM standards and criteria. They also have some other grades recently which are completely biodegradable.

Also the polylactide type of polymers are completely biodegradable and have been shown scientifically to biodegrade and disintegrate very fast, but require thermophilic conditions to initiate the hydrolysis of the polymer structure. Polylactide does not biodegrade fast enough in standard tests because of the test conditions, where the temperature is too low. In a natural composting process the polymer will be biodegraded and disintegrated very fast because of higher temperatures.

15.10.3 McDonald's Approves Degradable Container Product Design

EarthShell Corporation, Santa Barbara, California, announced that after more than two years in validation and testing, the product design of its hinged-lid containers has met performance targets and has received final approval by McDonald's Corporation, Oak Brook, Illinois. Introduced in Chicago locations in April 2000, earth compostable packaging has been used to serve more than 7 million Big Mac hamburgers.

15.10.4 Degradable Polymer Blends for Innovative Medical Devices

Polymers like polylactides, polyglycolides, their copolymers and blends have excellent properties and have thus been developed for the applications such as wound closure, osteosynthesis, nets for support while wounds are healing, degradable bandages and surgical cord for reinforcement, hollow fibres for treatment of damaged nerves as well as prostheses for anastomoses.

For example a biodegradable intravascular stent prototype is moulded from a blend of polylactide and trimethylene carbonate. Furthermore these polymers are being considered for the development of products for the treatment of paradontal diseases, fillers for tooth extraction, artificial vessels and drug carrier systems. Why would a medical practitioner want a material to degrade? There may be a variety of reasons, but the most basic begins with the physician's simple desire to have a device that can be used as an implant and will not require a second surgical intervention for removal. Besides eliminating the need for a second surgery, the biodegradation may offer other advantages. For example, a fractured bone that has been fixed with a rigid, non biodegradable stainless implant has a tendency for refracture upon removal of the implant whereas, an implant prepared from a biodegradable polymer can be engineered to degrade at a rate that will slowly transfer load to the healing bone.

15.10.5 Degradable Polymers for Synthetic Organs

A new method of post-mastectomy reconstruction uses bio-engineered polymers to, in effect, allow a breast to regenerate itself. Researchers at the Institute of Bioscience and Engineering at Rice University, USA, have come up with a 'tissue scaffold', a preformed mould made from a biodegradable polymer blend of polyactic and glycolic acid. This 'scaffold' is used along with long-term growth factor delivery systems that stimulate and guide cell differentiation, proliferation, and migration to produce soft tissue in the three-dimensional shape desired. As the new tissue grows, the polymer scaffolding slowly disintegrates, until eventually all that remains is a regenerated, completely natural-looking breast.

15.10.6 Green Games

The 2000 Summer Olympic Games in Sydney, Australia, provided the greenest Games ever thanks to Biocorp of the US, the leader in biodegradable industry. Biocorp was the sole supplier of biodegradable, compostable, plastic utensils including knives and forks, straws, cups, spoons, plates, and also biodegradable compostable garbage bags. Biocorp supplied some 45 million biodegradable items to the Games.

Unlike conventional plastic cutlery, Biocorp's products do not contain any PE or PS. The resin used to make the products is a mixture of corn starch, water and other biodegradable materials that can be broken down in the composting process.

15.10.7 Improved Polymer Blend for Agricultural Mulching Film Commercialisation

Nihon Gosei Kagaku has been developing loose packaging film, garbage bags and the other biodegradable plastic applications by importing Novamont's starch/polyester polymer blend for ten years. Agricultural mulching film was considered as one of its most promising and important applications. Sufficient initial strength of the film for furrow application by machine, and reliable soil coverage during several weeks of plant growth were the common and one of the most important challenges to be met by any kind of biodegradable plastic film.

Recently, Nihon Gosei announced that Novamont had improved the film strength by increasing the polyester content of Mater-Bi and is starting the final application tests in fields in north-eastern Japan from February. As farmers have come to recognise more the environmental and labour-saving values of biodegradable films in agriculture, the company has high expectations from the field test results. This product can command a price of up to 2.5 times conventional plastic prices for biodegradable mulching films, provided that the government gives incentives to farmers for the use of biodegradable agricultural films.

15.10.8 Biodegradable Fishing Line from Toray

Biodegradable fishing line 'Field Mate' was recognised as the most environment friendly fishing line in the Guinness Book of Records [84]. The product made from PLA by applying Toray's proprietary filament technology has been commercialised since 1996.

15.11 Developing World Markets for Biodegradable Plastics

The Asian market for biodegradable plastics is poised for massive growth, thanks to two new developments in Japanese Environmental Law. Japan has been in the forefront of early research in the field of state sponsored efforts over the past decade. Many Japanese companies were among the earliest developers of biodegradable plastics materials. In recent years, they have been outpaced by a US Company, Cargill Dow. Its 1,40,000 tonnes/year plant in Nebraska (Japan) will supply polylactic acid for biodegradable and other applications. In 2000 the Basic Law to create a Sustainable Society was passed, it will oblige all companies, who produce more than 50 tonnes/year of organic waste, to divert it, either to feed use or organic recycling. The result will be an explosion in composting and consequently a demand for biodegradable bags in which to do it.

The revised Plastic Packaging Recycling Law (Japan), passed in 2000, imposed extended responsibility on manufacturers, obliging them to pay for ultimate disposal of materials, they place on the market, with annual reporting and auditing.

The Japan Organic Recycling Association (JORA) was created in 2000 to co-ordinate the national organic recycling efforts relating to standardisation and technology exchange. Japan Biodegradable Plastics Society (JBPS), which belongs to JORA, has established the 'Green Pla' logo mark for biodegradable plastics materials, that meet its existing standards.

There is increasing public concern in Taiwan about dioxin contamination from incineration of plastics and other types of waste, and the problems of field burning of agricultural film residues, which may lead to a ban. The Taiwan Government is a strong supporter of recycling of all kinds. An Organic Recycling Association was formed in 2001. From 2002, there will be separate collections of organic and incinerable waste at street level. Korea has also come a long way. Food residuals and organic waste are to be banned from landfills by 2005 and instead aerobic digestion will be used. Use of biodegradable plastic bags that are 100% biodegradable is advised.

Demonstration projects for composting using biodegradable plastic bags are also being planned in Hong Kong and Shanghai, with the help of US Grains Council.

India, also has huge potential, Pakistan has not moved beyond the basic educational stage. Thailand is looking at producing tapioca starch for biodegradable plastics applications, while the Philippines is showing interest in the lactic acid route.

In the United States, the future of biodegradables will depend on developing applications with competitive pricing and performance that demonstrate the economics of composting organic residuals *versus* landfilling, from a systems perspective.

According to Frost and Sullivan [50, 85] the market for degradable plastics should reach 7575 metric tons in 2004, representing growth of 35%/year, equivalent market values of $185 M. This compares to market demand of 9072 metric tons ($23 M) in 1997. Biodegradable polyesters and PET are widely available. Products include polylactic acid, biodegradable PET, polybutylene succinate, polybutylene adipate, PHB, PHV, polyethylene succinate, polyester amide and PCL and starch blends. Growth is being driven by environmental concerns but high costs and established markets compete. Frost and Sullivan report that there are 29 competitors in the US market, the main players are: National Starch, Showa Highpolymer, Daicel, DuPont and Cargill-Dow.

15.12 Cost of EDP

15.12.1 Resin Cost

These new biopolymers are priced higher than the commodity polymers typically in use in plastics applications. However, producers are currently working toward bringing down the price of degradable polymers by increasing production capacity and improving process technology. Five years ago PLA and PHB sold for more than $55 per kilogram. Today PLA, depending on the quantity, is between $3.30 USD and $6.60 USD per kg and PHB, in large quantity is near $8.80 USD per kg.

A small toothbrush manufacturer, Fine Company, developed a biodegradable toothbrush stem technology replacing acrylonitrile-butadiene-styrene or PP resins with a PLA/wood pulp composite and started supplying the product to drug stores. It is important for mass marketing that the selling price of this 'Ecott' toothbrush was decided as 250 yen/piece, a little high but within the range of common commodities.

The starch used to create the biodegradable plastic, typically wheat gluten, costs about 35 cents a kg. The cheapest commercial plastics cost about $2.20 a kg. Thus, when this starch-based plastic becomes available to manufacturers, it could be the cheapest biodegradable plastic. Now, most environment-friendly plastics cost about $5.50 per kg. In the first stage of the project, researchers have brought the cost of their plastic down to about $3.30 per kg. The ecological efficiency of natural polymers is very high. NatureWorks PLA uses 30% to 40% less fossil fuel than is required to produce conventional plastic resins. Also, overall carbon dioxide emissions are lower compared with hydrocarbon-based resins because carbon dioxide is removed from the atmosphere by growing corn.

15.12.2 Injection Moulding

Although properties of the starch ester resin are comparable to PS, and can be injection-moulded in cycle times comparable to PS, there are two problems: cost, and in some applications, weight because of its higher density. If one could effectively foam the product in an extrusion or injection moulding operation, less material would be required and so the cost per article would be less. This would be true for all biodegradable plastics. Therefore, generic process technology to make biodegradable foam plastics would be very useful in the effort to successfully commercialise biodegradable plastic products.

US prices for biodegradable polymers range from $3.30 to $6.60/kg for PLA to $6.60 to $8.80/kg for PHA. High prices have limited the use of biodegradable polymers to special applications such as loose-fill packaging, compost bags, and medical devices. The total 1996 consumption of biodegradable polymers in the United States, Western Europe, and Japan is estimated to have been approximately 14,000 tons (excluding PVA).

The problem with pricing is that, at least in the US, prices are unlikely to drop until a larger market opens up, but mass market sales are hindered by high prices.

With all their advantages, why have Canada and the United States not used more bioplastics? In Europe, especially in Germany, northern Italy, the Netherlands and Scandinavia, the biodegradable bag is used extensively for the collection of organic wastes. Use of the traditional PE bag is largely prohibited.

The first reason may be cost. Currently, biodegradable bags and food service utensils are priced between two and three times the price of comparable products made from conventional PE and other plastics. This price differential is the result of the very different technologies and processes used for PE and for these biodegradable products.

When, however, the cost savings resulting from the elimination of debagging are taken into consideration, the overall cost differential between conventional PE bags and biodegradable bags vanishes. The biodegradable bag becomes as economical as the PE bag. Similarly, if all the waste generated in a fast-food restaurant were compostable, the restaurant's overall waste disposal costs would be expected to drop significantly, since it would no longer be necessary to separate the plastic and the Styrofoam from the food waste.

15.12.3 Improvements in PCL Resin Reduces the Extrusion Costs of Film to the Level of PE Film

Daicel has succeeded in introducing structural branching into the PCL resin. Recently, Daicel announced that an important effect of the technology was to raise the melt viscosity of the resin and thus reduce the extrusion costs of PCL film to the same level as PE. The low melt viscosity of PCL resin had prevented the application of high speed extrusion technology that is typical of PE film and resulted in comparatively higher converting costs. Besides improving extrusion efficiency, the quality improvements in uniformity of film thickness and transparency are so improved that truly commercial PCL film manufacturing has been realised. Even so the sales price of PCL film remains up to three times higher than that of PE due to very high production cost of the PCL resin.

With the help of this technology, Daicel intends to sell larger amounts of resin and to reduce production costs. The company said that it would soon increase the production capacity of PCL resin to an annual 5,000 tons from the current 1,000, targeting sales of three billion yen in year 2000.

15.12.4 Competitively Priced Ball Point Pen Made of Corn

A major Japanese pen and pencil supplier, Mitsubishi Pencils started selling the Ecocorn ball point pen from April 2000. It is said that each pen comprises up to forty grains of corn. The company set the sales price at a competitive 100 yen per piece with the expectation of selling five million pens in the first year.

15.12.5 Topy Green's Marketing Situation of Biodegradable Mulching Films

Musashino City, one of the important vegetable growing areas supplying the Tokyo Metropolis, is subsidising the farmers' purchase of biodegradable mulching films from last year. The subsidy from general tax revenue, totalled 380,500 yen in 1998, is rationalised primarily to prevent illegal field-burning of used films that may cause harmful dioxin emission (and draw strong complaints from surrounding residents).

Tables **15.5** and **15.6** summarise the target pricing of biodegradable polymers in USA, Germany and Japan.

15.13 Conclusions

Biopolymer technology has emerged because of the difficulties encountered in waste disposal and recyclability of conventional plastics. Bioplastics are environmentally friendly, can be extruded (as a film) or injection moulded. After serving their useful purpose, they disintegrate completely, leaving no residue. They are available commercially. PLA is the most produced polymer from renewable resources while PCL is the only synthetic polymer

Table 15.5 Global prices of biodegradable polymers			
	United States ($/kg)	Germany (Euros/kg)	Japan (Y/kg)
PLA	3.30-6.60	3.60-4.75	1,000
PET-based	4.40	NA	NA
Aliphatic polyesters	NA	NA	500-800
Aliphatic/aromatic copolyester	4.75	3.50-4.40	NA
Starch-based Z class	6.40	4.75	700
PCL	6.60	NA	1,000
PHBV	4.00-8.80	7.90	1,100-1,200
NA: *not available*			

Table 15.6 Long-term target pricing for biodegradable polymers			
	United States	Germany	Japan
Target pricing	$1.10/kg	25 Euros/kg ($2.75/kg)	500 yen/kg ($3.85/kg)
Pricing basis	Commodity polymer, i.e., price/performance. No market value for biodegradability	EU packaging Waste Directive DSD recycling fee for non-biodegradables = 13.75 euros/kg	Recycling laws and fees still uncertain Tied to PET pricing.
DSD: *Duales Systeme Deutschland*			

which is completely biodegradable. Blends of bioplastics lead to cost dilution, improved processability and property improvements. Successful applications of biodegradable plastics include; agricultural mulching films, seedling pots, garbage bags for food/kitchen wastes collection, loose fills, prepaid cards, etc. Woven and nonwoven fabrics, expanded mould packages for fragile electronic commodities, and food contact package applications are especially expected to contribute the tomorrow's market growth. PE-starch blends do not degrade completely and PE residues are not acceptable in the compost. To increase the rate of disintegration and the degree of fragmentation, metals such as cobalt, manganese and copper are added to these PE-starch blends. The compost is contaminated with these metals, which may impose significant long-term health risks as the compost is used in agricultural applications. By inserting PE, an artificial substance, into soil we are altering the fundamental property of soil. This alternation goes far beyond what human beings can afford. The qualities and originality of soils are and remain the basis of our planet. This repetitive alternation needs to come to an end until we are able to measure the consequences. PE added to the soil daily, even in infinitesimally small quantities, is unacceptable since it will affect ecology by altering the soil. Innovations in the development of materials from biopolymers have led to the preservation of fossil-based raw materials, the reduction in the volume of waste, complete biological degradability and compostibility in the natural cycle, protection of the climate through the reduction of CO_2 released, as well as the application possibilities of agricultural resources for the production of bioplastics.

Up to 80% of all waste at sporting venues is organic waste. The inability to collect this waste in an uncontaminated form has meant an enormous opportunity for recycling has been missed.

By using biodegradable utensils and disposing of them in biodegradable bags the organic waste stream will remain uncontaminated. This will result in thousands of tonnes of waste being diverted from landfill.

References

1.   J. Rose, *Chemistry and Industry (London)*, 1992, 8, 284.

2.   E. Chellini, Proceedings of the ICS-UNIDO International Workshop, Environmentally Degradable Plastics: Industrial Development and Application, Seoul, Korea, 2000, 7.

3.   E. Chellini, ICS-UNIDO International Workshop, 'Environmentally Degradable Plastics: Position of EDPs in Plastic Waste Management, Lodz-Pabianice, Poland, 2001, 9.

4.   S.H. Kim, J.Y. Lim, S. Lim and Y.H. Kim, Proceedings of the ICS-UNIDO International Workshop, 'Environmentally Degradable Plastics: Industrial Development and Application', Seoul, Korea, 2000, 146.

5.   *Chemical Market Reporter*, 1998, **254**, 19, FR 14.

6.   A.K. Kulshreshtha and S.K. Awasthi, *Popular Plastics and Packaging*, 1998, **43**, 11, 53.

7.   D.H. Kim, Proceedings of the ICS-UNIDO International Workshop, Environmentally Degradable Plastics: Industrial Development and Application, Seoul, Korea, 2000, 269.

8.   Y.H. Kim and S.H. Kim, Proceedings of the ICS-UNIDO International Workshop, Environmentally Degradable Plastics: Industrial Development and Application, Seoul, Korea, 2000, 361.

9.   K. Ohshima, Proceedings of the ICS-UNIDO International Workshop, Environmentally Degradable Plastics: Industrial Development and Application, Seoul, Korea, 2000, 343.

10. M. Ito, Proceedings of the ICS-UNIDO International Workshop, Environmentally Degradable Plastics: Industrial Development and Application, Seoul, Korea, 2000, 216.

11. G. Ghislandi and C. Bastioli, Proceedings of the ICS-UNIDO International Workshop, Environmentally Degradable Plastics: Industrial Development and Application, Seoul, Korea, 2000, 277.

12. S. Quan, Proceedings of the ICS-UNIDO International Workshop, Environmentally Degradable Plastics: Industrial Development and Application, Seoul, Korea, 2000, 304.

13. I. Tomka, inventor; I. Tomka, assignee; WO 9005161 A1, 1990.

14. I. Tomka, inventor; no assignee; US 5362777, 1994.

15. C. Cordes and W-D. Jeserich, inventors; Basf AG, assignee; EP 0000397 B1, 1980.

16. G. Lay, J. Rehm, R.F. Stepto, M. Thoma, J-P. Sachetto, D.J. Lentz and J. Silbiger, inventors; Warner-Lambert Company, assignee; US 5,095,054, 1992.

17. I. Tomka, inventor; Bio-tech Biologische Naturverpackungen GmbH, assignee; EP 0542155 B1, 1998.

18. I. Tomka, inventor; no assignee; US 5280055, 1994.

19. I. Tomka, inventor; Bio-tech Biologische Naturverpackungen GmbH, assignee; EP 0596437 B2, 2002.

20. J. Lorcks, W. Pommeranz and H. Schmidt, inventors; Bio-tech Biologische Naturverpackungen GmbH, assignee; WO 9619599 A1, 1996.

21. J. Lorcks, W. Pommeranz and H. Schmidt, inventors; Bio-tech Biologische Naturverpackungen GmbH, assignee; WO 9631561 A1, 1996.

22. H. Kibbutz, inventor and assignee; GB 1586344A, 1981.

23. R.P. Westoff, H.F. Otey, C.L. Mehltretter and C.R. Russell, *Industrial Engineering Chemistry: Product Research and Development*, 1974, **13**, 123.

24. G.J.L. Griffin in *Emerging Technologies for Materials and Chemicals*, Eds., R.M. Rowell, T.P. Schultz and R. Narayan, ACS Symposium Series No. 476, American Chemical Society, Washington, DC, USA, 1992, 203.

25. J.L. Willett, inventor; Fully Compounded Plastics, Inc., assignee; US 5087650, 1992.

26. U.R. Vaidya and M. Bhattacharya, *Journal of Applied Polymer Science*, 1994, **52**, 617.

27. L. Nie, R. Narayan and E.A. Grulke, *Polymer*, 1995, 36, 2227.

28. M. Bhattacharya, U.R Vaidya, D. Zhang and R. Narayan, *Journal of Applied Polymer Science*, 1995, **57**, 539.

29. Z. Yang, M. Bhattacharya and U.R. Vaidya, *Polymer*, 1996, **37**, 2137.

30. M. Kim and A.L. Pometto, *Journal of Food Protection*, 1994, **57**, 1007.

31. R.L. Shogren, *Carbohydrate Polymers*, 1993, **22**, 93.

32. M.F. Koenig and S.J. Huang, *Polymer*, 1995, **36**, 1877.

33. M.A. Kotnis, G.S. O'Brien and J.L Willett, *Journal of Environmental Polymer Degradation*, 1995, **3**, 97.

34. H.F. Otey and M.W. Doane in *Starch Chemistry and Technology*, 2nd Edition, Eds., R.L. Whistler, J.M. Bemiller and E.F. Paschall, Academic Press, New York, NY, USA, 1984, 389.

35. J.W Lawton and G.F. Fanta, *Carbohydrate Polymers*, 1994, **23**, 275.

36. C. Bastioli in *Degradable Polymers, Principles and Applications*, 1st Edition, Eds., G. Scott and D. Gilead, Chapman and Hall, London, UK, 1995, 112.

37. A.C. Albertsson, *Journal of Applied Polymer Science*, 1978, **22**, 3419.

38. A.C. Albertsson, *European Polymer Journal*, 1980, **16**, 623.

39. A.C. Albertsson, C. Barenstedt and S. Karlsson, *Journal of Applied Polymer Science*, 1994, **51**, 1097.

40. R. Narayan and W. Lafayette, *Kunstoffe German Plastics*, 1989, **79**, 92.

41. G.J.L. Griffin, inventor; Epron Industries Limited, assignee; US 4,983,651, 1991.

42. W. Sung and Z.L. Nikolov, *Industrial & Engineering Chemistry Research*, 1992, **31**, 2332.

43. D. Bikiaris, J. Prinos and C. Panayiotou, *Polymer Degradation and Stability*, 1997, **56**, 1.

44. G. Scott in *Degradable Polymers, Principles and Applications*, 1st Edition, Eds., G. Scott and D. Gilead, Chapman and Hall, London, UK, 1995, 169.

45. D.W. Wiles, J.F. Tung, G. Scott and G. Swift, Proceedings of the ICS-UNIDO International Workshop, Environmentally Degradable Plastics: Industrial Development and Application, Seoul, Korea, 2000, 298.

46. F.H. Otey, R.P. Westhoff and C.R. Russell, *Industrial Engineering Chemistry: Product Research and Development*, 1997, **16**, 305.

47. F.H. Otey, R.P. Westhoff and W.M. Doane, *Industrial Engineering Chemistry: Product Research and Development*, 1980, **19**, 1659.

48. B.Jasberg, C. Swanson, T. Nelsen and W. Doane, *Journal of Polymer Materials*, 1992, **9**, 153.

49. G.F. Fanta, C.L. Swanson and W. Doane, *Journal of Applied Polymer Science*, 1990, **40**, 811.

50. R.L. Shogren, R.V. Greene and Y.V. Wu, *Journal of Applied Polymer Science*, 1991, **42**, 1701.

51. R.L. Shogren, A.R. Thompson, R.V. Greene, S.Y. Gordon and G. Cote, *Journal of Applied Polymer Science*, 1991, **47**, 2279.

52. G.F. Fanta, C.L. Swanson and W.M Doane, *Carbohydrate Polymers*, 1992, **17**, 51.

53. M. Defosse, *Modern Plastics International*, 1999, **29**, 1, 31.

54. *Chemistry in Britain (London)*, 1990, **26**, 6, 525.

55. *Chemical Weekly*, 1998, **43**, 27, 81.

56. *Chemical Engineering News*, 1996, **74**, 31, 40.

57. *Japan Chemical Week*, 1996, 37, 1869, 1.

58. *ASTM Standards Pertaining to the Biodegradability and Compostability of Plastics*, ASTM, West Conshohocken, PA, USA, 1999

59. ASTM D5488, *Standard Terminology of Environmental Labeling of Packaging Materials and Packages*, 1994.

60. ASTM D5338, *Standard Test Method for Determining Aerobic Biodegradation of Plastic Materials Under Controlled Composting Conditions*, 1998.

61. *Biodegradable Polymers*, Report No.115, Stanford Research Institute, Menlo Park, CA, USA, 1998.

62. *FTC Guide for the Use of Environmental marketing Claims*, Federal Trade commission, Washington, DC, USA, 1998.

63. TC61/SC5/WG22, the technical committee TC (61) dealing with plastics, subcommittee (SC) 5, and working group (WG) 22 biodegradable plastics.

64. ASTM D2096, *Standard Test Method for Colorfastness and Transfer of Color in the Washing of Leather*, 2000.

65. DIN V 54900, *Testing of the Compostability of Plastics*, 1998.

66. TC20/SC4/WG2, the technical committee TC (20) aerospace fastener systems, subcommittee (SC) 4, working group (WG) 2 dealing with design parameters and procurement specification for bolts.

67. R. Narayan, Proceedings of the ICS-UNIDO International Workshop, Environmentally Degradable Plastics: Industrial Development and Application, Seoul, Korea, 2000, 24.

68. Proceedings of the ICS-UNIDO International Workshop, Environmentally Degradable Plastics: Position of EDPs in Plastic Waste Management, Lodz-Pabianice, Poland, 2001, 46.

69. H. Sawada, Proceedings of the ICS-UNIDO International Workshop, Environmentally Degradable Plastics: Industrial Development and Application, Seoul, Korea, 2000, 62.

70. C. Vasile, Proceedings of the ICS-UNIDO International Workshop, Environmentally Degradable Plastics: Position of EDPs in Plastic Waste Management, Lodz-Pabianice, Poland, 2001, 24.

71. C. Vasile, M.M. Macoveanu, L. Constantin, M.C. Pascu, G. Cazacu, *Cellulose Chemistry and Technology* (in press).

72. Y. Doi and Y. Kikkawa, Proceedings of the ICS-UNIDO International Workshop, Environmentally Degradable Plastics: Industrial Development and Application, Seoul, Korea, 2000, 160.

73. G. Braunegg, R. Bona, G. Haage, E. Schellauf and E. Wallner, Proceedings of the ICS-UNIDO International Workshop, Environmentally Degradable Plastics: Industrial Development and Application, Seoul, Korea, 2000, 181.

74. S.Y. Lee, Proceedings of the ICS-UNIDO International Workshop, Environmentally Degradable Plastics: Industrial Development and Application, Seoul, Korea, 2000, p.201.

75. G. Braunegg, R. Bona, G. Haage, E. Schellauf and E. Wallner, Proceedings of the ICS-UNIDO International Workshop, Environmentally Degradable Plastics: Position of EDPs in Plastic Waste Management, Lodz-Pabianice, Poland, 2001, 33 and 37.

76. S.S. Im and E.S. Yoo, Proceedings of the ICS-UNIDO International Workshop, Environmentally Degradable Plastics: Industrial Development and Application, Seoul, Korea, 2000, 228.

77. G.E. Agafitei, M.C. Pascu, G. Cazacu, M. Popa, R. Hogea and C. Vasile, *Macromolecular Materials Engineering*, 1999, **267**, 44.

78. G.E. Agafitei, M.C. Pascu, G. Cazacu, M. Popa, R. Hogea and C. Vasile, *Macromolecular Materials Engineering*, 2000, **283**, 93.

79. P.R. Gruber, Proceedings of the ICS-UNIDO International Workshop, Environmentally Degradable Plastics: Industrial Development and Application, Seoul, Korea, 2000, 91.

80. T. Yagi, K. Inabaam and N. Kawashima, Proceedings of the ICS-UNIDO International Workshop, Environmentally Degradable Plastics: Industrial Development and Application, Seoul, Korea, 2000, 117.

81. M.L. Tansengco and A. Jarerat, Proceedings of the ICS-UNIDO International Workshop, Environmentally Degradable Plastics: Industrial Development and Application, Seoul, Korea, 2000, 136.

82. X.H. Wang, S. Susan and P. Seib, Proceedings of the 85th American Association of Cereal Chemists Annual Meeting, Kansas City, MI, USA, 2000.

83. K. Tianyike and X.S. Sun, Proceedings of the 85th American Association of Cereal Chemists Annual Meeting, Kansas City, MI, USA, 2000.

84. *Guiness World Records 1999*, Ed., M. Young, Gullane Publishing, London, UK, 1999.

85. Schriftenreihe Umwelt No. 271/1, *Okobilanz Starkehaltiger Kunststoffe Bewertung mit Uberarbeiteter Wirkungsorientierter Methodik,* Volumes 1 and 2, BUWAL, Bern, Switzerland, 1996.

16 Liquid Crystalline Polymers in Polymer Blends

Dana-Ortansa Dorohoi

16.1 Introduction to Liquid Crystals

Liquid crystallinity [1-3] is a mesomorphic state, between the liquid and the crystalline states, characterised by a long-range orientational order and either by partial positional order or by complete positional disorder. Substances in the liquid crystalline state are in fact ordered fluids.

The mesomorphic behaviour depends directly on the molecular shape and group polarisability. A mesomorphic compound contains anisotropic units having a large axial ratio, such as (a) rigid rods, (b) discs-shape, (c) planks or (d) helices (**Figure 16.1a-d**), small molecules or macromolecules with anisotropic properties. These molecules must also resist entropic forces which would otherwise tend, for example, to turn a rod into a random coil. The result of this is that a substance exists as a mesophase only under suitable conditions of temperature, pressure or concentration.

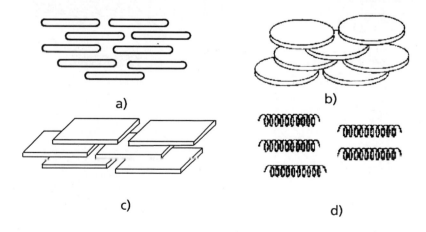

a)

b)

c)

d)

Figure 16.1 Possible components of liquid crystal: a) rods; b) discs; c) planks; d) helices

The first detailed description of the nature of liquid crystals [4], was done in 1888 by Reinitzer, for a melt of natural cholesteryl acetate and cholesteryl benzoate. Today many examples of natural liquid crystals, such as DNA, cholesterol, or tobacco mosaic virus are known.

There are also a lot of synthetic liquid crystals used in technical applications such as: cyano-biphenyls, *para*-linked benzene rings stilbene, azo-links. Kelker and Hatz [3] list many of the chemical substances that exhibit liquid crystallinity, in certain conditions.

The earliest classification of liquid crystals was made by Friedel [5]. He distinguished between nematics, cholesterics and smectics mainly on the basis of optical studies. The nematic mesophase is the least ordered liquid crystalline phase. One dimensional order, shown in **Figure 16.2a**, characterises the nematic mesophase. Cholesterics [6] (**Figure 16.2b**) are twisted structures usually due to the optically active molecules. The smectic mesophase is two dimensional ordered, in layers (**Figure 16.2c**), hence the Greek term 'smectic'. However, Friedel argued that the cholesterics should be regarded as twisted nematics, and so, one can consider that there are only two basically different types of liquid crystals: nematics and smectics [5].

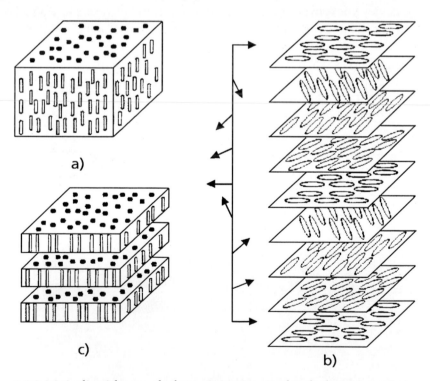

a)

c)

b)

Figure 16.2 Main liquid crystal phases: a) nematic; b) cholesteric and c) smectic mesophases

The liquid crystals that become ordered by heating are called 'thermotropic' [2, 3] and those that become ordered by dissolution are called 'lyotropic' [7, 8].

It has been found that many compounds show [4, 6, 9] first-ordered phase transitions within the smectic range, suggesting the existence of different types of smectics. Based on the aspect of microscopic textures as well as of X-ray patterns and on the order in which different phases appear in compounds with more than one smectic phase, Sackmann and co-workers [10] introduced the identification of different smectic mesophases by capital letters A, B, C, D E, F or H. The H-phase has been identified by de Vries and Fishel [11]. So, for example, smectics A, C and F are characterised by a disordered arrangement of the molecular axes within the smectic layer, B_n have a hexagonal lattice in the smectic layer, while a pseudohexagonal lattice in the plane perpendicular to the molecules characterises B_t smectic. D smectics are characterised by a cubic arrangement of spherical molecular aggregates.

Within each smectic layer, the molecules are arranged in a rather random manner and have considerable freedom of translation and rotation around their long axis. The loose organisation of the molecules makes the layers rather flexible and they often adopt complex curved arrangements. The flat layers have been observed to slide very easily over each other, but the viscosity perpendicular to the layers is much higher due to a strong tendency to maintain the integrity of the layer.

The importance of this very interesting phase of matter is pointed out by various applications in technics, in oil, food, detergent, or soap industries, in medicine, in biology or environment science.

The great interest in the study of liquid crystals is justified by practical applications in: numerical displays, information store, colour television, surface thermography of microcircuits, in detection or visualisation of electromagnetic waves, ultrasounds or vibrational fields, in aerodynamic tests, in chromatography, in the infrared interferometry. Besides their technical applications, liquid crystals represent an exciting and challenging area of research in biology, revealing the effect of molecular organisation on chemical reactivity.

In the last decades, the liquid crystalline polymers included in blends were used to induce or to improve the anisotropy of some materials and as '*in situ*' reinforcements.

16.2 Liquid Crystalline Polymers and Their Properties

The polymers exhibiting one or several mesomorphic phases are called liquid crystalline polymers. The first findings in liquid crystalline polymers are attributed to Blumstein

[12] and Finkelmann [13]. Polymer molecules having a stiff main-chain or a flexible main-chain with mesogenic side groups (**Figure 16.3**) are known to form mesomorphic phases as main-chain [14], or as side–chain types [15].

One can obtain main-chain systems by blending bifunctional monomers containing two or more complementary groups, i.e., electron or proton donor and acceptor groups [16, 17]. Main-chain liquid crystals are known for their ability to form high tensile strength fibres and moulding.

The principal approach in the design of the supramolecular side-chain liquid crystalline polymers consists in the introduction of the mesogenic units via the interaction of the polymers containing pendant binding side groups as complementary low molar mass compounds [18-20]. Some new side-chain liquid crystalline polymers in which mesogenic units are attached to the polymer backbones as pendant groups through flexible spacers are reported in [15, 21].

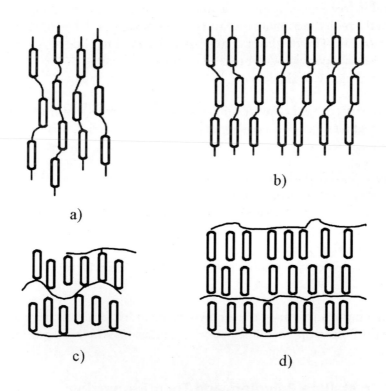

a)

b)

c)

d)

Figure 16.3 Liquid crystalline polymer: a) main-chain nematics; b) main-chain smectics; c) side-chain nematics; d) side-chain smectics

Side-chain liquid crystalline polymers have applications in optical components such as brightness enhancement films for display devices. Anisotropic networks, elastomers and gels also contain side-chain liquid crystalline polymers [22].

Anisotropic aromatic building blocks can be copolymerised with flexible monomers to get partially ordered samples composed both from ordered regions created by liquid crystal groups and random coils of flexible units. Placed between the rigid polymers and the liquid crystalline side groups, the flexible units decouple the motion of liquid crystalline groups between the polymer main-chains, increasing the anisotropy of the samples. Incorporated in elastomers, liquid crystalline polymers change their colour on stretching, being used as pressure or temperature sensors.

The chemistry of the commercial liquid crystalline polymers has been dominated by aromatic polyamides, aromatic polyesters, polycarbonates, polyphosphazines, siloxanes, or by mixtures of such compounds.

The main liquid crystalline polymers commercially available are reviewed by Isayev [23]. As is shown in [23], thermotropic liquid crystalline polymers were firstly synthesised by the DuPont Chemical Company in 1972 [24, 25] and commercialised as Ekkcel 1-2000. This was an aromatic copolyester consisting of *p*-hydroxybenzoic acid (HBA), terephthalic acid (TA) and 4,4′-dioxydiphenol (DODP) with a high melting point. Liquid crystal polymer Ekkcel 1-2000 became the originator of a new family of liquid crystalline polymers called Xydar, firstly manufactured in 1985 by Dartco Manufacturing Company and then by Amoco Chemical Company.

In 1973 the copolyester X7G, based on HBA and polyethyleneterephthalate (PET) was commercialised by Tennessee Eastman Chemical Company [26, 27].

Unitika Chemical Company presently commercialises Rodrun, a modified version of X7G. Celanese Research Company [28-31] synthesised a family of thermotropic liquid crystalline polymers, based on HBA and hydroxynaphthoic acid (HNA), firstly commercialised in 1985 under a trade name Vectra.

Since 1985 a number of other companies introduced new liquid crystalline polymers [23] such as: Rodrun by Unitika (1986); KU by Bayer AG; Ultrax by BASF; Victrex by ICI Company; Granlar by Granmont Inc. (1988); Zenite by DuPont (1995); Novacor by Mitsubishi; Econol by Sumitomo and Rhodester by Rhone-Poulenc.

Although the statistics indicate a significant growth of liquid crystalline polymer consumption, a number of companies such as BASF, ICI and Bayer have announced that they are moving out of the liquid crystalline polymer business.

Let us mention some commercially available main-chain liquid crystalline polymers and their properties in **Tables 16.1-16.3** [32]. Some of the main properties [32] of liquid crystalline polymers are:

- The liquid crystalline polymers formed by highly aromatic monomers are very stiff. For example, Kevlar fibres have a specific modulus significantly higher than the one of steel [33, 34] or carbon fibres produced from a liquid crystal pitch [35] which have moduli up to 700 GPa.

- The highly aromatic liquid crystalline polymers are very brittle, have very low failure strains, but can be very strong [32].

- The viscosity of liquid crystalline polymers aligned in the flow direction increases with the molecular weight to the power 1, while measured under the same conditions, the viscosity of normal flexible polymer depends on the superior powers of the molecular weight. Thus, for equal molecular weights, liquid crystalline polymers have a much lower viscosity compared with the normal flexible polymers. The most ordered liquid crystalline polymers have the highest values of the viscosity and a pronounced smectics anisotropy. It results that liquid crystalline polymers alter the rheology of bulk polymers, reducing their viscosity.

- The mechanical properties of liquid crystalline polymers are highly anisotropic. Injection moulded specimens [36] can show anisotropies in modulus of up to 20.

- Liquid crystalline polymers have low thermal expansion coefficients. This results in low mould shrinkage and hence very high tolerance components can be moulded. This is one of the most exploited property of liquid crystalline polymers. For example, intricate electronic components that could not be moulded from traditional polymers were achieved using liquid crystalline polymers.

- Although the general tendency of polymer viscosity is to be reduced with temperature, the viscosity of liquid crystalline polymers can even increase with temperature, near isotropic fluid point [32].

- Anisotropy of the electro-optical properties of the liquid crystalline polymers [37-42] and of their blends [43] has also been reported.

Liquid crystalline polymers have been extensively studied from the point of view of texture types, phase transitions [44, 45], cholesteric pitches [46, 47], electric or magnetic field effects [48].

Table 16.1 Some commercially available liquid crystalline polymers [32]
Reproduced with permission from C.S. Brown and P.T. Alder in Polymer Blends and Alloys, Chapter 8, Eds., M.J. Folkes and P.S. Hope, published by Blackie, 1993

Nr	LCP name	Manufacturer	Content
1	Vectra A	Celanese	HNA(27%), BA(73%)
2	Vectra B	Celanese	HNA(58%), TA(21%), 4AP(21%)
3	X7-range	Eastman Kodak	PET/HBA (HBA 40, 60 or 80%)
4	Rodrun	Unitika	PET/HBA based materials
5	LX-scries	DuPont	TA, HQ, PHQ
6	Ultrax KR	BASF	-
7	Xydar	AMOCO	BP, HBA, TA
8	Victrex SRP	ICI	HBA, HQ, IA
9	KU-9211	Bayer	HBA, TA, HQ, IA, BP

HNA: *2-hydroxy-6-naphthoic acid*   HBA: *4-hydroxybenzoic acid*
4AP: *4-aminophenol*   PET: *poly(ethylene terephthalate)*
TA: *terephthalic acid*   HQ: *hydroquinone*
IA: *isophthalic acid*   PHQ: *phenyl hydroquinone*
BP: *4,4 ˝dihydroxybiphenyl*

16.3 The Effect of Liquid Crystalline Polymers on the Processing and on the Physical Properties of Polymer Blends

Liquid crystalline polymers offer the possibility of improving the melt processability and of enhancing the mechanical properties of the flexible isotropic polymers. In order to obtain high-performance materials, many mixtures of thermotropic liquid crystals with isotropic polymers or with thermoplasts were prepared and carefully studied. Although the addition of a small amount of liquid crystalline polymer into the isotropic polymer has produced a significant reinforcement effect, a further improvement in mechanical properties has been hampered by a poor interfacial adhesion between the liquid crystalline polymer and the isotropic ones. Therefore the study of the miscibility as well as isothermal phase separation processes are usually described in the specialised literature.

Table 16.2 Physical properties of some commercially available liquid crystalline polymers [32]
Reproduced with permission from C.S. Brown and P.T. Alder in Polymer Blends and Alloys, Chapter 8, Eds., M.J. Folkes and P.S. Hope, published by Blackie, 1993

No.	LCP	MP (°C)	T_g (°C)	$T_{proc.}$(°C)
1	Vectra A	283	105	320-340
2	Vectra B	280	110	300-330
3	X7-range	250-300	70,143	270-350
4	Rodrun	250	70,143	270-350
5	LX-series	310	180	340-360
6	Ultrax KR	-	195	270-32-
7	Xydar	412	-	>420
8	Victrex SRP	280	-	300-350
9	KU-9211	-	198	330-360

MP: *melting point* T_g: *glass transition temperature*

Table 16.3 Mechanical properties of some commercially available liquid crystalline polymers [32]
Reproduced with permission from C.S. Brown and P.T. Alder in Polymer Blends and Alloys, Chapter 8, Eds., M.J. Folkes and P.S. Hope, published by Blackie, 1993

No.	Property	Vectra A950	Vectra B950	Ultrax KR-4002	KU 1-9211	LX-2000
1	Tensile strength (MPa)	165	188	160	160	117
2	Tensile modulus (GPa)	9.71	9.3	8.2	9.95-16.0	28.3
3	Tensile elongation (%)	3.0	1.3	2.8	1.7-2.8	0.6
4	Flexural modulus (GPa)	9	15.2	-	6.2-10.0	18
5	Notched Izod impact strength (J/m) (KJ/m²)	520 -	415 -	- 59	- 32	197 -
6	Heat deflection temperature (°C)	180	-	118	156	185

Some commercial main chain thermotropic liquid crystalline polymers have been mixed with some polymers to obtain engineering plastics with improved mechanical properties. Relationships between morphology and physical properties of the blends obtained were extensively analysed. Two phase structures were shown by the morphological studies, showing the immiscibility of the liquid crystalline polymers with flexible chains. Obviously, investigations such as phase separation and control of the supermolecular assembly are very important for practical applications in order to prepare high-performance engineering plastics.

Side-chain liquid crystalline polymers have been used as matrix resins in polymer dispersed liquid crystals in order to improve the wide angle viewability and the optical transmittance of the obtained blends [49].

Polymer dispersed liquid crystal is an nonhomogeneous composite film comprising a low molecular weight liquid crystal dispersed in a polymer matrix, prepared by polymerisation or by thermally induced phase separation from an initially homogeneous state. The refractive index of the mesogens of low molecular weight liquid crystals is usually matched with that of the mesogenic side-groups of the side-chain liquid crystalline polymers so that, the refractive index appears the same from the all viewing angles [50]. The electro-optical performances of these materials depends strongly on the uniformity, size and shape domains, as well as on the topology of the liquid crystal within the dispersed domains. The size and shape of the liquid crystal domains are generally determined by both thermodynamical and kinetic characteristics of the phase separation during preparation process.

Lately, polymer dispersed liquid crystals have contributed to sheets consisting of droplets of nematic liquid crystals in an amorphous polymer with technical applications. The orientation in the nematic droplets can be influenced by electric or magnetic fields, that change the aspect of the sheet from cloudy to clear, opening up the possibility for electrically switchable windows and panels and for a large area signs and advertising boards [1].

In switchable panels the liquid crystal is constrained in a confined geometry [51] and the ratio of surface contact area to bulk volume is high. The important aspects of the confinement, such as porous polymer network assemblies in nematic liquid crystals, polymer stabilised cholesterics with their implications for reflexive cholesteric displays, filled nematics or anisotropic gels have been also reported in [51].

Initially, the liquid crystallinity has been controlled via specific interactions in binary solutions of low molecular mass mesogens having unlike groups [52]. The clearing temperature of these solutions was usually higher than the clearing temperatures of each single component and often smectic behaviour was strongly enhanced or induced. Then, the liquid crystallinity of polymers or of polymer blends has been induced or modified via specific noncovalent interactions [53, 54]. In the majority of these systems, liquid crystallinity is induced or increased by charge transfer, such as electron donor-acceptor interactions [54].

The balance between the inter- and intra- molecular interactions is responsible for the morphology as well as for degree of order of the miscible blends. The dipolar interactions may also be responsible for liquid crystallinity in the case of some mixtures [55].

Along with improved processing, the presence of the liquid crystalline polymers in blends can determine an enhanced rigidity and strength or a viscosity variation versus temperature. For example, when a sample was cooled from the isotropic phase, the viscosity firstly increased and then it suddenly dropped at the isotropic-anisotropic transition [32]. The decreasing in the viscosity was explained by the increase in the order degree of the texture.

By reducing the viscosity of mixtures, liquid crystalline polymers have the ability to alter the rheology of the bulk polymers. Properties of mixtures are usually modified when compared to those of the individual polymers. For example the rheological properties of polymer mixtures are not additive [56].

The rheological properties of the mixtures, consisting of an amorphous polymer and a liquid crystalline polymer of the droplet type, depend strongly on the following factors:

- size and morphology of the droplets

- direction of flow that deforms droplets

- interface interactions between components.

Studying the rheology-morphology relationships for several polymer mixtures, Han and Kim [57] pointed out the influence of the shape of the droplets on the rheological behaviour.

The viscosity of the polymer mixtures [58, 59] and their rheological behaviour [60] as a function of the concentration were also studied by several authors. They pointed out positive or negative deviations from the additivity rule, attributed to homogeneous and to heterogeneous mixtures, respectively.

Decreasing the viscosity also improves the processability of the liquid crystalline polymer mixtures [61] made for obtaining materials with high mechanical strength.

16.4 Specific Interactions in Polymer Blends Containing Liquid Crystalline Polymers

The liquid crystallinity of the side-chains can be controlled by molecular structure (mesogenic units are attached to the polymer backbones as pendant groups through flexible spacers [15, 21]) as well as by specific noncovalent interactions (intermolecular hydrogen bonds [62, 63], ionic interactions [64, 65] or electron donor-acceptor interactions [66, 67]).

Specific interactions assist not only the polymer miscibility, improving their blend processability, but also influence the structural, mechanical and thermal properties of the blends. The specific charge transfer interactions themselves, or accompanied by dipolar interactions, are responsible for induced liquid crystallinity in the majority of the systems containing side-chains. This approach has been exploited for many years in the binary mixtures of electron rich and electron poor low molecular mass mesogens. Recently, similar effects have been observed for side-chain liquid crystalline copolymers, containing electron donor and electron acceptor mesogenic groups. The same behaviour as in the copolymers were shown in the majority of the corresponding homopolymer blends in which liquid crystallinity is induced by specific interactions.

16.4.1 Electron Donor-Acceptor Interactions

Extended studies on the specific interactions in polymers, copolymers and polymer blends containing electron donor and electron acceptor side-chain groups were made by Kosaka and Uryu [68-70]. A remarkable effect of the central linking groups on the thermal properties of poly 4{[6-(methacryloyloxy)hexyl]oxy}-*N*-(carbazolylmethylene) aniline (PM6Cz)/PM6XNO$_2$ polymer blends, with the chemical structures of the components from **Scheme 16.1,** has been reported [68].

Analysing the miscibility in PM6Cz/PM6XNO$_2$ blends containing a carbazolyl (Cz) and nitrophenyl with various central linking groups, Kosaka and Uryu showed that the miscibility is influenced both by specific interactions and by molecular structures of the mesogenic groups. The increased temperature of mesophase-isotropic transition and the high entropy change on clearing indicated the formation of 1:1 electron donor-acceptor complex with a high orientation of the mesogenic side-groups in the miscible blends: 1:1 PM6Cz/polymethacryloyloxyhexyloxynitrostilbene (PM6SBNO$_2$) (T$_g$ = 73 °C and the mesophase-isotropic transition temperature T$_i$ =207 °C) and 1:1 PM6Cz/PM6AzNO$_2$ (T$_g$ = 68 °C and T$_i$ =185 °C).

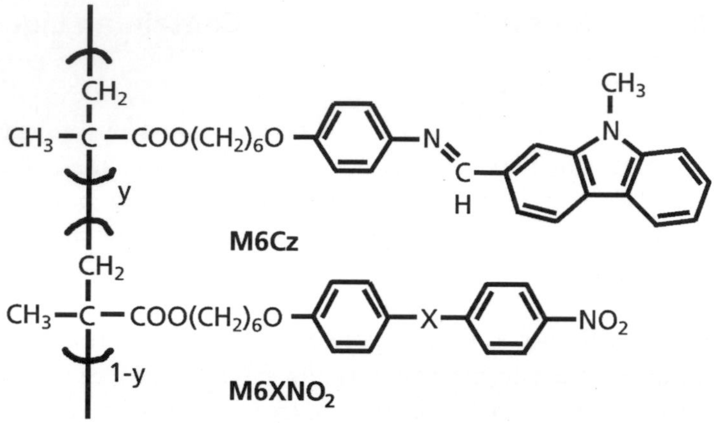

CH_2

$CH_3-\overset{|}{\underset{|}{C}}-COO(CH_2)_6O$ —〇— N≡C—H ... carbazole, N—CH_3

M6Cz

CH_2

$CH_3-\overset{|}{\underset{|}{C}}-COO(CH_2)_6O$ —〇—X—〇—NO_2

1-y   **M6XNO₂**

Scheme 16.1 Molecular structures of copolymers M6Cz-M6XNO₂. {X = -N=CH in M6BANO₂; X = -CH=CH in M6SBNO₂; X = -N=N in M6AzNO₂; -COO in M6PBNO₂}

Although their components have electron donor and electron acceptor groups, the PM6Cz/PM6PBNO₂ blend exhibited, over the entire composition range, two glass transitions (T_{g1} = 38 °C and T_{g2} = 75 °C) and two mesophase-isotropic transitions (T_{i1} = 100 °C and T_{i2} = 165 °C), corresponding to those of respective homopolymers. The immiscibility of this blend has been explained by the steric effects of the carboxyl groups that disturbed the overlapping between the mesogenic groups, the energy of the specific interactions being smaller than those corresponding to the steric effects.

The phase diagrams of PM6Cz/PM6XNO₂ blends suggest a dependence of the phase transition temperatures on the proportions of the PM6Cz units in the mixture. In the case of PM6Cz/PM6SBNO₂ blends, as it is shown in **Figure 16.4**, for proportions of PM6Cz bigger than 0.7, coexist two separated phases (smectic + nematic at temperatures lower than 150 °C and smectic + isotropic in 150-200 °C temperature range). For proportions of PM6Cz between 0.3-0.6, blends exhibited a smectic behaviour, and for proportions of PM6Cz smaller than 0.3 a nematic one. The presence of a smectic phase, for the proportions of PM6Cz centred on 0.5, supports that the electron donor acceptor complex formed in PM6Cz/PM6SBNO₂ blends is of the type 1:1.

X-ray diffraction showed smectic phases for the miscible blends [68]. Measured from the X ray pattern of quenched PM6Cz/PM6SBNO₂ blends, the smectic layer spacing was d = 31.5A (2θ = 2.80°) and the distance between the mesogenic side groups was d = 4.41A (2θ = 20.14°).

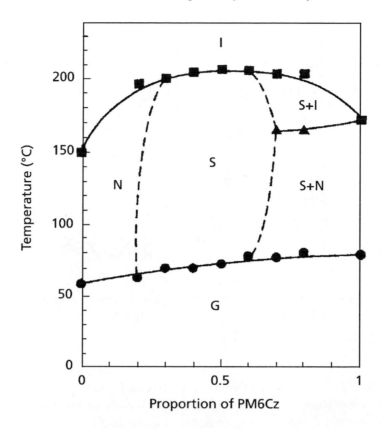

Figure 16.4 Phase transition temperatures *versus* PM6Cz proportion in
PM6Cz/PM6SBNO₂ blends [68]

Reproduced with permission from Y. Kosaka and T. Uryu, Journal of Polymer Science;
Polymer Chemistry Edition, 1995, 13, 2221. Copyright 1995, John Wiley & Sons Inc

The three times increase of the isotropisation enthalpy for 1:1 PM6Cz/PM6SBNO₂ blend, compared with those of the individual homopolymers, proved an increasing degree of order, induced in blends by specific interactions.

The blends made by homopolymers with similar central linking groups have remarkable thermal stability. To demonstrate this assertion, Kosaka and Uryu [69] used the homopolymers from **Scheme 16.2**. All PM6Cz/PM6BA-Y blends with Y = -NO₂; -CN; -OCH₃ exhibited homogeneous smectic phase when the proportion of the mesogenic carbazolyl units was in the 0.3-0.6 range, though the smectic phase did not appear in both homopolymers.

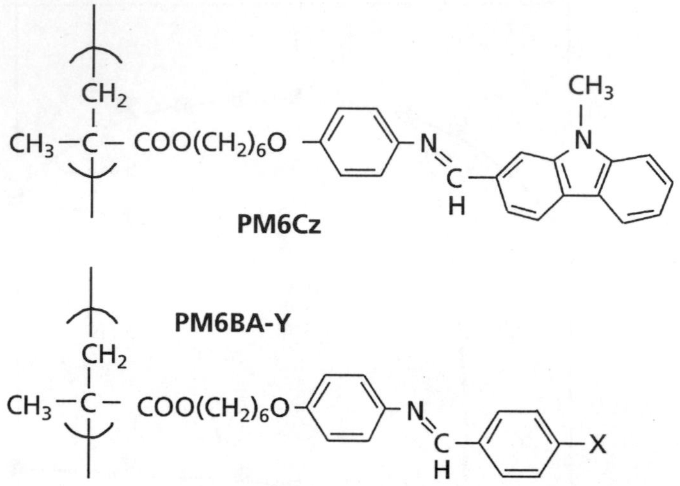

Scheme 16.2 Homopolymers: PM6Cz and PM6BA-Y with Y = NO$_2$ (PM6BA-NO$_2$); CN (PM6BA-CN); OCH$_3$ (PM6BA-OCH$_3$)

The 1:1 PM6Cz/PM6BA-NO$_2$ miscible blend had a single glass transition at 69 °C and a single mesophase-isotropic transition at 190 °C. This blend exhibited a focal conic texture smectic phase by optical polarising microscopy (**Figure 16.5**). X-ray pattern [69] gave a thickness of the smectic layer d = 30.9A (2θ = 2.86°) and a distance between the mesogenic groups d = 4.43A (2θ = 20.0°), smaller than those in the pure homopolymers (d = 4.64A for PM6Cz and d = 4.63A for PM6BA-NO$_2$).

The thickness of the smectic layer determined in the blend was equivalent to the maximum length of the monomeric units in the mesogenic copolymers, showing that electron donor (carbazolylmethylene)aniline groups overlapped the electron acceptor (nitrobenzylidene)aniline groups, as it is illustrated in **Figure 16.6**.

Kosaka and Uryu [69] studied the influence of specific interactions on the blend miscibility by replacing nitro group from PM6BA-NO$_2$ (**Scheme 16.2**) with cyano, or methoxy groups. While PM6Cz/PM6BA-CN was a miscible blend, phase separation occurred in PM6Cz/ Poly 4{[6-(methacryloyloxy)hexyl]oxy}-*N*-(4′-methyloxybenzilidene) aniline (PM6BA-OMe) blend (characterised by two glass transitions at T$_{g1}$ = 63 °C and T$_{g2}$ = 80 °C and two mesophase-isotropic transitions at T$_{i1}$ = 139 °C and T$_{i2}$ = 175 °C, corresponding to the respective polymers). So, using substituents with different capacity to accept electrons, Kosaka and Uryu provided that the blend miscibility was caused by intermolecular electron donor-acceptor interactions. For the systems formed with stronger

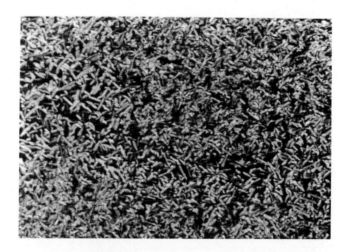

Figure 16.5 Optical polarised microphotographs of the 1:1 PM6Cz/PM6BA-NO$_2$ polymer blend at 180 °C [69]

Reproduced with permission from Y. Kosaka and T. Uryu, Macromolecules, 1994, 27, 6286. Copyright 1994, American Chemical Society

interactions, smectic phase was shown, showing the capacity of these interactions to induce a superior degree of order. They suggested that electron interactions between electron rich and electron poor units is a weak charge transfer between the electron donor carbazolyl groups and the electron acceptor nitrophenyl groups.

Kosaka and Uryu [70] also examined the structural and thermal properties of the side-chain liquid crystalline polymers containing both (2- or 3-quinolinylmethylene)aniline and (4-nitrobenzylidine)aniline groups as well as of their blends. They prepared 4-{[6-(methacryloyloxy)hexyl]oxy}-N-(4-nitrobenzylidene)aniline (M6NO$_2$) as an electron acceptor monomer and methacrylate monomers containing (quinolinylmethylene)aniline groups such as: (3-quinolinylmethylene)aniline (M6Q3); (2-quinolinylmethylene)aniline (M6Q2); (2-naphthylmethylene)aniline (M6N); (6-methoxy-2-naphthylmethylene)aniline (M6NOMe) and (4-methoxybenzylidene)aniline (M6OMe) as electron donor monomers. Concomitantly, the corresponding copolymers and homopolymers [Poly 4{[6-(methacryloyloxy)hexyl]oxy] N (1′ R-benzilidene) aniline (PM6R)] having: R = -NO$_2$; -Q$_3$, -Q$_2$, -N –NOMe; -OMe, were prepared in order to study the specific interactions in their mixtures and blends.

The PM6R homopolymers with average molecular weights ranging between 2.42 x 10^4 to 5.23 x 10^4 were obtained in high yields, all of them exhibiting liquid crystallinity between the glass- and the isotropic-temperatures. The isotropic temperature of the

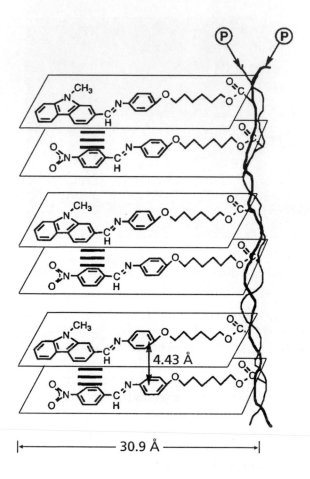

Figure 16.6 Schematic illustration of a proposed mesophase structure in the 1:1 PM6Cz/PM6BA-NO$_2$ blend [69]

Reproduced with permission from Y. Kosaka and T. Uryu, Journal of Macromolecules, 1994, 22, 6286. Copyright 1994, American Chemical Society

polymers with methoxy group was higher, compared with that of the other homopolymers, showing that the methoxy substituent stabilises the mesophase.

Liquid crystallinity of pure polymers depends on the R- structure. So, the Poly 4{[6-(methacryloyloxy)hexyl]oxy}-N-(2-quinolinylmethylene) aniline (PM6Q2) homopolymer exhibited a nematic phase from 49 °C to 72 °C, whereas PM6Q3 gave a smectic phase from 55 °C to 143 °C.

The morphology of the binary PM6R/PM6NO$_2$ blends was shown by differential scanning calorimetry (DSC) curves and by optical polarising microscopy. So, DSC curves of PM6OMe/PM6NO$_2$ blends exhibit three endothermic peaks at 121 °C, 140 °C and 162 °C, corresponding to the isotropic transitions of PM6NO$_2$, PM6OMe and to the 1:1 PM6OMe/PM6NO$_2$ miscible blend, while the PM6Q3/PM6NO$_2$ miscible blend had a single glass transition, at 52 °C and a single crystal liquid isotropic phase transition at 166 °C. These results suggest that the electron donor – acceptor interactions acted between equimolar electron donor and acceptor groups.

PM6Q3, being a structure with a stronger electron donor nature as compared PM6Q2, it forms miscible blends with PM6NO$_2$, while PM6Q2 does not. The miscible PM6Q3/PM6NO$_2$ blend has a smectic focal-conic fan texture (**Figure 16.7**).

As is shown in **Table 16.4**, for PM6Q3/PM6NO$_2$ blend, the smectic layer spacing was 28.9 A, almost the same as that of the (M6Q3- M6NO$_2$) ([M6Q3] = 0.55) copolymer, indicating that the polymer blend has a mesophase structure similar to that of the copolymer containing the same mesogens.

The phase diagram of the PM6Q3/PM6NO$_2$ blends is given in **Figure 16.8**. A single smectic phase appeared in the polymer blends with approximately equimolar compositions. In the blends containing PM6Q3 in the proportions smaller than 0.3 as

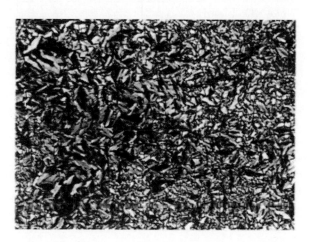

Figure 16.7 Optical polarised microphotograph of 1:1 PM6Q3/PM6NO$_2$ blend [70]

Reproduced with permission from Y. Kosaka and T. Uryu, Macromolecules, 1995, 28, 24, 8295. © 1995, American Chemical Society

Table 16.4 Results of X-Ray measurements of homopolymers, and polymer systems containing electron donor and acceptor groups. [2θ value (deg.) in parentheses; D_1^s the smectic spacing and D_2^g the distance between side-groups] [70]

Reproduced with permission from Y. Kosaka and T. Uryu, Macromolecules, 1995, 28, 8295. Copyright 1995, American Chemical Society

Nr.	Sample	D_1^s (2θ)	D_2^g (2θ)
1	PM6Q3	27.8 (3.18)	4.44 (19.98)
2	PM6Q2		4.60 (19.30)
3	PM6NOMe	29.6 (2.98)	4.49 (19.74)
4	PM6OMe		4.51 (19.66)
5	PM6NO$_2$		4.63 (19.17)
6	(M6Q3-M6NO$_2$) copolymer	29.2 (3.02)	4.46 (19.80)
7	(M6Q2-M6NO$_2$) copolymer	28.5 (3.10)	4.43 (20.04)
8	(M6N-M6NO$_2$) copolymer	29.2 (3.02)	4.45 (19.92)
9	(M6NOMe-M6NO$_2$) copolymer	29.6 (2.98)	4.41 (20.12)
10	(M6OMe-M6NO$_2$) copolymer	29.0 (3.04)	4.36 (20.34)
11	1:1 PM6Q3/ M6NO$_2$ blend	28.9 (3.06)	4.45 (19.92)

well as higher than 0.7, the phase separation occurred. For the molar fractions of PM6Q3 outside the range 0.3-0.7, optical polarising microscopy showed the existence of mixtures consisting from PM6NO$_2$, PM6Q3 and the PM6Q3/PM6NO$_2$ (1:1) miscible blend.

Studying the effect of specific interactions on the thermal behaviour of the polymers P_1 and P_2 as well as of the copolymers P_3 containing the same mesogenic groups (**Scheme 16.3**), and of their P_1/P_2 blends, Imrie and Paterson [53] observed that the relative modest interactions between two unlike mesogenic species should increase the clearing temperatures above the glass temperature, revealing liquid crystalline behaviour.

The significant difference between the colour of the substances P_1 and P_2, permitted optical polarising microscopy studies referring to phase separation in the P_1/P_2 blends (with different proportions of the polymer P_1) *versus* temperature. Thus, when cooling the isotropic phase of each blend at 0.2 °C/min, from 10 degrees above the clearing

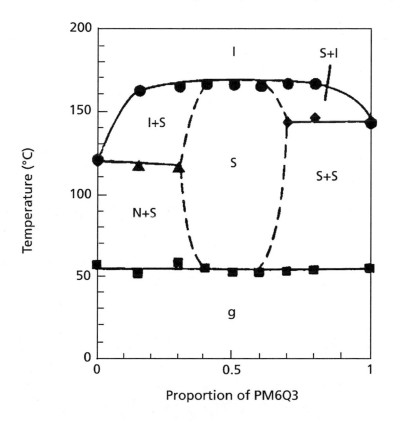

Figure 16.8 Phase transition temperatures on the proportion of the PM6Q3 units in the PM6Q3/PM6NO$_2$ blend [70]

Reproduced with permission from Y. Kosaka and T. Uryu, Macromolecules, 1995, 24, 8295. Copyright 1995, American Chemical Society

temperature, clear, characteristic focal conic fan textures appeared, indicating the presence of a smectic A phase. The phase separation was not complete, but there were regions rich or poor in each component, the separation being less pronounced for the equimolar blends. The smectic A-isotropic transitions were essentially independent on the blend composition.

These aspects suggest that the specific interactions between the unlike mesogenic units can assist the miscibility of the polymers.

The manner in which the liquid crystalline phase develops in the blends also states that a specific interaction between the unlike mesogenic groups leads to the mesophase formation.

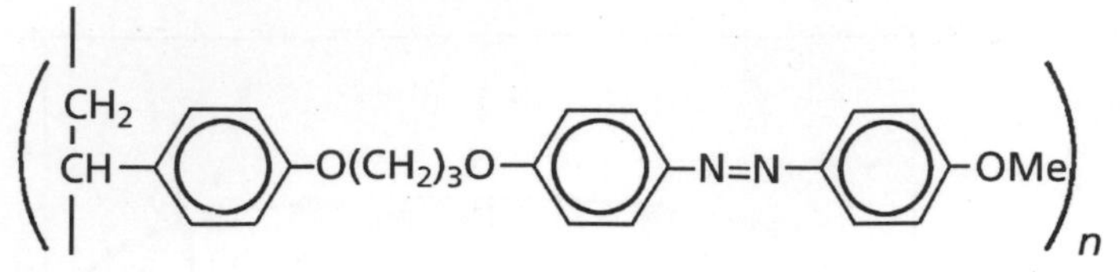

P₁

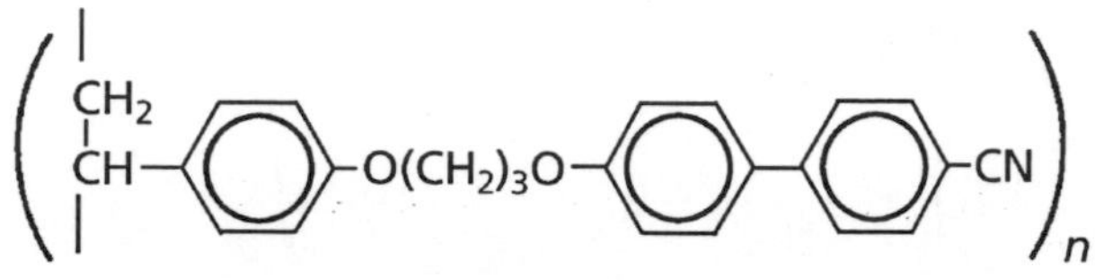

P₂

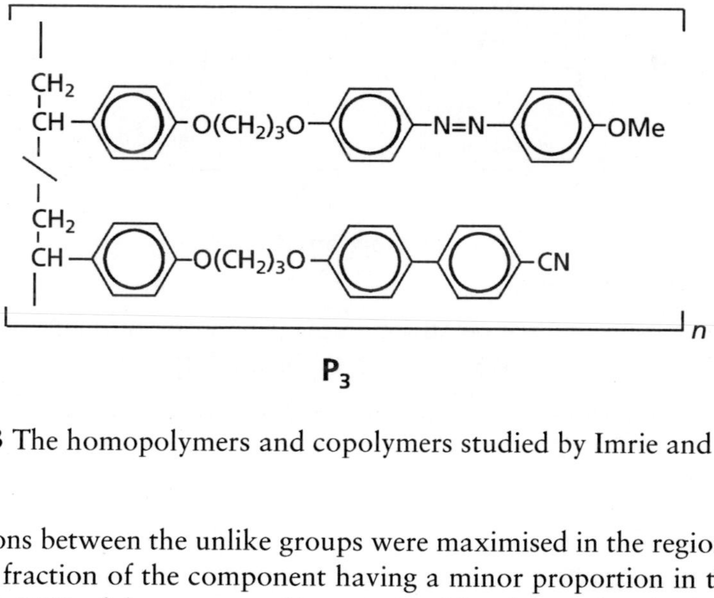

P₃

Scheme 16.3 The homopolymers and copolymers studied by Imrie and Paterson [53]

The interactions between the unlike groups were maximised in the regions containing a large relative fraction of the component having a minor proportion in the blend. Thus the blend with 25% of the compound P_1 contained focal conic fans in the regions rich in the component P_1, whereas for 75 wt% of the compound P_1, the focal conic fans developed initially from the domains rich in component P_2.

16.4.2 Hydrogen Bond Interactions

Phase behaviour and phase separation in blends containing a liquid crystalline polymer and an amorphous polymer, interacting by hydrogen-bonds were analysed by Chen and co-workers [71] in the case of poly{styrene-*co*-[*p*-2,2,2-trifluoro-1-trifluoromethyl)ethyl-α-methylstyrene]} (PS(OH)) and a thermotropic liquid crystalline polymer poly{*p*-(*tert*-butyl)phenylene fumarate} (PBPF) soluble in common solvents. For the PS(OH)/PBPF blends, the miscibility was found only at high PBPF compositions and low temperatures, due to the intermolecular hydrogen bonding [71, 72] between -C(CF$_3$)$_2$OH groups of PS(OH) and the phenolic –OH-end groups of PBPF, as shown by Fourier transform infra red spectroscopy (FTIR) analyses.

Even PS/PBPF blends, containing pure polystyrene (PS), were phases totally separated over all composition and temperature ranges, the incorporation of -C(CF$_3$)$_2$OH groups into polystyrene led to a partial miscibility of PS(OH) in PBPF.

After drying at 60 °C under vacuum, the cast films of 10/90 and 30/70 (wt/wt) PS(OH)/PBPF were transparent and homogeneous. Upon annealing at a temperature above the glass temperature, the sample showed a faint anisotropy under crossed polarisers. After annealing at very high temperatures, phase separation took place with formation both of an anisotropic phase and an isotropic one. Two-phase morphology with a periodic distance of about 10 μm and a bicontinuous feature, corresponding to a spinodal decomposition, has been shown in the micrographs of the annealed PS(OH)/PBPF blends. In the anisotropic domains [71], a certain nematic texture of liquid crystalline polymer has been shown.

On the contrary, the cast films with a larger weight fraction of PS(OH) were translucid. An irregular two-phase structure with anisotropic and isotropic phases was observed distinctly under optical polarising microscope. Upon annealing at high temperatures above glass temperature, the domains of anisotropic and isotropic phases grew in size and a certain nematic texture of liquid crystalline polymer was observed in the anisotropic phase. By temperature increasing above glass temperature, the PS(OH)/PBPF blends with 90/10 and 70/30 (wt/wt) compositions showed a transition of the nematic phase into an isotropic one.

In **Figure 16.9** the observations described previously are summarised. An arbitrarily drawn line delimitates the single- and two-phase regions. Similar results were obtained for the PS(OH)-4/PBPF and PS(OH)-8/PBPF blends containing 4 mol% and 8 mol% of -C(CH$_2$)$_3$OH in polystyrene, as shown in **Figure 9A** and **9B**. Although the miscibility increased with increasing numbers of C(CF$_3$)$_2$OH groups in polystyrene, a total miscibility was not obtained over the entire composition range.

The phase boundary for PS(OH)-4/PBPF and PS(OH)-8/PBPF blends, shown in the phase diagrams from **Figures 9A** and **9B**, suggests a separation into two phases from miscible blend, when the weight molar fraction increases from 0.3 to 0.5. This result was also shown by the optical polarising microscopy.

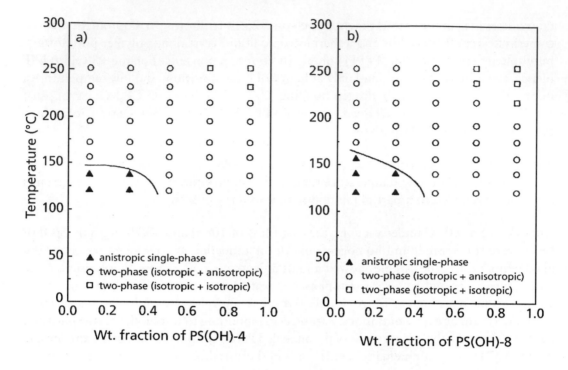

Figure 16.9 Phase diagrams of a) PS(OH)-4/PBPF and b) PS(OH)-8/PBPF blends [71]

Reproduced with permission from W. Chen, J. Wu and M. Jiang, Macromolecular Chemistry & Physics, 1998, 199, 8, 1683. Copyright 1998, Wiley-VCH

In order to study the molecular origin of the miscibility in the PS(OH)/PBPF blends, FTIR technique was used. As is shown in [71, 72], in such blends, the most probable are intermolecular hydrogen bonds between the ester carbonyl groups of PBPF and –OH groups of PS(OH), i.e., $CF_3)_2OH..O=C-$. Hydrogen bonds are also expected between:

- $-C(CF_3)_2OH$ and $>(CH_3)_2OH$ end groups of PBPF;

- $-C(CF_3)_2OH$ and $-C(CF_3)_2OH$ within the pure PS(OH);

- $>(CH_3)_2OH$ and $>(CH_3)_2OH$ within the pure PBPF.

Increasing PS(OH)-8 concentration into the PS(OH)-8/PBPF blends, at the room temperature, the $3487 \ cm^{-1}$ IR band corresponding to the self-associated -OH end groups of pure PBPF (phenolic –OH... phenolic –OH dimers), is progressively shifted to higher frequency (up to $3512 \ cm^{-1}$). IR band $3512 \ cm^{-1}$ has been assigned to the –OH stretching vibrations of -OH..-OH dimer within the $-C(CF_3)_2OH$ groups. The shifted peak to the corresponding frequency of pure components instead of the split of the peaks by the modification of the blend composition, implies the intermolecular hydrogen bonding between –OH groups of PS(OH) and phenolic –OH end-groups of PBPF.

The presence of an additional very broad band centred at 3362 cm^{-1}, assignable to the –OH stretching vibration of OH..OH multimers within $-C(CF_3)_2OH$ groups in the 50/50, 70/30, 90/10 PS(OH)-8/PBPF blends, as well as in the pure PS(OH)-8, indicated that there are voluminous aggregations of the PS(OH)-8 component to form PS(OH) –rich phase. Concomitantly, this band did not emerge in the spectra of the single-phase blends (10/90 and 30/70 PS(OH)-8/PBPF blends). In other words, the miscibility in the PS(OH)/PBPF has been explained by a balance between the intermolecular hydrogen bonds between –OH from PS(OH) and phenolic -OH groups from PBPF (favourable to the miscibility) and intramolecular hydrogen bonds within PS(OH) or PBPF (unfavourable to the miscibility).

The changes from the 1150-1050 cm^{-1} and 1300-1100 cm^{-1} ranges of FTIR spectra, induced by modifying the PS(OH) content in blends, argued the replacement of the intramolecular hydrogen bonds between C-OH groups in PS(OH)-8 by the intermolecular hydrogen bonds between $-C(CF_3)_2OH$ and phenolic –OH end groups of PBPF.

With the increase of PBPF content, the intermolecular hydrogen bonds between the phenolic –OH end-groups of PBPF and the –OH groups of PS(OH) increased, while the intramolecular hydrogen bonds within PS(OH) decreased. So, at high composition in PBPF, single-phase blends were obtained. Also, as the degree of intermolecular hydrogen bonding in the PS(OH)/PBPF system is dependent on the concentration of phenolic OH end-groups, the further increase in OH content of PS(OH) might be ineffective for an additional improvement of the miscibility.

The blends of the mesogenic 4-*n*-octyloxybenzoic acid and polystyrene, poly(4-vinyl pyridine) or poly(2-vinyl pyridine) were analysed [73] from point of view of thermal behaviour using DSC and optical polarising microscopy. The acid is essentially immiscible with polystyrene over the entire composition range. However, molecular mixing was observed for blends containing about 0.3 of either poly(4-vinyl pyridine) or poly(2-vinyl pyridine). The authors explained the miscibility by hydrogen bond formation between the acid and pyridine groups. At high concentrations of the acid, phase separation occurred although the presence of hydrogen bonding was shown in FTIR spectra.

No pronounced differences are observed between the blends containing the different isomers of poly(vinyl pyridine). Two competing processes can influence the miscibility and structural features of the above mentioned blends [73]:

- the self association of the acid, forming dimers

- the specific interaction between the acid and the pyridine groups, yielding hydrogen bonded complexes.

Studies on the polymer systems have shown, however, that the equilibrium constant for the self-association of carboxylic acids is greater than that for complex formation, despite the fact that the hydrogen bond between the unlike components is a stronger interaction. This fact suggested that the entropic contribution to hydrogen bond formation could play a decisive part to determine the phase behaviour. The occurrence of hydrogen bond formation for all the blend compositions indicates that the interphase region is stabilised via hydrogen bonding and hence a high compatibility is observed.

Alder and co-workers [73] suggested that in order to use hydrogen bonding in the construction of supramolecular side-chain liquid crystalline polymers, in which the side-chain is noncovalently bonded to the backbone, the strength of the hydrogen bond should be increased. However, simply increasing it, a proton transfer can result. The authors recommended an alternative approach, using complementary units containing multiple binding sites, as it has been used to construct mesogenic units [74].

16.5 Rheology of the Blends Containing Liquid Crystalline Polymers

Liquid crystalline polymers offer to the blend technologist considerable possibilities for improving the processability as well as the mechanical and thermal properties of the polymer blends, by controlling their viscosity.

The factors influencing the viscosity of the blends containing liquid crystalline polymers are:

- the morphology (droplets, ellipsoids, fibres or layers) of the melt, determined by the previously applied processing;

- the processing history (previous processing, previous stress applied to the melt, or thermal history);

- the chemical structure of the components, the content in liquid crystalline polymer, as well as the temperature.

The processability improvement can be realised either by temperature reducing [75, 76] or/and by lessening the die swell [77, 78] or the mould shrinkage [79], using a suitable liquid crystalline polymer in corresponding proportions.

16.5.1 Experimental Data on the Blend Viscosity

Blizard and Baird [80] studied the viscosity reduction for the blends made of polycarbonate (PC) and a liquid crystalline copolyester 60-HBA-PET, containing 60 wt% HBA and 40 wt% PET. They measured, at 260 °C, a viscosity reduction from 1000 Pa-s corresponding to neat PC, to 20 Pa-s at 20 wt% liquid crystalline polymer. The increase over 20 wt% of the liquid crystalline polymer content in PC/60-HBA-PET blend determined a slow decreasing of the blend viscosity until 6 Pa-s, corresponding to the measured viscosity of the neat liquid crystalline copolyester.

Studying the viscosity of amorphous polyamides (PA) blended with liquid crystalline HBA-HNA, at 260 °C and different shear rates, Siegmann and co-workers [81] observed a steep viscosity reduction at small content of copolymer; the greatest reduction of viscosity being seen at 5 wt% (See **Figure 16.10**). The viscosity increased with the growth of liquid crystalline copolymer concentration, the greatest increasing has been reported for the PA/(HBA-HNA) blends of a high content in liquid crystalline copolymer. **Figure 16.10** illustrates the same behaviour of PA/HBA-HNA blend for four shear rates (54 s^{-1}; 135 s^{-1}; 540 s^{-1} and 2700 s^{-1}).

When the viscosity of liquid crystalline polymer (LCP) was almost the same of the pure PA, the PA/LCP blends exhibited a viscosity of an order of magnitude lower than those of the components. Examples of viscosity reduction can also be found in the review of Dutta and co-workers [75].

Viscosity increasing has been reported for some liquid crystalline polymer blends [32], such as for the blend containing 20 wt% of HBA-HNA liquid crystalline copolymer in polyethylene terephthalate-12 (PET12) that exhibited an increased viscosity in the 270-290 °C temperature range.

Weiss and co-workers [82] also reported a rise in viscosity of a blend obtained by mixing 4,4′-dihydroxydimethylbenzalazine liquid crystal with polystyrene, but only at low shear rates (of about 1 s^{-1}).

For the blends obtained from polysulfone Amoco's-3500 (PS) and a liquid crystalline polyester BASF's Ultrax-4002 (PE), Kulichikhin and co-workers [83] noticed the dependence of viscosity on the liquid crystalline polymer content, at three temperature values, as plotted in **Figure 16.11**. The measured viscosity at 280 °C, gave a linear fall with liquid crystalline polymer content throughout the composition range. At 260 °C, the fall was initially faster, but increased at higher liquid crystalline polymer concentrations, giving a minimum value at about 50 wt%. At 240 °C, two viscosity minima appeared near 30 wt% and 70 wt% content in liquid crystalline polymer.

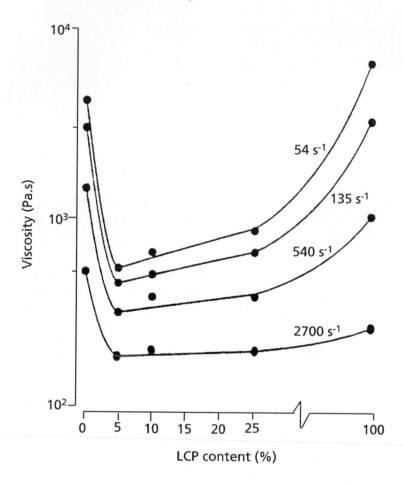

Figure 16.10 The PA/(HBA-HNA) blends viscosity *versus* the HBA-HNA content at 260 °C at four shear rates [81]

Reproduced with permission from A. Siegmann, A. Dagan and S. Kening, Polymer, 1985, 26, 1325. Copyright 1985, Elsevier

At the same temperature, different shear rates can induce increasing or decreasing of the blend viscosity, as it was shown by Nobile and co-workers [84] in the case of the BPPC/PET-OB blend containing bulk bisphenol-A polycarbonate (BPPC) and a liquid crystalline copolyester (PET-OB) with ethylene terephthalate (PET) and oxybenzoate (OB) units (see **Figure 16.12**).

At very low shear rate (10^{-2} s^{-1}), the viscosity was found to increase with liquid crystalline polymer content, at a shear rate of 0.17 s^{-1}, the viscosity seems to be a constant over all

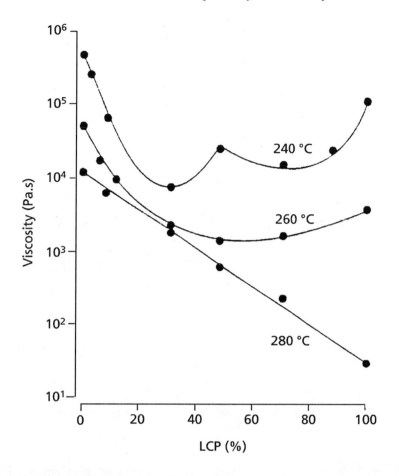

Figure 16.11 Viscosity *versus* composition for polysulfone blended with a liquid crystalline polyester (Ultrax). After Kulichikhin and co-workers [83]

Reproduced with permission from V.G. Kulichikhin, O.V. Vasil'eva, I.A. Litinov, E.M. Antopov, I.L. Parsamyan and N.A. Plate, Journal of Applied Polymer Science, 1991, 42, 363, Figure 2. Copyright 1991, Wiley

concentrations, whereas at higher shear rate (700 s⁻¹), a significant drop appeared (about a factor of 10% at 50 wt% liquid crystalline polymer content).

16.5.2 Theoretical Expressions for the Blend Viscosity

Theories have been developed to explain the behaviour of the viscosity of the blends containing a liquid crystalline polymer.

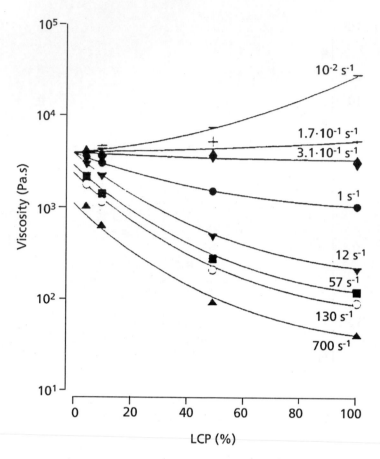

Figure 16.12 Viscosity *versus* PET-OB copolyester content in (bisphenol-A-PC)/(PET-OB) blend at 240 °C, at different shear rates [84]

Reproduced with permission from M.R. Nobile, E. Amendola, L. Nicolais, D. Acierno and C. Carfarna, Polymer Engineering Science, 1989, 29, 244, Figure 7. Copyright 1989, Society of Plastics Engineers

In the case of deformable droplets of one fluid in another, when no chemical interactions occur between the dispersed phases, the viscosity has been expressed by Schowalter [85], using the extended Einstein equation:

$$\eta = \eta_1 \left[1 + \left(5\eta_1 / \eta_2 + 2 \right) \left(2\eta_1 / \eta_2 + 1 \right) \varphi_2 \right] \tag{16.1}$$

where η_1, η_2 are the viscosities of the components 1 and 2 and φ_1, φ_2 are the relative volume fractions. From Equation (16.1) it results only an increase of the blend viscosity for all compositions and all viscosity ratios.

Taylor [86, 87] extended the concept of shear flow behaviour for the rigid spheres in Newtonian fluids, giving for the blends viscosity η_b, in the case of component 2 dispersed in component 1, the following equation:

$$\eta_b = \eta_1\left[1 + \left\{(\eta_1 + 2.5\eta_2)/(\eta_1 + \eta_2)\right\}\varphi_2\right] \tag{16.2}$$

Equation (16.2) was obtained when the droplets retain their spherical shape in the shear flow field (the dispersed phase forms uniform spherical droplets). The effect of the droplet size on the viscosity is not reflected in Equations (16.1) and (16.2).

Equations (16.1) and (16.2) do not explain the experimentally observed reduction in the blend viscosity at different liquid crystalline polymer content.

According to Chuang and Han [88], if no chemical interactions occur between the phases, the experimental viscosities would generally be lower than those theoretically predicted. They compared Taylor's prediction (16.2) with the observed shear viscosity of poly(amide imide) (PAI)/LCP system [89], at two shear rates (3.93 s^{-1} and 118.1 s^{-1}), both when PAI is taken as component 2 (predicted values though low, closely followed the experimental values) and when PAI is taken as component 1 (the predicted values were higher and are different from the observed values.

Heitmiller and co-workers [90] predicted a dispersion for the viscosity of mixtures based on 'inverse volume-weighted' rule and assuming concentric layers of component 1 in the component 2. For a large number of layers, the viscosity is given by:

$$\frac{1}{\eta} = \frac{\varphi_1}{\eta_1} + \frac{\varphi_2}{\eta_2} \tag{16.3}$$

with the same significance of the parameters as in Equations (16.1) and (16.2).

A similar relation has been established by Lees [91] for the viscosity of perfect cylindrical layers following through a circular die. The possibility of forming such layers with decreased viscosity at high liquid crystalline polymer concentrations is mentioned by Lee [77, 78].

Equation (16.3) can in fact explain large viscosity reduction with small addition levels, if $\eta_1 \gg \eta_2$. According with Equation (16.3), the blend viscosity, η_b, varies monotonically with the volume fractions φ_1 and φ_2. Equation (16.3) under-predicts the viscosity values, but the predicted values are closer to experimental ones at higher shear rates, than at lower shear rates.

Hashin [92] extended the model for the prediction of the upper and lower bound for the elastic modulus of a composite material and obtained a viscosity envelope for polymer blends.

In the case of a Newtonian fluid, the viscosity equation is:

a) upper bound:

$$\eta_b = \eta_2 + \left[\varphi_1 / \left\{ 1/(\eta_1 - \eta_2) + \frac{2}{5} \frac{\varphi_2}{\eta_2} \right\} \right]$$

b) lower bound:

$$\eta_b = \eta_1 + \left[\varphi_2 / \left\{ 1/(\eta_2 - \eta_1) + \frac{2}{5} \frac{\varphi_1}{\eta_1} \right\} \right]$$

For non-Newtonian fluids the above equations become:

a) upper bound:

$$\eta_b = \eta_2 + \left[\varphi_1 / \left\{ 1/(\eta_1 - \eta_2) + \frac{1}{2} \frac{\varphi_2}{\eta_2} \right\} \right]$$

b) lower bound:

$$\eta_b = \eta_1 + \left[\varphi_2 / \left\{ 1/(\eta_2 - \eta_1) + \frac{1}{2} \frac{\varphi_1}{\eta_1} \right\} \right]$$

The experimental values for the viscosity of the PAI/LCP blends as function of PAI concentration superpose above the predicted results according to the Hashin model for two shear rates, all the values being closed to the lower bound of the envelope with the higher shear rate [89].

De Meuse and Jaffe [93] applied a linear dependence to express natural logarithm of the viscosity for miscible blends containing two liquid crystalline polymers of different composition, using the following equation:

$$\ln \eta = \varphi_1 \ln \eta_1 + \varphi_2 \ln \eta_2 \tag{16.4}$$

The reduced viscosity of the bulk polymer blended with a liquid crystalline polymer characterised by a low viscosity, could qualitatively be explained by involving polymer blend theory developed by treating the two fluids as isotropic fluids approaching Lees layers [91]. However the blend viscosity which is much smaller than the viscosities of its both components, can not be explained by the traditional theories discussed previously.

The modifications produced in the blend morphology by various phenomena (concentration change, flow, heating or cooling) as well as the flowing type must be known, in order to explain the different kind of the viscosity behaviour.

When a blend melt containing droplets begins to flow several events occur. The droplet can be deformed into a fibre, the fibres could break into droplets again, as described Utracki [94], or droplets could coalesce. It is even possible that fibres coalesce to form platelets. These phenomena could influence the viscosity behaviour.

During the flow, a continuous velocity profiles within a liquid crystalline blend are more probable, but, as it has been shown by Kulichikhin and co-workers [83], these profiles might be highly discontinuous.

The problem is complicated by the shear thinning shown by liquid crystalline molecules, that become oriented in the flow direction and hence the decrease of the viscosity occurs. When the applied shear is realised, the molecules will remain oriented in the low viscosity state. This shear thinning memory effect may well be accentuated in a blend, self perpetuating the viscosity reducing.

Beery and co-workers [95] and Berry [96] pointed out that the elongational viscosity of liquid crystalline polymers and their blends, that can be 100 to 400 times greater than their shear viscosity, was not systematically studied. Most practical polymer processing involves a considerable elongational component to the flow, e.g., at the entry in an extruder die. Berry concluded that elongational flow, rather than the shear component dominates the die entry pressure drop. Thus, the measured shear viscosities could be in fact the result of differences in elongational viscosity. The importance of elongational flow producing a desirable morphology was also noted by Dutta and co-workers [75]. They considered that the extensional deformation of the liquid crystalline polymer domains, during the flow in the entrance region of the viscosimeter, could explain the decreasing of the viscosity at high shear rates.

16.5.3 Model Describing Rheological Behaviour of Immiscible Blends

The mechanical properties of the blends depend strongly on the morphological characteristics of the dispersed phase. Adhesion between the two phases strongly influences the mechanical properties of the blends. The rheology of the phase separated fluids with complex interfaces, created by shearing of two immiscible fluids, was studied from point of view of the modifications in the area and the orientation of the dispersed phase. A review is given in [96].

A model has been developed [97, 98] to describe the rheological behaviour of a liquid in which the dispersed phase changes its area as well as its orientation under the flow. The system considered comprised two isotropic, non viscoelastic components, with equal viscosities. The flow field enlarges and orients the interface between the phases, driving the minor phase from an otherwise spherical shape, whereas the interfacial tension opposes to these tendencies. Under these conditions, the stress tensor s and the total interfacial area per unit volume Q, for a flow with velocity gradient tensor k(t), are related to the same properties for a flow with velocity gradient ck(ct) by the expressions:

$$\sigma\big[t, ck(ct)\big] = c\sigma\big[ct, k(ct)\big] \qquad (16.5)$$

$$Q\big[t, ck(ct)\big] = cQ\big[ct, k(ct)\big] \qquad (16.6)$$

where c is a constant.

Thus, the stress tensor and the interfacial area at the time t under a velocity gradient ck(ct) are c times higher than their values at the moment ct, under a velocity gradient k(ct). From equation (16.5) for a velocity gradient independent on time, the steady-state stress is given by:

$$\sigma(ck) = c\sigma(k) \qquad (16.7)$$

Under a steady shearing flow, the transient stress growth is given by:

$$\sigma(t, ck) = c\sigma(ct, k) \qquad (16.8)$$

Rheological studies [99, 100] on immiscible blends of two flexible chain polymers gave results in accordance with the equation mentioned above, showing that the ratio of the transient to steady-state stress depends only on kt, and not on the individual values t and k.

In this model, the rheological behaviour is traced to the dependence on k of the structural feature and the consequent lack of characteristic time or length scales. For example, the dependence on k of the size and shape of the disperse phase in an immiscible blend can be described using the results of this model. Rheological models similar to that predicted for immiscible blend of isotropic fluids might be applied to an immiscible liquid crystalline phase dispersed into an isotropic matrix, with the flow tending to elongate and orient the liquid crystalline phase. But, an extensional flow that orients the dispersed phase more effectively does not verify the relationships in Equations (16.5)-(16.8). The deviations [97, 98] could be explained by the component viscoelastic behaviour, neglected in this model.

16.5.4 Factors Influencing Rheological Behaviour

The rheological behaviour of a polymer mixture is correlated to thermodynamic state and structural features of the melt or of the solution. This assertion can be illustrated by the results from [56] regarding the rheological behaviour of mixtures made of two immiscible, semirigid, thermotropic and lyotropic, showing cholesteric anisotropic phases, liquid crystalline polymers hydroxypropyl cellulose (HPC) and ethyl cellulose (EC) in acetic acid solutions.

At the room temperature, the ternary phase diagram exhibits large biphasic regions with complex zones of: isotropic-isotropic; isotropic-anisotropic and anisotropic-anisotropic phase equilibrium [103, 104].

Large and complex biphasic regions as a function of temperature for a polymer concentration of 40 wt% in acetic acid are seen in the phase diagram of HPC/EC/acetic acid system. Optical polarising microscopy also showed anisotropic, isotropic and isotropic-anisotropic zones, changing their aspects depending on the content of two polymers in HPC/EC/acetic acid blends.

Figure 16.13 illustrates the viscosity dependence on the shear rate in the case of HPC/EC/acetic acid mixture with a total polymer concentration of 40 wt% in acetic acid.

Depending on HPC content in HPC/EC polymer, three morphological types were seen [56] in HPC/EC/acetic acid solutions:

- A monophase anisotropic structure for HPC/EC with high content in HPC (100-90 wt%);

- An anisotropic-isotropic biphasic structure for the intermediate concentrations of HPC (90-10 wt%). Rheological measurements showed decreasing deviations from the additive rule with the shear rate increasing.

- An isotropic monophase at low content (10-0 wt%) of HPC in HPC/EC polymer. In this region a positive excess viscosity, compared with the value predicted by the additivity rule, has been measured.

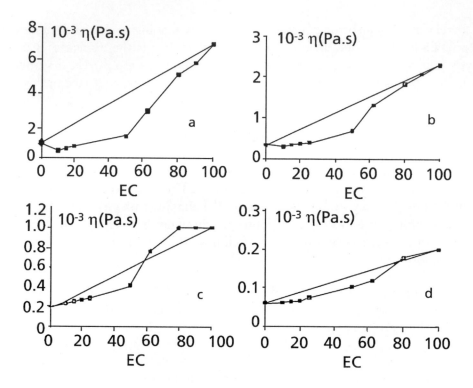

Figure 16.13 Viscosity as function of EC content in HPC/EC mixture (total polymer concentration 40 wt% in acetic acid) for various shear rates: a) 1.25×10^{-2}; b) 3.95×10^{-2}; c) 1.25×10^{-1}; d) 12.5 s^{-1} [56]

Reproduced with permission from S. Ambrosino and P. Sixou, Journal of Polymer Science, Part B, 1994, 32, 1, 77. Copyright 1994, John Wiley & Sons

The decrease of viscosity from the predicted values by the additivity rule has been explained by phase separation, while the increased viscosity values, with very small deviations from the additive rule, characterised the homogeneous phase. Near the isotropic-anisotropic transitions, the solutions exhibited an anisotropy induced by shear.

Lee and co-workers [105] studied the phase behaviour and rheology of binary blends of PC and a semiflexible liquid crystalline polyester (LCPE) having the structure from **Scheme 16.4**. A distinctive nematic- to isotropic-transition in the pure polymers and in LCPE/PC blends has been seen.

The phase diagram of LCPE/PC binary blends, obtained by DSC and optical polarising microscopy, exhibited isotropic (I), nematic (N) and glass (G) phases, depending on the temperature and on the LCPE content in the blends.

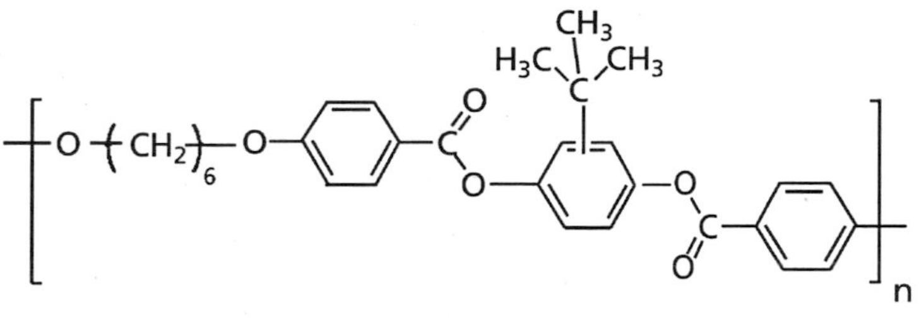

Scheme 16.4 Liquid crystalline polyester (LCPE) studied in [105] by Lee and co-workers

Dynamic oscillatory measurements show that there are some interactions between the separate isotropic and anisotropic phases. The complex viscosity of the blends is between those of pure components and show a significant deviation from the logarithmic rule of mixtures expressed by Equation (16.4). In **Figure 16.14a** and **16.14b**, the storage and loss moduli versus frequency for various blend compositions at 180 °C and 200 °C respectively, are plotted.

From **Figure 16.14** it results that the loss moduli $G''(\omega)$ in the pure liquid crystalline polymer are about two to five times higher than the storage moduli, $G'(\omega)$ over the entire frequency range. In the case of the LCPE/PC blends, $G''(\omega)$ was found depending on the frequency as, $G''(\omega) \approx \omega^{0.8}$, while the storage modulus $G'(\omega)$ changed slope over the entire frequency range as well as in other thermotropic melts [106, 107].

The blend with a weight fraction, ω_{LCP}, (of about 0.1) exhibits values for $G'(\omega)$ and $G''(\omega)$ lower than those in the pure LCPE. The cross point of $G'(\omega)$ and $G''(\omega)$ shifts to a higher frequency in comparison to that for neat PC, indicating a decrease in the longest relaxation time of the blend. When the weight fraction of LCPE exceeds 0.5, moduli decrease monotonically with increasing ω_{LCP}, but do not follow a simple mixing rule, suggesting an influence of the interfacial tension on the rheological properties. At 200 °C (**Figure 16.14b**) as the LCPE concentration increases, the slope of $G''(\omega)$ becomes smaller than in the pure PC, but still larger than in pure LCPE, proving the partial miscibility of the two components.

Figure 16.15 shows a plot of complex viscosity $\eta^*(\omega)$, $\omega=1$ rad/s, as a function of temperature, during cooling between 265-180 °C with a rate of cooling of 0.1 °C/min. For pure PC, the viscosity increased with the decreasing temperature, in a similar manner as for the most liquids. The LCPE viscosity begins to fall with decreasing temperature near 240 °C, corresponding to temperature of the isotropic to nematic phase transition. The viscosity continues to decline until about 220 °C, then the phase transition is complete

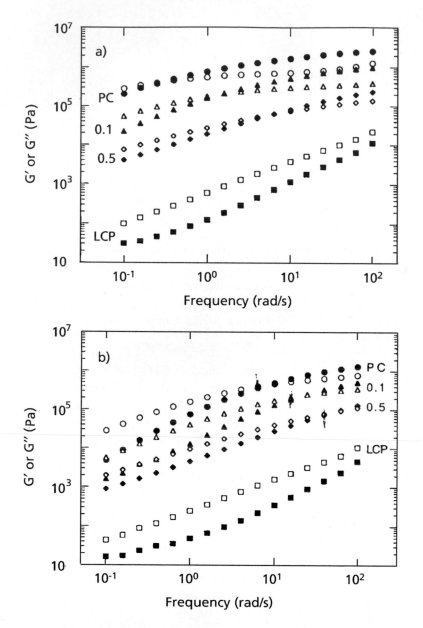

Figure 16.14 Storage and loss moduli at the various composition of LCPE/PC blends measured in biphasic region at (a) 180 °C and (b) 200 °C. Numbers of the figure denote the LCPE weight fraction. Black and white symbols denote the G′ and G″, respectively, at each composition [105]

Reproduced with permission from S. Lee, P.T. Mather and D.S. Pearson, Journal of Applied Polymer Science, 1996, 58, 2, 243, Figure 6. Copyright 1996, Wiley

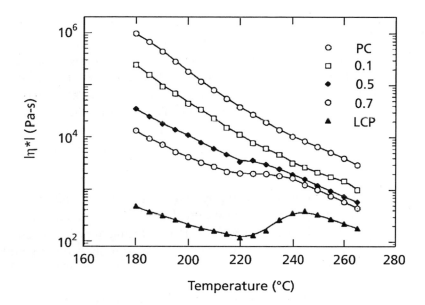

Figure 16.15 Dynamic viscosity at the various blend composition measured at 1 rad/s during cooling after annealing 20 minutes at 270 °C. Numbers on the figure denote the LCPE weight fraction [105]

Reproduced with permission from S. Lee, P.T. Mather and D.S. Pearson, Journal of Applied Polymer Science, 1996, 58, 2, 243, Figure 7. Copyright Wiley

and the viscosity begins to rise with decreasing temperature. The finite temperature range over which viscosity decreases during cooling is due to the polydispersity of the liquid crystalline polymer.

In the case of LCPE with a weight fraction is about 0.7, a similar, but less dramatic viscosity change was observed. This experimental fact was in accordance with the optical polarising microscope observations which showed that, at cooling, the isotropic-nematic phase separation occurs in the blends, below the isotropic-nematic transition temperature of the pure liquid crystalline polymer. With temperature decreasing, the reduction in viscosity of the LCPE-rich phase is compensated by an increasing viscosity of the PC-rich phase. In these conditions, the viscosity of LCPE/PC blends appears to remain unchanged for a certain range of temperatures.

This behaviour is diminished as the LCPE content decreases, since the viscosity of the PC-rich phase dominates at low LCPE composition. So, the temperature dependence of viscosity for the weight fraction of about 0.1 LCPE in the LCPE/PC blend is similar to that of pure PC, but shows considerably reduced values.

The complex viscosity as a function of the blend composition, at 180 °C, showed a continuous decrease with the addition of LCPE. The values are between those corresponding to the pure components. For small LCPE compositions, the viscosity gradually decreases with increasing weight fraction of LCPE. It seems that LCPE molecules dissolved in the PC-rich phase within the miscible composition range act to plasticise the blends. The larger LCPE compositions, for which phase separation occurs, show a smaller negative slope of the viscosity-composition curves. Finally, a large drop in viscosity at high LCPE compositions (bigger than 0.8 wt%) was observed for all oscillation frequencies.

Based on a vast material concerning the implications of rheology modification caused by the nature of liquid crystalline polymer used in the blend obtaining, some processing advantages arose, such as processing improvement; or reducing of processing temperature. Concomitantly, die swell or mould shrinkage can be less pronounced if a suitable liquid crystalline polymer is chosen.

Lee [77, 78] has injection moulded a blend of chlorinated poly(vinyl) alcohol (PVC) and HBA-PET as liquid crystalline copolymer. He found that the liquid crystalline polymer increased the spiral mould flow length and was able to model the improvement in injection moulding processing using a simple power law. He also noted that extrudate die swell was reduced by addition of the liquid crystalline polymer.

In order to establish optimum gas barrier performance from injection moulded parts, Toshikazu [108] investigated the liquid crystalline polymer/ethylene vinyl alcohol copolymer blend in a 75 ton injection machine with a 20 cm diameter extruder, observing that screw torque and melt pressure decreased in all blend systems and established a relationship between morphology, viscosity and surface tension.

Dutta and co-workers [75] have also discussed the implications of rheology modifications arising from liquid crystalline polymers incorporation. They highlighted the benefit of reduced melt temperature which may imply reduced energy costs or less degradation.

16.6 Liquid Crystalline Polymers as Reinforcements

The rod-like molecular conformation and chain stiffness give their 'self-reinforcing' properties to the liquid crystalline polymers.

The thermotropic liquid crystals have become available because they can be produced in the melt to give highly orientated structures that are largely retained on cooling.

The good, but anisotropic, mechanical properties of the relatively expensive liquid crystalline polymers recommend their use to reinforce the less expensive, isotropic, but mechanically weaker engineering plastics, such as polyethylene, polypropylene, PS, PC or polyether sulfones.

The most common reinforcing commercial liquid crystalline polymers are Vectra, Econol, Rodrun, Sumicosuper. Composites containing Kevlar are examples of liquid crystalline blends in commercial use. Kevlar itself is a liquid crystalline polymer and a solution spun in its lyotropic state. Another available thermotropic melt spun, manufactured by Celanese, the heat treated Vectra [109] have a tensile strength comparable with that of Kevlar 49.

The advantages of the liquid crystalline polymers used in reinforcement are:

- the high *para*-linked aromatic content gives samples with good moduli;

- the very low melt viscosity allows good flow properties and processability;

- the high chemical resistance;

- the low mould shrinkage determining accurate mouldings;

- they readily form fibrous structures, for example injection moulded samples of a neat liquid crystalline polymers resins having an appearance similar to wood.

Because of the incompatibility of isotropic and liquid crystalline materials, the reinforcing is difficult to realise with commercial engineering plastics and liquid crystalline polymers. The molecular reinforcement needs a perfect molecular dispersion of the liquid crystalline polymer in the matrix [110, 111].

Another approach close to the ideal molecular reinforcement, despite a total incompatibility of matrix and liquid crystalline polymer, has been described by Taesler and co-workers [112, 113]. They obtained 'lyotropic blends' based on the coprecipitation of a substituted liquid crystalline polyester with various matrix polymers.

The generation of reinforcing species *in situ* offers advantages over the addition of solid fibres and fillers. The use of techniques such as *in situ* crystallisation and polymerisation have been reviewed by Kiss [114].

The mechanical properties of *in situ* composites, were listed by Crevecoeur and Groeninckx [115].

16.6.1 Reinforcing Action of Hydroxybenzoic Acid Based Liquid Crystalline Polymer Blends

Some results in the reinforcement of the blends using liquid crystalline polymers will be presented next.

Siegmann and co-workers [81] firstly reported *in situ* composite formation in the systems of amorphous polyamide and Vectra with improvements in processing and reinforcement as well.

A fibrillar structure was observed, by scanning electron microscopy (SEM) characterisation of the etched samples, when the liquid crystalline polymer concentration exceeded 25 wt%. For the liquid crystalline polymer concentration smaller than 25 wt%, some improvement in elastic modulus and tensile strength have been reported, but ultimate elongation sharply decreased as it is shown in **Figure 16.16**.

Chung [116] obtained an extruded blend displaying an anomalous skin-core structure from polyamide-12 (PA-12) and (HNA) - HBA Vectra type liquid crystalline polymers. Improvements in mechanical properties were observed up to 80 wt% HNA-HBA concentrations in PA/(HNA-HBA) blend, when phase segregation occurred. In the case of 1:1 PA/(HNA-HBA) blend, the coefficient of thermal expansion was found to be null.

Kiss [117] investigated the effect of mixing Vectra A or Vectra B grades with wide variety of amorphous and crystalline polymers, including polyether sulfones, polyetherimide (PEI), polyarylate (PAr), polyacetal (PAc), polyamide-6 (PA-6), polybutylene terephthalate (PBT), PC, poly(chlorotrifluoroethylene) (PCFE) and poly(etheretherketone) (PEEK). The blends were obtained by compounding techniques, by extrusion into strands, or by injection moulding. Differences in adhesion of the liquid crystalline polymer phase in the various systems determined the morphology of the studied blends and the extension of the elongated liquid crystalline domains resulting from extended droplets to fibres [117].

Dramatic increases in the tensile and flexural moduli (two to three times), with smaller increases in the flexural and tensile strength, accompanied by a substantial decrease in elongation to break were observed for all blends studied by Kiss.

Swaminathan and Isayev [118] showed that the extent of property improvement is dependent on the method of compounding of PEI/Vectra A959 blends. They compared the PEI/Vectra A959 blends obtained by using a static mixer and a co-rotating twin-screw extruder as well. The blends obtained at static mixer had a higher degree of fibrillation and corresponding improvements in Young's modulus over those produced using the twin-screw extruder.

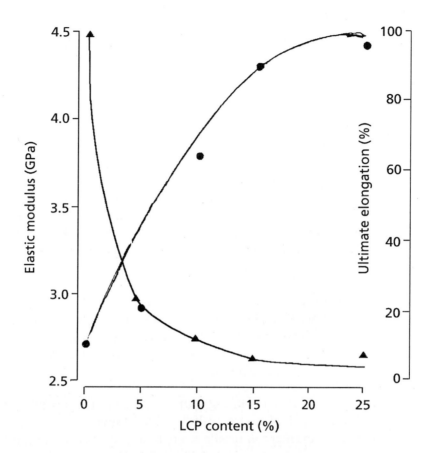

Figure 16.16 Elastic modulus and ultimate elongation of injection moulded LCP/PA blends *versus* LCP content [81]

Reproduced with permission from A. Siegmann, A. Dagan and S. Kening, Polymer, 1985, 26, 1325. Copyright 1985, Elsevier Science

The interphase adhesion between the PEI and liquid crystalline polymer was found to be poor, skin-core morphology being in injection moulded specimens. SEM of PEI/Vectra A959 blends showed droplets or elongated structures of the liquid crystalline phase at concentrations below 10 wt% and only fibrils (typically 2-5 μm diameter) at concentrations of the liquid crystalline polymer greater than 20 wt%.

The PEI/Vectra A959 extrudates from a capillary die with shorter length:diameter (L/D) ratios had the modulus and strength higher than those obtained with longer L/D ratios.

Blizard and Baird [80] studied the extent of fibrillation in the PA-6/(PET-HBA;40:60) blends prepared by extrusion through capillary dies of various L/D ratios and immediate quenching. They obtained blends with higher mechanical properties in which fibrils appeared when the liquid crystalline concentration exceeded 30 wt%.

The influence of L/D ratio on the fibril formation was studied using a capillary die with a short band [80]. Capillary rheometry showed that fibrils are readily produced with a die of L/D = 7.82, but not with one having L/D = 21.4.

La Mantia and co-workers [119] obtained similar results for the (PA-6)/Vectra B950 blends. SEM analyses of a 20 wt% Vectra B950 blend, extruded at a shear rate of 1200 s^{-1} through a capillary die (L/D = 40), showed no fibrillation, but when the blend passed, under identical conditions, through a die having L/D = 0 and conical inlet, fibrils were produced. These results showed that the fibrils can be lost during the flow in a long capillary, when the orientational relaxation time of the liquid crystalline polymer is less than the time necessary to flow through the capillary.

The importance of extensional flow on fibril formation for the PC/(PET-HBA) blends has been underlined by Jung and Kim [120]. SEM analyses for strands of blends containing 30 wt% liquid crystalline polymer produced through capillary rheometry, showed the lack of fibrillation, although a 5 wt% liquid crystalline polymer blend, having a draw ratio of 15, gave an extruded fibril morphology.

For the PC/Vectra A900 and PC/(PET-HBA, 40:60) blends, as well as for PPS/(PET-HBA 40:60 or 20:80) blend having polyphenylene sulfide (PPS), Ramanathan and co-workers [121, 122] reported an improvement in tensile strength and modulus, but did find that the fibril formation for all cases, except the 1:1 PPS/(PET-HBA) for both ratios of PET-HBA (40:60 or 20:80) extruded with a draw ratio of up to 9. The lack of fibrillation in the case of Vectra A900/(PET-HBA) blend may be explained by the higher temperature of processing (260 °C) at which the viscosity of liquid crystalline polymer component is likely to be higher than those of PC. For the PPS/Vectra A900 blend, the lack of fibrillation was attributed to a possible chemical reaction between the two components under processing conditions.

Lee and Dibenedetto [123] used the chemical reactions to improve the adhesion between incompatible aromatic fibres and thermoplastic matrix. In composites LCP/LCP such as KU-9211(BayerA6)/(PET-HBA; 40:60), KU-9211 acted as reinforcing phase. Extruded and drawn samples properties were influenced by chemical interactions between the two thermotropic liquid crystalline polymers.

The PET-HBA (40:60) also bonds well with conventional thermoplastics [124, 125], such as PET.

Zhang and co-workers [126] studied the correlation between the miscibility of the blend components, the liquid crystalline polymer concentration and the morphology for the PET-HBA (40:60) with PS, PC and PET blends. By DSC, SEM and dielectric thermal analysis methods, they demonstrated the dependence of the blend morphology on the component miscibility. So, liquid crystalline polymer was immiscible with PS and gave the most coarse and defined morphology, while for the most miscible system PET/(PET-HBA; 40:60) the obtained blend had a less distinctive, but the finest morphology. Transition from ellipsoidal to fibrous liquid crystalline polymer domains with the increasing shear rate in capillary extrudates was also show [126].

Compatibility between PET and a LCP, such as PET-HBA (40:60), was investigated by Brostow and co-workers [127]. In the case of the PET/LCP blend with high content of liquid crystalline polymer, they observed a limited degree of solubility of LCP in PET. Adding a small amount of LCP (2-5 wt%) the observed mechanical properties decreased, then increased to maximum and fall again as the LCP content increased.

Friedrich and co-workers [128] studied the blends PC/(PET-HBA) that usually had two values for T_g (the lower associated with the PET-rich phase of liquid crystalline polymer and the higher with HBA-rich phase), indicating the immiscibility. Concomitantly, DSC measurements gave a single T_g value on annealing of the PC/(PET-HBA) blend, interpreted as a partial miscibility through transesterification.

For the same system, PC/(PET-HBA; 40:60), Jung and Kim [120] reported that T_g of PC decreased by a small addition of PET-HBA (40:60) although the glass transitions of the liquid crystalline polymer remain unchanged. They interpreted this fact as a possible exclusion of the PC from liquid crystalline polymer phase, while some partial mixing in the PC phase occurred. Good interface adhesion was also observed between the two components in PC/(PET-HBA; 40:60) blends [104].

Changes in crystallisation behaviour have been reported in some systems by Minkova and co-workers [129] that found an increase in temperature of non-isothermal crystallisation by adding Vectra B to PPS; the nucleation density of the blend increased independently on the concentration.

Shin and Chung [130] showed that PET was nucleated by an unusual liquid crystalline polymer [melting temperature (T_m) = 272 °C] containing long methylene spacer units.

While the blend components were found to be immiscible, the strong interface adhesion between liquid crystalline fibrils and PET determined a good fibre reinforcement.

16.6.2 Reinforcement by Rigid Rod Polyester Blends

An important finding of Kricheldorf and co-workers [131] was the formation lyotropic blends from rigid-rod polyesters (RRP), which is not liquid crystalline as a neat materials, and commercial poly(ε-caprolactone) (PCL), acting as solvent. The RRP studied in [131] were those from **Scheme 16.5**.

Despite the non-symmetrical substitution patterns, most polyesters, such as RRP derived from monosubstituted terephthalic acid [132], are semicrystalline materials. The glass transition temperatures of RRP are so low, that drying of the freshly precipitated samples at 115 °C has an annealing effect.

The melts of RRP showed a Schlieren texture and mobility typical for a nematic liquid crystal at optical polarising microscopy. Only the RRP derived from phenylhydroquinone do not form a liquid crystalline melt because their very low melting temperatures.

In order to obtain a mono-molecular 'solution' [131] of the RRP in a matrix of PCL or at least a blend close to this ideal case, both the PCL and the RRP were dissolved in a co-solvent, precipitated in methanol, and dried at 60 °C. Neat CH_2Cl_2, or CH_2Cl_2 with small amounts of trifluoroacetic acid (TFA) were used as co-solvent. Systematic mechanical measurements for blends of 8 wt% RRP-2a in PCL, obtained from either CH_2Cl_2 or CH_2Cl_2 with small amounts of TFA, showed that TFA did not influence the blend properties as can be seen in **Table 16.5**.

Table 16.5 Influence of trifluoroacetic acid (TFA) in the solvent mixture on the properties of the RRP-2a/PCL blend with 8 wt% of RRP-2a				
No.	Vol.% of TFA	E-modulus[a] (MPa)	Max. stress[a] (MPa)	η_{inh}[b] (dL/g)
1	0	780	15.5	1.14
2	2	810	16.7	1.11
3	5	820	17.0	1.0
4	10	785	15.7	0.96
5	20	920	16.5	0.94
[a] *The given values are averaged over five measurements. Measured from the neat matrix at 20 °C with c = 2 g/L in CH_2Cl_2/TFA (volume ratio 4:1)*				

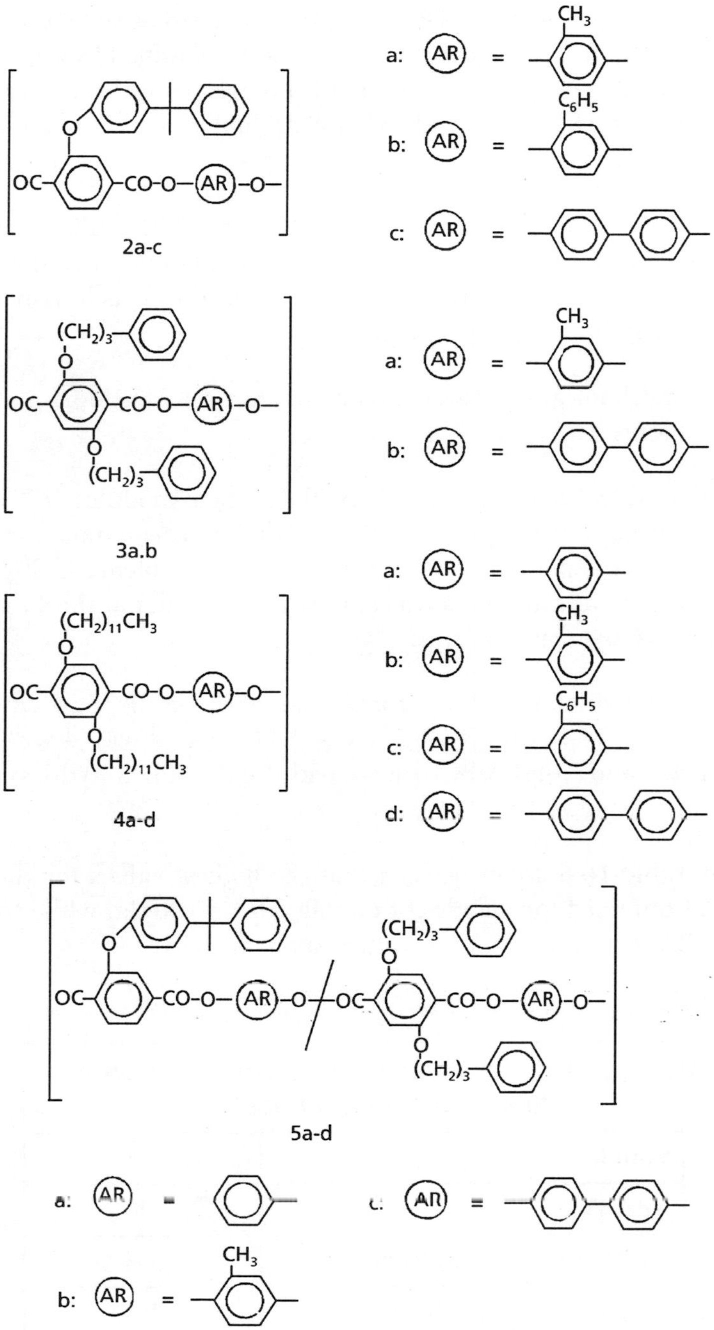

Scheme 16.5 Rigid-rod polymers (RRP) used by Kricheldorf and co-workers [131] to obtain RRP/LCP blends

The lyotropic blends of 2, 3, 4 and 5 RRP in PCL (PCL acted as solvent) that could not be achieved by blending in an extruder or kneader, were obtained by coprecipitation of both components from the co-solvent. A significant mechanical reinforcement effect, was shown in the case of RRP/PCL blends when RRP had aromatic substituents, such as 2c, 3b, 4d and 5c.

Mechanical measurements of the blends of 8 wt% 1, 2, or 4 RRP in PCL were conducted in two ways: by dynamical mechanical analyses (DMA) of hot pressed films, achieved at a frequency of 1 Hz in the temperature domain [–110, 40]°C such as by stress-strain measurements of doggy-bone type specimens.

The results of the mechanical measurements achieved by Kricheldorf and co-workers [131] are given in **Figure 16.17**.

The elastic modulus (E) reflecting the elasticity, the storage modulus (E') reflecting the transformation of mechanical energy into heat, and their quotient, tan δ, were recorded. **Figure 16.17a** displays the measurement results for RRP-2a blend. Compared to neat PCL (**Figure 16.17b**), the loss modulus was almost unchanged, but the storage modulus showed a 80% gain at low temperatures.

The normalised storage moduli E'/E'_{PLC} for the samples: neat PCL; 4 wt% in RRP in RRP/PCL blends with RRP-2a and RRP-1; PMMA/PCL blend with 4 wt% polymethyl methacrylate (PMMA) and P3-HBA/PCL blend, with 4 wt% poly(3-hydroxybenzoic acid) (P3-HBA) in PCL are listed in **Table 16.6**.

From the data in **Table 16.6** it can be seen that the highest values for the normalised storage moduli are obtained for blends containing RRP 1 and 2a, while for those with PMMA and P3-HBA, the corresponding values are near unity.

Table 16.6 Normalised storage moduli of 4 wt% RRPs in RRP/PCL blends		
No.	Sample	E'_B / E'_{PCL}
1	Neat PCL	1
2	4 wt% RRP-2a/PCL blend	1.6
3	4 wt% RRP-1/PCL blend	1.8
4	4 wt% PMMA/PCL blend	0.8
	4 wt% P-3HBA/PCL blend	1.02

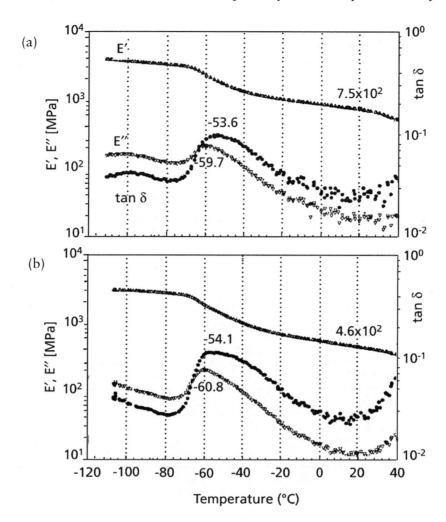

Figure 16.17 DMA measurement of a blend containing 4 wt% of RRP 2a. Neat PCL measured under identical conditions gave a storage modulus (E') of 4.6 x 10^2 MPa (E'' = loss modulus) [131]

Reproduced with permission from H.R. Kricheldorf, L.H. Wahlen, C. Friedrich and T.J. Menke, Macromolecules, 1997, 30, 9, 2642. Copyright 1997, American Chemical Society

RRP/PCL blends, containing increasing concentrations of RRP, have higher normalised elastic moduli values, showing that the stiffening effect of the RRP is additive for concentrations below 10%. From the data in **Table 16.7**, it also shows that the reinforcing effect of 2b is nearly as high as that of 2a when both polyesters are compared on the basis of equimolar amounts of the repeating units.

The normalised elastic moduli (stress-chain measurements) for some RRP/PCL blends of 8 wt% RRP in PCL are listed in **Table 16.8**. The highest values from this table correspond to the blends containing the -5c, -2a, and –1 RRP.

Between the 8 wt% RRP/PCL blends with RRP derived from methylhydroquinone (-2a, -1, -3a, -5a, -4b), the most effective from point of view of normalised elastic are those containing 2a, -1 and –3a RRP (see **Table 16.8**).

All RRP-4/PCL blends have normalised elastic moduli near the unity, excepting RRP-4c/PCL blend that exhibited a value below the unity.

In the 8 wt% RRP-5/PCL blends, the high normalised elastic moduli are induced by a high rigidity of the repeating units (see **Scheme 5** and **Table 16.8**). The highest value of the normalised elastic moduli, of about 2.40, has been obtained for RRP-5c/PCL blend.

Table 16.7 Normalised elastic moduli (stress-strain measurements) *versus* RRP concentration in RRPs/PCL blends			
EB/E_{PCL}			
RRP-2a/PCL		RRP-2b/PCL	
Neat PCL	1	Neat PCL	1
1 wt%	1.3	1 wt%	1.2
2 wt%	1.6	2 wt%	1.3
4 wt%	1.52	4 wt%	1.4
8 wt%	1.92	8 wt%	1.6

Table 16.8 Normalised elastic moduli (stress-strain measurements) for the blends RRP/PCL, containing 8 wt% RRP					
No.	RRP/PCL Blend	EB/E_{PCL}	No.	RRP/PCL Blend	EB/E_{PCL}
1	Neat PCL	1	6	RRP-4b	1.05
2	RRP-1	1.90	7	RRP-4c	0.95
3	RRP-2a	1.98	8	RRP-5a	1.79
4	RRP-3a	1.80	9	RRP-5b	1.45
5	RRP-4a	1.12	10	RRP-5c	2.40

The reinforcing effect of some RRP in blends with PCL was explained by electronic interactions between the aromatic π-electrons and the polar ester groups of the PCL.

16.6.2.1 Rheology

From the rheological measurements it resulted that the time sweeps at a frequency of 1 Hz generally show a slow increase of both G' and G'' and reached constant values after 2 hours. The increase of G' for 1 wt% RRP in PCL, at 80 °C was 30% from the initial value. The comparable long relaxation times of RRP/PCL blends could even indicate that the lyotropic blends posses a distinct anisotropy, characteristic to a nematic liquid crystalline phase. As it can be seen in **Figure 16.18**, the 1 wt% RRP-4b/PCL blend has a viscoelastic behaviour, with G'' closed to that of the neat PCL. At lower frequencies (w_{at} <1 Hz), the storage modulus (G'') considerably deviates from that of PCL, reaching a secondary plateau at around 100 Pa.

Inducing morphological changes, the shear significantly influences the properties of these materials. Shearing of the samples at elevated temperatures (time sweep for 2 h at 180 °C, 1 Hz and 10% strain) prior to rheological investigation leads to significant changes in the rheological response of the blends.

16.6.2.2 Morphology

Connections between the transition temperature and the breakdown of orientation and thus, the reinforcement in the RRP/PCL blends are shown in [131]. PCL is a semicrystalline polymer above its melting point T_m, but its blends containing 1 wt% of a RRP showed birefringence only in 15-25% of the total volume, while the concentrations of 4 wt% or 8 wt% of RRP in RRP/PCL blends induced birefringence in the entire volume. The birefringent phase can be sheared and oriented above T_m of the PCL and thus, it must be considered as a mobile anisotropic phase [131].

The homogeneous birefringence is stable at temperature below 150 °C for many hours (its final disappearance has never been observed), but slowly vanishes when the temperature approaches the T_m of the RRP. Above T_m of the neat RRP the Schlieren texture of the neat RRP may become detectable.

It results that an initially homogeneous finely dispersed phase of RRP in PCL, undergoes an irreversible phase separation at temperatures around the T_m of RRP.

For all the blends the phase separation was irreversible if the cooling rate was very high or extremely slow (10 °C/min). The existence of a rather stable mobile birefringent

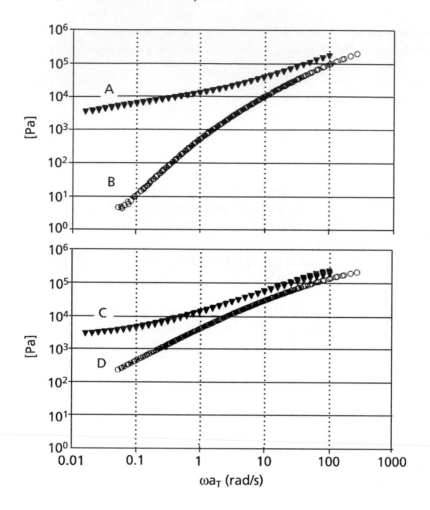

Figure 16.18 Rheological measurements (80 °C) of the neat PCL and the blends containing 1 wt% of 2a: (A) storage modulus of the blend; (B) storage modulus of PCL; (C) loss modulus of the blend; (D) loss modulus of PCL [131]

Reproduced with permission from H.R. Kricheldorf, L.H. Wahlen, C. Friedrich and T.J. Menke, Macromolecules, 1997, 30, 9, 2642. Copyright 1997, American Chemical Society

phase below the T_m of RRP but above T_m of the polymer matrix, is characteristic of these blends.

The formation of the anisotropic phase is dependent on the electronic interactions between π-electrons and the polar ester groups of the PCL. Only the blends containing aromatic substituents show a significant reinforcing effect.

Kircheldorf and co-workers [131] proposed a schematic representation of the order in RRP/PCL blends showing double stacks (**Figure 16.19A**) when the terephthalic acid is monosubstituted or monostacks (**Figure 16.19B**), when the terephthalic acid is disubstituted.

The coprecipitation of PCL with an RRP having more or fewer aromatic substituents yields a blend with a nearly molecular dispersion of the RRP. The individual RRP chains of the stacks of few chains are solvated by PCL chains which are more or less aligned parallel along the RRP

The anisotropy in the matrix surrounding the RRP chains (or stacks) is induced by the orientation of the PCL chains along the RRP chains. The induced anisotropy of the matrix contributes to the mechanical reinforcement. The birefringent mobile phase in RRP/PCL blends encompasses a much larger volume than the neat RRP itself, allowing shearing above T_m of PCL.

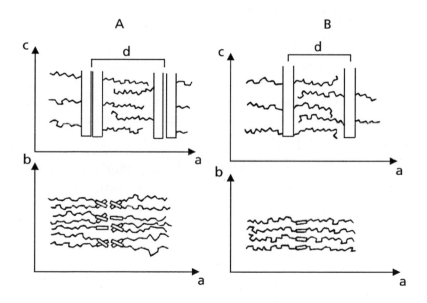

Figure 16.19 (A) Schematic illustration of the double stacks formed by polyester derived from monosubstituted terephthalic acids. (B) Schematic illustration of the stacks and layer structure of polymers derived from 2,5-disubstituted terephthalic acids [131]

Reproduced with permission from H.R. Kricheldorf, L.H. Wahlen, C. Friedrich and T.J. Menke, Macromolecules, 1997, 30, 9, 2642. Copyright 1997, American Chemical Society

The electronic interaction contributes to the increasing of anisotropy and to a transfer of the mechanical forces between the matrix and RRP, significantly reinforcing it. So, aromatic substituents seem to have a compatibility role in RRP/PCL blends. This interaction determines the appearance of lyotropic blends from two isotropic components, such as PCL and RRP-2c.

Because the compatibilising role of aromatic substituents for various matrix, a lyotropic blend can be obtained even from two isotropic components, for example the RRP-2c/PCL blend reported in [131].

The Ballauff and Flory theory [133] predicts that the formation of the mesophase strongly depends on the free volume of the neat liquid crystalline polymer. The melt of RRP-2c containing too much free volume above the T_g, as well as the intensive solvation by more or less parallel aligned PCL chains improve the stiffness of this aggregate and reduce its free volume. So, from two isotropic components a nematic blend can appear.

16.6.3 Aromatic Liquid Crystalline Polymers as Reinforcements

La Mantia and co-workers [134-136] measured the melt viscosity, the spinnability evaluated through the melt strength and the breaking stretching ratio (BSR), such as the mechanical properties for several blends of flexible polymers with wholly aromatic and semi-aromatic liquid crystalline polymers spun under different conditions. The thermoplastic polymers and liquid crystalline polymers listed in **Table 16.9** were used to obtain blends with improved mechanical properties.

Blends containing 0.5 wt%, 10 wt%, 20 wt% LCP and a thermoplastic polymer were prepared with a laboratory single screw extruder (D = 19 mm, L/D = 25, Brabender, Germany) equipped with a die assembly for ribbon extrusion. The shear viscosities of some investigated polymer/LCP blends, measured at shear rate of 24 s^{-1} generally decreased with the growth of LCP content.

The blends PET/Va and PA/VB have minimum of shear viscosity, that usually can appear [133, 134] when the viscosity of LCP is similar or higher than that of the matrix. Concomitantly, the addition of LCP in the studied blends [134] leads to an increase of elongational viscosity as it is shown in **Figure 16.20**.

Figure 16.20 contains the values of the elongational viscosity as a function of extrusion rate for the VB, PA, and PA/VB (90:10) blend. The elongational viscosity of the 90:10 PA/VB blend is considerably higher than that of the pure matrix. Similar behaviour, reported in [95, 137], has been attributed to the considerable amount of energy required for deforming

Table 16.9 Characteristics of the polymers used for the blend preparation [134]

Reproduced with permission from F.P. La Mantia, A. Roggero, U. Pedretti and P.L. Magagnini, Liquid Crystalline Systems, Technological Advances, Eds., A.I. Isayev, T. Kyu and S.Z.D. Cheng, American Chemical Society, Washington, DC, 1996, Chapter 8, 110

Sample	M_w	Name	Manufacturer
PP	680000	D60P	Himont
PC	36000	Sinvet 301	Enichem Polimeri
PET			Enichem Polimeri
PA	62000	ADS40	SNIA
VA		Vectra A-900	Technical Polymers Ticoria
VB		Vectra B-950	Technical Polymers Ticoria
SBH		SBH 1:1:2	Eni Tecnologie
SBHN		SBHN 1:1:3:5	Eni Tecnologie

Chemical composition of the liquid crystalline polymers is: Va=[(4-hydroxybenzoic acid (HBA) (73%), 2-hydroxy-6-naphthoic acid (HNA) (27%)]: VB= [HNA (60%), terephthalic acid (TA) (20%), 4-aminophenol (4AP) (20%)]; SBH= [sebacic acid (SA) (25%), 4,4'-dihydroxybiphenyl (DHBP) (25%), HBA (50%)], SBHN = [SA (10%), DHBP (10%), HBA (30%), HNA (50%)]

and orienting the liquid crystalline polymer domains in the convergent flow at the die entrance.

Melt strength and BSR measurements (carried out under non-isothermal conditions of the spinning line), were made by La Mantia and co-workers in order to obtain information on the spinnability of the studied polymer/LCP blends. The results are shown in **Figure 16.21** for melt strength and in **Table 16.10** for BSR.

Decrease in melt strength for some blends such as PC/VB, PC/SBH, PET/SBH, PET/SBHN and PA/SBH concomitantly with the increases in melt strength for the other were reported. Under non-isothermal conditions, the viscosity of some blends may be higher than that of the neat matrices, in some temperature intervals, if the viscosity of the two components has different activation energies.

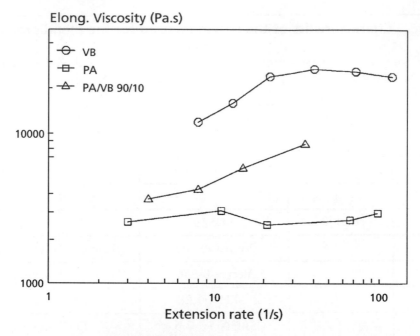

Figure 16.20 Elongational viscosity of VB, PA and PA/VB 90/10 [135]

Reproduced with permission from F.P. La Mantia and A. Valenza, Makromolekulare Chemie, Macromolecular Symposia, 1992, 56, 151. Copyright 1992, Wiley-VCH

The BSR values, i.e., the ratios of the wind up speeds to the extrusion speeds to filament breaking, measured at a shear rate 24 s^{-1}, are listed in **Table 16.10** for some of the blends studied.

By adding LCP to the polymer/LCP blends, the BRS decreased for all investigated blends, indicating the two phases nature of these blends. The extent of BSR reduction significantly depends on the chemical structure of the two blended components. The decrease of BSR is very small in the case of PET blended with semi-crystalline LCP such as SBH or SBHN, whereas it significantly drops when PET is blended with wholly aromatic ones, such as Va. The same reduction of BSR exhibited the PA/LCP and PC/LCP blends with aromatic or semi-aromatic LCP. The spinning behaviour of the PP/SBH blend is characterised by a marked increase of melt strength and a fairly small BSR reduction with the increase of LCP content.

The strongest modulus improvement is found for PA especially when a wholly aromatic copolyesteramide (VB) is added into it (E_b/E_{LCP} = 2.75, for 20 wt% LCP). Good reinforcement was also obtained for the blends PA/SBH, containing a semi-aromatic copolyester.

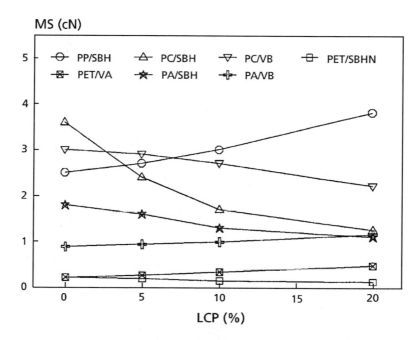

Figure 16.21 Melt strength of some LCP/polymer blends as a function of the LCP content [117]

Reproduced with permission from C. Kiss, Polymer Engineering and Science, 1987, 27, 410. Copyright 1987, Society of Plastics Engineers

Table 16.10 Breaking stretching ratios (BSR) of the polymer/LCP blends, measured at a shear rate of 24 s^{-1} [134]

Reproduced with permission from F.P. La Mantia, A. Roggero, U. Pedretti and P.L. Magagnini, Liquid Crystalline Systems, Technological Advances, Edited by A.I. Isayev, T. Kyu and S.Z.D. Cheng, American Chemical Society, Washington, DC, 1996, Chapter 8, 110

LCP %	PP/SBH	PC/SBH	PC/VB	PET/ SBH	PET/ SBHN	PET/Va	PA/SBH	PA/VB
0	650	730	1400	710	710	710	1400	2000
5	610	520	1100	710	700	620	950	1000
10	500	400	650	700	660	400	670	850
20	450	200	400	660	620	300	540	750

PA: polyamide 6 SBH; Sebacic acid 25%, 4, 4-dihydroxybiphenyl 25%, 4-hydroxybenzoic acid 50% VB; 2-hydroxy-6-naphthoic acid 60%, terephthalic acid 20%, aminophenol 20%

An enhancement of mechanical properties was obtained for the PC-based blends but E_b/E_{LCP} was smaller comparatively with PA based blends. The results are explained by the difference between the tensile modulus of PC fibres and PA fibres prepared under comparable conditions (60 GPa for PC and 0.8 GPa for PA).

For PET-based blends, addition of a semi-aromatic LCP, such as SBH or SBHN, leads to remarkable modulus enhancement, whereas a stiffer, wholly aromatic, LCP such as Vectra A, does not influence the value of the elastic modulus of the blends. On the contrary, for compression moulded blends, the modulus improvement was found to be higher for PET/Va blends than for PET/SBH blends. Although PP/SBH blends show good values of melt strength and BSR, the improvement in mechanical properties was negligible. The authors explained this fact by a strongly depressed interphase adhesion between PP and SBH in the solid state compared with molten state.

Despite the fact that the LCP usually reduce the shear viscosity of molten flexible polymers, the melt strength and BRS of the polymer/LCP blends may still be high enough to grant good spinnability, indicating the fibrillation of the LCP particles and the orientation of the matrix macromolecules. High values of melt strength and BRS indicate a good phase compatibility in the molten state.

A considerable reinforcing effect has been shown only for the blends with good spinnability and good interphase adhesion.

16.6.4 Poly (amide imide) Blends

Isayev and Varma [89] studied the blends of PAI, high performance polymer having high viscosity and melt reactivity, with a LCP having easy processability and self-reinforcing capacity. The blends were obtained using two installations; one containing a co-rotating twin screw extruder and another one with a single screw extruder attached to a static mixer. The blends were injection moulded and characterised both before and after heat treatment. The thermal, flow and mechanical properties of the blends and of the pure components were studied. The PAI used were Torlon 4000Tf-40 (PAI-1) and Torlon 4203L (PAI-2) supplied by Amoco Performance Products. The thermotropic LCP was Vectra A950 a random HBA-HNA copolyester from Celanese Company.

The DMTA studies indicated that PAI-1/LCP and PAI-2/LCP are immiscible blends. After heat treatment, the molecular stiffness and T_g increased concomitantly with a decrease of the loss tangent. Viscosity studies [89] revealed a complete segregation of the components, the lower viscosity component (LCP) forming the outer layer.

After heat treatment, the tensile modulus and elongation of the blends and of the pure components increased concomitantly, with an increasing in the flexibility and toughness of the liquid crystalline polymer, due to the loss of the orientation. Tensile strength modulus of LCP decreased on the heat treatment. The water loss by heating leads to the decrease of flexibility of the PAI chains.

A decrease in toughness on blending, due to the mutual incompatibility of the components, such as a general increase in toughness on curing were reported [89] for PAI-1/LCP and PAI-2/LCP blends.

The general tendency after heating of PAI/LCP blends with different proportions of PAI, is an increasing of the break stress and of the secant modulus. No significant improvement in the modulus and other tensile properties were observed on curing.

Mechanical studies of the blends PAI/LCP show that the properties of the pure components are better than those of the blends, hence no reinforcement due to the fibre formation is evident by morphological studies. Voids and destruction of fibres on heat treatment were observed. Agglomerations and segregation, indicating a poor mixture, combined with the lack of interaction between the phases, lack of reinforcing fibres and the inhibition of the imidisation reaction of PAI in the blends contributed to the poor mechanical properties of the blends.

16.7 Concluding Remarks

Because the high-performance polymers, such as LCP, are often quite costly, there is an incentive to obtain materials with relative small amounts of liquid crystalline polymers into commercially available polymers.

Main-chain or side-chain liquid crystalline polymers can be mixed with flexible engineering polymers in order to obtain systems as polymer-dispersed liquid crystalline polymer, or polymer-stabilised liquid crystalline polymer, used in electro-optical applications (liquid displays, light shutters, etc).

When a liquid crystalline polymer is characterised by a stiff conjugated π-electrons, the blends obtained are electrically conductive and can be used as electrodes in plastic light-emitting diode, such as those containing 3 wt% polyaniline in PMMA.

Increasing applications of liquid crystalline polymers in many fields are based on their capacity to modify compatibilisation, morphology, as well as mechanical and thermal performances of the blends. These results are very useful to those industries using liquid crystalline polymer blends for electronics, telecommunications, fibre optics, consumer goods, automotive, marine or aerospace applications.

At present, consumption of liquid crystalline polymers in blending is rapidly expanding and is expected to increase in the coming years, offering possibilities and directions for future research and industrial applications.

Acknowledgments

I gratefully acknowledge Professor Cornelia Vasile, who read this chapter and offered many helpful suggestions to improve it.

References

1. *Liquid Crystals: The Fourth State of Matter*, Ed., F.D. Saeda, Marcel Dekker, New York, NY, USA, 1979.

2. *Handbook of Liquid Crystals*, Eds., D. Demus, J.W. Goodby, G.W. Gray, H.W. Spiess and V. Vill, Wiley-VCH, Weinheim, Germany, 1998.

3. *Handbook of Liquid Crystals*, Eds., H. Kelker and R. Hatz, Oxford University Press, Oxford, UK, 1980.

4. A. de Vries in *Liquid Crystals: The Fourth State of Matter*, Ed., F.D. Saeda, Marcel Dekker, New York, NY, USA, 1979, Chapter 1, 1.

5. G. Friedel, *Anales de Physique*, 1922, **18**, 273.

6. H.W. Gibson in *Liquid Crystals: The Fourth State of Matter*, Ed., F.D. Saeda, Marcel Dekker, New York, NY, USA, 1979, Chapter 3, 99.

7. S. Friberg in *Lyotropic Liquid Crystals*, Ed., R.F. Gould, Advances in Chemistry Series, American Chemical Society, Washington, DC, USA, 1976, 114.

8. K. Hiltrop in *Liquid Crystals*, Eds., H. Baumgartel, E.U. Frauk and W. Grunbein; Guest Ed., H. Stegemeyer, Springer Verlag, New York, NY, USA, 1994, Chapter 4, 143.

9. S.E.B. Petrie in *Liquid Crystals: The Fourth State of Matter*, Ed., F.D. Saeda, Marcel Dekker, New York, NY, USA, 1979, Chapter 4, 163.

10. H. Sackmann and D. Demus, *Molecular Crystals and Liquid Crystals*, 1973, **21**, 239.

11. A. de Vries and D.L. Fishel, *Molecular Crystals and Liquid Crystals*, 1972, **16**, 311.

12. A. Blumstein, G. Maret and S. Vilasagar, *Macromolecules*, 1981, **14**, 1543.

13. H. Finkelmann, *Angewandte Chemie, International Edition*, 1987, **26**, 816.

14. *Liquid Crystalline Polymers*, Ed., C. Carfagna, Pergamon Press, Oxford, UK, 1994.

15. *Side Chain Liquid Crystals Polymers*, Ed., C.B. McArdle, Blackie, Glasgow, UK, 1989.

16. M. Kotera, J.M. Lehn and J.P. Vigneron, *Chemical Communications*, 1994, 197.

17. C. Alexander, C.P. Jariwala, C.M. Lee and A.C. Griffin, *Die Makromolekulare Chemie - Macromolecular Symposia*, 1994, 77, 283.

18. T. Kato and J.M.J. Frechet, *Macromolecules*, 1989, **22**, 9, 3818.

19. T. Kato and J.M.J. Frechet, *Macromolecules*, 1990, **23**, 9, 360.

20. U. Kumar, J.M.J. Frechet, T. Cato, S. Ujiie and K. Timura, *Angewandte Chemie, International Edition*, 1992, **31**, 1531.

21. *Recent Advances in Liquid Crystalline Polymers*, Ed., L.L. Chapoy, Elsevier, London, UK, 1985.

22. S.M. Kelly, *Liquid Crystals Today*, 1996, 6, 4,1.

23. A.I. Isayev in *Liquid Crystalline Polymer Systems Technological Advances*, Eds., A.I. Isayev, T. Kyu and S.Z.D. Cheng, American Chemical Society, Wasighton, DC, USA, 1996, Chapter1, 1.

24. M. Jaffe, P. Choe, I.S. Chung and S. Makhija, *Advances in Polymer Science*, 1994, **117**, 297.

25. S.G. Cottis, J. Economy and L.C. Wohrer, inventors; The Carborundum Company, assignee; US Patent, 3,975,487, 1976.

26. W.J. Jackson, Jr. and H.F. Kuhfuss, inventors; Eastman Kodak, assignee; US Patent 3,778,410, 1973.

27. W.J. Jackson, Jr. and H.F. Kuhfuss, *Journal of Polymer Science, Chemistry Edition*, 1976, **14**, 2043.

28. G.W. Calundann, inventor; Celanese Corporation, assignee; US Patent, 4,067,852, 1978.

29. A.J. East and G.W. Calundann, inventors; Celanese Corporation, assignee; US Patent, 4,318,841, 1982.

30. A.J. East and G.W. Calundann, inventors; Celanese Corporation, assignee; US Patent, 4,318,842, 1982.

31. G.W. Calundann and M. Jaffe in Proceedings of the Robert A. Welch Foundation Conference on Chemical Research: Synthesis of Polymers, 1982, **36**, 247.

32. C.S. Brown and P.T. Alder in *Polymer Blends and Alloys*, Eds., M.J. Folkes and P.S. Hope, Blackie, Glasgow, UK, 1993, Chapter 8, 195.

33. M.G. Dobb and J.E. McIntyre, *Advances in Polymer Science*, Springer, New York, NY, USA, 1984, 61.

34. C.B. Hansson, *Physical Data Book*, Pergamon Press, Oxford, UK, 1972.

35. A.A. Bright and L.S. Singer, *Carbon* 1979, **17**, 57.

36. P.G. Hedmark, J.M.R. Lopez, M. Westdahl, P. Werner, J. Janseen and U.W. Gedde, *Polymer Engineering and Science*, 1988, **28**, 19, 1248.

37. F. Roussel, J.M. Buisine, U. Maschke and X. Coqueret, *Molecular Crystals and Liquid Crystals*, 1997, **299**, 321.

38. L. Nasta, S. Frunza, L. Cianga, O. Tonita M. Postolache and D. Dorohoi, *Analele Stiintifice Universite 'Al .I. Cuza' Iasi, s. Fizica*, 1997-1998, **43-44**, 121.

39. D. Dorohoi, L. Nasta, M. Cotlet, S. Frunza and O. Tonita, *Analele Stiintifice Universite 'Al. I. Cuza' Iasi, s.Chimie*, 1999, 7, 1, 121.

40. D. Dorohoi, *Analele Stiintifice Universite 'Al .I. Cuza' Iasi, s.Chimie*, 2000, 8, 1, 35.

41. D. Dorohoi, M.I. Postolache and M. Postolache, *Journal of Macromolecular Science, Part B, Physics*, 2001, 40, 239.

42. V.N. Tsvetkov, E.J. Riumtsev, I.N. Shtennikova, E.V. Korneeva, B.A. Krentsel and Y.B. Amerik, *European Polymer Journal*, 1973, **9**, 6, 481.

43. H.R. Kricheldorf, L.H. Wahlen, C. Friedrich and T.J. Menke, *Macromolecules*, 1997, **30**, 9, 2642.

44. A. Blumstein, S. Vilasagar, S. Ponrathnam, S.B. Clough and R.B. Blumstein, *Journal of Polymer Science: Polymer Physics Edition*, 1982, **20**, 5, 877.

45. M.J. Seurin, A.T. Bosch and P. Sixou, *Polymer Bulletin*, 1983, **9**, 8-9, 450.

46. F. Fried, J.M. Gilli and P. Sixou, *Molecular Crystals and Liquid Crystals*, 1983, **98**, 209.

47. R.S. Werbowyj and D.G. Gray, *Macromolecules*, 1984, **17**, 8, 1512.

48. J.M. Gilli, J.F. Pinton, P. Sixou, A. Blumstein and O. Thomas, *Molecular Crystals and Liquid Crystals*, 1985, **1**, 123.

49. J. West in *Liquid-Crystalline Polymers*, Eds., R.A. Weiss and C.K. Ober, ACS Symposium Series, No. 435, American Chemical Society, Washington, DC, USA, 1990, Chapter 32, 475.

50. J.W. Doane, in *Liquid Crystals: Applications and Uses*, Ed., B. Bahadur, World Scientific, River Edge, NJ, USA, 1990, 361.

51. *Liquid Crystals in Complex Geometries*, Eds., G.P. Crawford and S. Zumer, Taylor and Francis, London, UK, 1996, Chapter 2, 21.

52. Y. Kosaka, T. Kato and T. Uryu, *Macromolecules* 1994, **27**, 2658.

53. C.T. Imrie and B.J.A. Paterson, *Macromolecules*, 1994, **27**, 22, 6673.

54. G. Peltz, M. Novac, W. Weissflog and D. Demus, *Crystal Research and Technology*, 1987, **22**, 125.

55. S. Takenaka, T. Hirohata, I. Simohara and S.J. Kusabayashi, *Journal of the Chemical Society, Perkin Transactions*, 1986, **2**, 1121.

56. S. Ambrosino and P. Sixou, *Journal of Polymer Science, Part B*, 1994, **32**, 1, 77.

57. C.D. Han and Y.W. Kim, *Transactions of the Society of Rheology*, 1975, **19**, 245.

58. L.A. Utracki, A.M. Catani, G.L. Bata, M.R. Kamal and V. Tan, *Journal of Applied Polymer Science*, 1982, **27**, 6, 1913.

59. G.V. Vinogradov, B.V. Yarlykov, M.V. Tsebrenko, A.V. Yudin and T.I. Ablazova, *Polymer*, 1975, **16**, 7-8, 609.

60. Y.S. Lipatov, A.F. Nesterov, V.F. Shumsky, T.D. Ignatova and A.N. Gorbatenko, *European Polymer Journal*, 1982, **18**, 11, 981.

61. *Thermotropic Liquid Crystal Polymer Blends*, Ed., F.P. La Mantia, Technomic Publishing Company, Inc., Lancaster, PA, USA, 1993.

62. U. Kumar, T. Kato and J.M. J.Frechet, *Journal of the American Chemical Society*, 1992, **114**, 6630.

63. T. Kato, H. Kihara, T. Uryu, A. Fujishima and J.M.J. Frechet, *Macromolecules*, 1992, **25**, 25, 6836.

64. S. Ujiie and K. Iimura, *Macromolecules*, 1992, **25**, 12, 3174.

65. D.N. Rodriguez, D. Guillon, A. Skouilos, Y. Frere and P. Gramair, *Makromolekulare Chemie*, 1992, **193**, 12, 3117.

66. M. Portugall, H. Ringsdorf and R. Zentel, *Die Makromolekulare Chemie*, 1982, **183**, 10, 2311.

67. T. Schleeh, C.T. Imrie, D.M. Rice, F.E. Karasz and G.S. Attard, *Journal of Polymer Science, Part A: Polymer Chemistry*, 1993, **31**, 7, 1859.

68. Y. Kosaka and T. Uryu, *Journal of Polymer Science, Polymer Chemistry Edition*, 1995, **33**, 13, 2221.

69. Y. Kosaka and T. Uryu, *Macromolecules*, 1994, **27**, 22, 6286.

70. Y. Kosaka and T. Uryu, *Macromolecules*, 1995, **28**, 24, 8295.

71. W. Chen, J. Wu and M. Jiang, *Macromolecular Chemistry and Physics*, 1998, **199**, 8, 1683.

72. M. Jiang, W. Chen and T. Yu, *Polymer*, 1991, **32**, 6, 984.

73. K.I. Alder, D. Stewart and C.T. Imrie, *Journal of Materials Chemistry*, 1995, 51, 12, 2225.

74. C. Fouquey, J.M. Lehn and A-M. Levelut, *Advanced Materials Journal*, 1990, **2**, 5, 254.

75. D. Dutta, H. Fruitwala, A. Kohli and R.A. Weiss, *Polymer Engineering and Science*, 1990, **30**, 17, 1005.

76. F.N. Cogswell, B.P. Griffin and J.B. Rose, inventors; Imperial Chemical Industries, assignee; US Patent 4,433,083, 1984.

77. B. Lee, *Polymer Engineering and Science,* 1988, **17**, 1107.

78. B. Lee, Proceedings of Antec '88, Atlanta, GA, USA, 1988, 1088.

79. A. Apicella, P. Iannelli, L. Nicodemo, L. Nicolais, A. Roviello and A. Sirigu, *Polymer Engineering and Science*, 1986, **26**, 9, 600.

80. K.G. Blizard and D.G. Baird, *Polymer Engineering Science,* 1987, **27**, 9, 653.

81. A. Siegmann, A. Dagan and S. Kenig, *Polymer,* 1985, **26**, 9, 1325.

82. R.A. Weiss, W. Huh and L. Nicolais, *Polymer Engineering and Science*, 1987, **27**, 9, 684.

83. V.G. Kulichikhin, O.V. Vasil'eva, I.A. Litinov, E.M. Antopov, I.L. Parsamyan and N.A. Plate, *Journal of Applied Polymer Science*, 1991, **42**, 2, 363.

84. M.R. Nobile, E. Amendola, L. Nicolais, D. Acierno and C. Carfagna, *Polymer Engineering and Science*, 1989, **29**, 4, 244.

85. W.R. Schowalter, C.E. Chaffey and H. Brenner, *Journal of Colloid and Interface Science*, 1968, **26**, 152.

86. G.I. Taylor, *Proceedings of the Royal Society, London,* 1932, **A138**, 431.

87. G.I. Taylor, *Proceedings of the Royal Society, London,* 1934, **A146**, 501.

88. H.K. Chuang and C.D. Han, *Journal of the American Chemical Society*, 1984, **106**, 171.

89. A.I. Isayev and T.R. Varma in *Liquid-Crystalline Polymer Systems*, Eds., A.I. Isayev, T. Kyu and S.Z.D. Cheng, ACS Symposium Series, No. 632, American Chemical Society, Washington, DC, USA, 1996, Chapter 10, 142.

90. R.F. Heitmiller, R.Z. Maar and H.H. Zabusky, *Journal of Applied Polymer Science*, 1964, 8, 873.

91. C. Lees, *Proceedings of the Physical Society, London,* 1990, **17**, 460.

92. Z. Hashin in *Second-Order Effects in Elasticity, Plasticity and Fluid Dynamics*, Eds., M. Reiner and D. Abir, McMillan Press, New York, NY, USA, 1964.

93. M.T. De Meuse and M. Jaffe, *Molecular Crystals and Liquid Crystals*, 1988, **157**, 535.

94. L.A. Utracki, *Polymer Alloys and Blends*, Carl Hanser Verlag, Munich, Germany, 1989, 192.

95. D. Beery, S. Kenig, A. Siegmann and M. Narkis, *Polymer Engineering and Science*, 1992, **32**, 1, 14.

96. G.C. Berry, *Trends in Polymer Science*, 1996, **4**, 9, 289.

97. M. Doi and T. Ohta, *Journal of Chemical Physics*, 1991, **95**, 1242.

98. H.M. Lee and O.O. Park, *Journal of Rheology*, 1994, **38**, 5, 1405.

99. G.K. Guenther and D.G. Baird, *Journal of Rheology*, 1996, **40**, 1, 1.

100. Y. Takahashi, N. Kurashima, I. Noda and M. Doi, *Journal of Rheology*, 1994, **38**, 3, 699.

101. F. Fried and P. Sixou, *Journal of Polymer Science, Polymer Chemistry Edition*, 1984, **22**, 1, 239.

102. P. Navard, *Journal of Polymer Science, Polymer Physics Edition*, 1986, **24**, 2, 435.

103. S. Ambrosino, T. Khallala, M.J. Seurin, A. Ten Bosch, F. Fried, P. Maissa and P. Sixou, *Journal of Polymer Science, Polymer Letters*, 1987, **25**, 8, 351.

104. S. Ambrosino and P. Sixou in *Polymer Association Structures: Microemulsions and Liquid Crystals*, Ed., M.A. El-Nokaly, ACS Symposium Series, No.384, American Chemical Society, Washington, DC, USA, 1989, Chapter 10, 142.

105. S. Lee, P.T. Mather and D.S. Pearson, *Journal of Applied Polymer Science*, 1996, **59**, 2, 243.

106. D.S. Kalika, L. Nuel and M.M. Denn, *Journal of Rheology*, 1989, **33**, 7, 1059.

107. D.S. Kalika, D.W. Giles and M.M. Denn, *Journal of Rheology*, 1990, **34**, 21, 139.

108. K. Toshikazu, *Processing of Liquid Crystalline Polymer and Engineering Plastics Blends*, University of Lowell, 1991. [PhD Thesis]

109. R.L. Brady, and R.S. Porter, *Journal Thermoplastic Composite Materials*, 1990, **3**, 253.

110. T.E. Helminiak, *Contemporary Topics in Polymer Science*, 1989, **6**, 17.

111. H.H. Chuah, T. Kyu and T.E. Helminiak, *Polymer*, 1987, **28**, 12, 2130.

112. C. Taesler and H.R. Kricheldorf and J. Petermann, *Journal of Materials Science*, 1994, **29**, 11, 3017.

113. C. Taesler, J. Petermann and H.R. Kricheldorf, *Innovations in Materials Research*, 1996, **1**, 89.

114. G. Kiss, A.J. Kovacs and J.C. Wittmann, *Journal of Applied Polymer Science*, 1981, **26**, 8, 2665.

115. G. Crevecoeur and G. Groeninckx, *Bulletin de la Societe Chimique de Belgique*, 1990, **99**, 1031.

116. T-S. Chung, *Plastics Engineering*, 1987, **43**, 10, 39.

117. G. Kiss, *Polymer Engineering and Science*, 1987, **27**, 6, 410.

118. S. Swaminathan and A.I. Isayev, *Polymer Materials Science and Engineering*, 1987, **57**, 330.

119. F.P. La Mantia, A. Valenza, M. Paci and P.L. Magagnini, *Polymer Engineering and Science*, 1990, **30**, 1, 22.

120. S.H. Jung and S.C. Kim, *Polymer Journal (Japan)*, 1988, **20**, 1, 73.

121. R. Ramanathan, K.G. Blizard and D.G. Baird, Proceedings of Antec '87, Los Angeles, CA, USA, 1987, 1399.

122. R. Ramanathan, K.G. Blizard and D.G. Baird, Proceedings of Antec '88, Atlanta, GA, USA, 1988, 1123.

123. W-C. Lee and A.T. Dibenedetto, *Polymer Engineering and Science*, 1992, **32**, 6, 400.

124. E.G. Joseph, G.L. Wilkes and D.G. Baird in *Polymeric Liquid Crystals*, Ed., A. Blumstein, Plenum Press, New York, USA, 1985, 197.

125. J-F. Croteau and G.V. Laivins, *Journal of Applied Polymer Science*, 1990, **39**, 11-12, 2377.

126. P. Zhang, T. Kyu and J.L. White, *Polymer Engineering and Science*, 1988, **28**, 17, 1095.

127. W. Brostow, T.S. Dziemianowicz, J. Romanski and W. Werber, *Polymer Engineering and Science*, 1988, **28**, 12, 785.

128. K. Friedrich, M. Hess and R. Kosfeld, *Die Macromolecular Chemie - Makromolekulare Symposia*, 1988, **16**, 251.

129. L.I. Minkova, M. Paci, M. Pracella and P. Magagnini, *Polymer Engineering and Science*, 1992, **32**, 1, 57.

130. B.Y. Shin and I.J. Chung, *Polymer Engineering and Science*, 1990, **30**, 1, 13.

131. H.R. Kricheldorf, L.H. Wahlen, C. Friedrich and T.J. Menke, *Macromolecules*, 1997, **30**, 9, 2642.

132. H.R. Kricheldorf and J. Engelhart *Journal of Polymer Science, Part A: Polymer Chemistry*, 1990, **28**, 9, 2335.

133. M. Ballauff, *Angewandte Chemie*, 1989, **101**, 261.

134. F.P. La Mantia, A. Roggero, U. Pedretti and P.L. Magagnini in *Liquid Crystalline Polymer Systems*, Eds., A.I. Isayev, T. Kyu and S.Z.D. Cheng, ACS Symposium Series, No. 632, American Chemical Society, Washington, DC, USA, 1996, Chapter 8, 110.

135. F.P. La Mantia and A. Valenza, *Die Makromolekulare Chemie - Macromolecular Symposia*, 1992, **56**, 151.

136. F.P. La Mantia and A. Valenza, *Polymer Networks and Blends*, 1993, **3**, 125.

137. D.E. Turek, G.P. Simion, F. Smejkal, M. Grosso, L. Incarnato and D. Acierno, *Polymer Communications*, 1993, **34**, 204.

Abbreviations

ρ	Density
σ	Tensile strength
φ	Volume fraction
α	Volume fraction of amorphous phase
φ_1	Volume fraction of PVA in the amorphous phase
χ_{12}	Thermodynamic interaction parameter
ε_b	Elongation at break
β-CD	β-Cyclodextrin
$\alpha_{diff.}$	Diffusion factor
σ_e	Electrical conductivity
τ_g	Swelling ratio
φ_p	Percolation threshold
(PP-MA)-g-PEO	Maleated polypropylene grafted with polyethylene oxide
(R)SF	(Regenerated) Silk Fibroin
α_{sep}	Separation factor
$\alpha_{sorp.}$	Sorption factor
1, 2 EPB	Epoxidated 1,2-polybutadiene
1, 4 EPB	1,4-Polybutadiene
12DA	Dodecane diamine
2DA	Ethylene diamine
3HV	3-Hydroxyvalerate
4AP	4-Amino phenol
4VP	4-Vinyl pyridine
6DA	Hexamethylene diamine
6F	Hexafluoroisopropylidene
6F-PC	Bis(hydroxyphenyl)-hexafluoropropane - PC copolymer

AA	Acrylic acid
AAM	Acrylamide
AAS	Acrylonitrile-acrylate (ester)-styrene copolymer
ABR	Acrylate (ester)-butadiene rubber
ABS	Acrylonitrile-butadiene-styrene terpolymer
ACN	Acrylonitrile
ADC	Allyl diglycol carbonate
AEP	1-(2-Aminoethyl) piperazine
AFM	Atomic force microscopy
AHL	Acid hydrolysis lignin
AIBN	2,2´-Azobis-isobutyronitrile
AMSMMA	α-Methylstyrene-methyl methacrylate
ANR	Allyl novolac resin
aPHB	Atactic PHB
aPMMA	Atactic PMMA
aPP	Atactic polypropylene
aPS	Atactic polystyrene
APU	Aqueous polyurethane
aPVA	Atactic PVA
ASA	Acrylate-styrene-acrylonitrile terpolymers
ASTM	American Society for Testing and Materials
ATBN	Amine Terminated Acrylonitrile Butadiene copolymer
BA	Butylacrylate
BD	1,4-Butane diol
BEPD	2-Butyl-2-ethyl-1,3-propanediol
BES	Bis(2-hydroxyethyl)-2-aminoethanesulfonic acid
BIE	Benzoin isobutyl ether
BisA-PSF	Bisphenol-A polysulfone
BMA	Butyl methacrylate
BMI	Bismaleinimide
BOD	Biochemical oxygen demand

BPO	Benzoyl peroxide
BPPC	Bisphenol-A polycarbonate
BR	Butadiene rubber
BS	Butadiene-styrene block copolymers
BSR	Breaking-stretching ratio
BTDA	Benzophenone tetracarboxylic dianhydride
BTX	Benzene-toluene-xylene
BuMA	Butylmethacrylate
C	Collagen
CA	Cellulose acetate
CCl4	Carbon tetrachloride
c_E	Exchange capacity
CELL	Cellulose
CFRP	Carbon Fibre Reinforced Plastic(s)
CHCL3	Chloroform
Cl-PD	3-Chloro-1,2-propanediol
CMC	Critical micelle concentration
CNBR	Acrylonitrile-butadiene rubber
COMPL	Complexed specimen
COPO	Ethylene-propylene-carbon monoxide copolymer
CoPP	Poly (propylene-co-ethylene);
c_p	Molar caloric capacity at constant pressure
CPE	Chlorinated PE
CP-MAS NMR	Cross-polarisation-magic angle spinning NMR
CPN	Copolymer networks(s)
CPS	Carboxylated polystyrene
CPVC	Chlorinated polyvinyl chloride
CR	Polychloroprene
CSR	Core-shell rubber
CTBN	Carboxyl terminated acrylonitrile butadiene copolymer
CTE	Coefficient of thermal expansion

CTNBR	Carboxy-terminated nitrile rubber
CZ	Carbazoyl
D	Diameter
DAA	*N, N*´-Di(2-hydroxyethyl)-3-aminopropionc acid
DAD	Donor-acceptor-donor
DAR	Dispersed acrylate rubber(s)
DAT	2,4-Diamino-1,3,5-triazine
DBTDL	Dibutyl tin dilaurate
DC	Deacetylated chitin
DCP	Dicumyl peroxide
DD	Deacetylation degree
DDS	Diamino dimethyl sulfone
DETA	Diethylene triamine
DGEBA	Di glycidyl ether of bisphenol-A
DGE-PPO	Diglycidyl ether of polypropylene oxide
DGTBBA	Diglycidyl ether of tetrabromo bisphenol A
DHBP	4, 4´-Dihydroxybiphenyl
DHT	Dehydrothermal treatment
DMA	Dynamic mechanical analysis
DMAc	Dimethylacetamide
DMAPAA	Dimethylamino propyl acrylamide
DMF	Dimethylformamide
DMS	Dynamic mechanical spectroscopy
DMSO	Dimethyl sulfoxide
DMTA	Dynamic mechanical thermal analysis
DNA	Deoxyribonucleic acid
DODP	4, 4´-Dioxyphenol
DOE	US Department of Energy
DOP	Dioctyl phthalate
DP	Degree of polymerisation
d-PDMS	Deuterated polydimethylsiloxane

DS	Degree of substitution
DSC	Differential scanning calorimetry
DSD	Duales System Deutchsland
DSDA	3,3´,4,4´-Diphenylsulfone tetracarboxylic dianhydride
DTA	Dynamic thermal analysis
DVB	Divinyl benzene
E	Ethylene
EA	Ethyl acrylate
EAA	Ethylene-acrylic acid
EAM	Ethylene-maleic anhydride copolymer
EB	Ethylene-butene rubbers
EBA	Ethylene butyl acrylate
EC	Ethyl cellulose
ECA	Ethylene-carbonate
ECO	Epichlorohydrin
EDMA	Ethylene dimethacrylate
EDP	Environmentally degradable polymers
EEA	Ethylene-ethyl acrylate
EG	Ethylene glycol
EGDMA	Ethylene glycol dimethacrylate
EGMA	Ethylene-glycidyl methacrylate copolymer
EHEM	Ethylene-hydroxyethyl methacrylate
EM4VP	Poly(ethyl methacrylate-*co*-4-vinyl-pyridine)
EMA	Ethylene-methacrylic acid random copolymer
EMMA	Ethylene methyl methacrylate copolymer
ENR	Epoxidated natural rubber
EO	Ethylene oxide
EOS	Equations of state theories
EP	Ethylene-propylene rubber
EPB	Ethylene-propylene-butylene terpolymer
EPDM	Ethylene-propylene terpolymer

EPR	Ethylene-propylene rubbers
EPS	Expanded PS
ER	Epoxy resin
ES	Ethylene-styrene copolymer (s)
ESCR	Environmental stress crack resistance
ESO	Epoxidated soya oil
ESR	Electron spin resonance
ETBN	Epoxy terminated acrylonitrile butadiene copolymer
ETE	Ether thio ether
ETP	Engineering thermoplastics
EVA	Ethylene-vinyl acetate copolymer
EVA-CO	EVA-carbon monoxide copolymers
EVA-VC	EVA copolymer grafted with vinyl chloride
EVOH	Ethylene vinyl alcohol copolymer
EVOH-COOH	Carboxyl-modified EVOH
FDA	Food & Drug Administration, USA
f-HDPE	Styrene and maleic anhydride modified HDPE
FR-HDPE	Flame retardant HDPE
FTC	The Federal Trade Commission
FTIR	Fourier-transform IR spectroscopy
GAL	Glutaraldehyde
GFRP	Glass fibre reinforced plastic
GFRsPS	Glass fibre reinforced sPS
GMA	Glycidyl methacrylate
GOX	Glucose oxidase
GRAS	Generally recognised as safe
GRC	Graft rubber concentrate(s)
GTF	Glass fibres
$H_{12}MDI$	Hydrogenated 4,4´-methylenebis(phenyl isocyanate)
HA	Hydroxylapatite
HB	3-Hydroxybutyric acid

HBA	*p*-Hydroxybenzoic acid
HD	High density
HDI	1,6 Hexamethyl diisocyanate
HDPE	High density polyethylene(s)
HDT	Heat deflection temperature
HEMA	2-Hydroxyethyl methacrylate
HFPC	Hexafluoropolycarbonate
HFPSF	Hexafluoro bisphenol A polysulfone
HIPS	High impact polystyrene
HMBIPSF	Hexamethyl biphenol polysulfone
HMW	High molecular weight
HNA	Hydroxynaphthoic acid
HNBR	Hydrogenated nitrile rubber
HP	High processability
HPA	High viscosity PA6
HPB	Hydrogenated polybutadiene
HPC	Hydroxy propyl cellulose
HPC	Hydroxypropyl cellulose
HPP	Homopolymer PP
HQ	Hydroxquinone
HS	Hard segment
HTBN	Hydroxyl terminated acrylonitrile butadiene copolymer
HTPB	Hydroxyl terminated poly butadiene
HV	High viscosity
HVA	Hydrovaleric acid
HVEM	High voltage electron microscope
HY	Hylon VII- high amylose starch
I	Isotropic
IA	Isophthalic acid
IBM4VP	Poly(isobutyl methacrylate-co-4-vinyl pyridine)
IFR	Imbedded fibre retraction method

IIR	Butyl rubber
ILSS	Interlaminar shear strength
iPMMA	Isotactic PMMA
IPN	Interpeneterating network(s)
ipp	Isobutylene/isoprene (butyl) rubber
iPP	Isotactic PP(s)
iPrOH	*i*-Propanol
iPVA	Isotactic PVA
IR	Infra-red
ISR	The Institute for Standards Research
ITO	Indium-tin oxide
JBPS	Japan Biodegradable Plastics Society
JORA	The Japan Organic Recycling Association
L	Lignin
L/D	Length:diameter ratio
LCP	Liquid crystal polymer
LCPE	Liquid crystalline polyester
LCST	Lower critical solution temperature
LDPE	Low density polyethylene
LED	Light emitting diodes
LLDPE	Linear low-density polyethylene(s)
LLDPE-G-MA	Linear low density polyethylene graft maleic anhydride
LMW	Low molecular weight
LMWPE	Low molecular weight polyethylene
LPA	Low viscosity PA
LPA6	PA6 with low viscosity
LPF	Lignin phenol formaldehyde
LPO	Lauroil peroxide
LV	Low viscosity
M6N	(2-Naphthylmethylene)aniline
M6NO2	4-[[6-(Methacryloyloxy)hexyl]oxy]-*N*-(4-nitro-benzylidene)aniline

M6NOMe	(6-Methoxy-2-naphthylmethylene)aniline
M6Ome	(4-Methoxybenzylidene)aniline
M6Q2	(2-Quinolinylmethylene)aniline
M6Q3	(3-Quinolinylmethylene)aniline
MA	Maleic anhydride
MAA	Methacrylic acid
MABS	Methacrylate-acrylonitrile-butadiene-styrene
MA-MI	Poly(methyl acrylate-*co*-maleimide)
MAO	Methylaluminoxane
MAP	MA grafted PP
MBC	Metallocene-based catalysts
MBS	Methacrylate-butadiene-styrene
MDA	Methylene dianiline
MDEA	*N*-methyldietanol amine
MDI	4,4′-Methylenebis(phenyl isocyanate)
MDPE	Medium density polyethylene
ME	Entanglement molecular weight
MEK	Methyl ether ketone
MFI	Melt flow index
MFR	Melt flow rate
MGE	MMA-GMA-ethyl acrylate terpolymer
MI	Melt index
MIT	Massachusetts Instituet of Technology
MLL	Metallocene low linear
MLLD	LDPE prepared using a metallocene catalyst
mLLDPE	Metallocene catalysed LDPE
MMA	Methyl methacrylate
MMAS	Methyl methacrylate-styrene
MMDI	Modified 4,4′ diphenylmethane diisocyanate
MMW	Medium molecular weight
M_n	Number average molecular weight

MNDA	1, 8-Diamino-*p*-menthane
m-PE	Metallocene polyethylene
MPS-5	5 mol% hydroxystyrene
MPW	Municipal plastic waste
MRGT	Multi catalyst reactor granule technology
MS	Microphase separation
M_w	Weight average molecular weight
MWD	Molecular weight distribution
MWL	Mechanical wood milling
MZCR	Multi-one circulating reactor technology
N	Nematic
NaPET	Sulfonic natrium-neutralised
NBR	Acrylonitrile butadiene rubber (nitrile rubber)
NC	Native corn (72% amylopectin)
N-MAm	*N*-methylolacrylamide
NMP	*N*-methylpyrolidone
NMR	Nuclear magnetic resonance
n-PrOH	*n*-Propanol
NR	Natural Rubber
OB	Oxybenzoate
ODA	4,4´-Oxy-dianiline
ODMS	Oligodimethylsiloxane
OS	Oligostyrene
OSL	Organosolv lignin
OXA	Oxazoline
P	Polydispersity
P((B-co-S)-g-AN)	Poly((butadiene-*co*-styrene)-g-AN)
P(α-MSAN)	Poly(α-methyl styrene-*co*-acrylonitrile)
P(AN-AM-AcAc)	Poly(acrylonitrile-acrylamide-acrylic acid)
P(CHMA-co-MMA)	Poly(cyclohexyl methacrylate-*co*-methyl methacrylate)
P(DMS-DPhS-VyMS)	Poly(dimethyl-diphenyl-vinylmethyl)siloxane

P(MA-co-AA)	Poly(methyl acrylate-*co*-acrylic acid)
P(MA-co-HEA))	Poly(methyl acrylate-*co*-hexyl acrylate
P(MMA-co-BMA)	Poly(methyl methacrylate-*co*-butyl methacrylate)
P(MMA-co-EA)	Poly(methyl methacrylate-*co*-ethyl acrylate)
PaMS	Poly (a-methyl styrene)
P(S-B-S)	Poly(styrene-butadiene-styrene)
P2CS	Poly(2-chlorostyrene)
P3-HBA	Poly(3-hydroxybenzoic acid)
P3HBE	Poly(3 hydroxybutyrate)
P3TESH	Poly[2-(3´-thienyl)ethansulfonic acid]
P3TESNa	Sodium salt of poly[2-(3´-thienyl) ethansulfonic acid]
PA	Polyamide(s)
PA12	Polyamide 12
PA6	Polyamide 6
PA6,12	Polyamide 6, 12
PA610	Polyamide 6, 10
PA66	Polyamide 6, 6
PA666	Polyamide 6, 6, 6
PAA	Polyacrylic acid
PAAM	Poly acrylamide
PAAm	Poly(allyl amine)
PAB	Poly(allylbiguanido-*co*-allylamine)
Pac	Polyacetal
PADC	Polyallyl diglycol carbonate
PAE	Polyamic acid di-ethylester
PAEBI	Poly(arylene-ether-benzimidazole)
PAI	Poly(amide-imide)
PAN	Polyacrylonitrile
PANI	Polyaniline
PAPSAH	Poly(aniline-*co*-N-propansulfonic acid aniline)
PAr	Polyarylate

PARA	Amorphous polyamide
PASM	Poly(acrylonitrile-styrene-methyl methacrylate)
PB	Polybutadiene rubber
PBA	Polybutyl adipate
PBA-2000	Poly(butylene adipate) with Mn2000
PBAA	Polybutyl acrylate-co-acrylic acid
PBI	Poly [2,2'-(*m*-phenylene-5,5'-benzimidazole)]
PBMA	Poly(butyl methacrylate)
PBMA	Polybutyl methacrylate
PBN	Polybutylene naphthalene
PBO	Oligobutylene oxide
PBPF	Poly{*p*-(*tert*-butyl)phenylene fumarate}
PBS	Polybutylene succinate
PBT	Polybutylene terephthalate
PBT/I	PBT-isophthalate
PBuMA	Poly(butylmethacrylate)
PBzMA	Poly(benzyl methacrylate)
PC	Polycarbonate (s)
PCB	Printed circuit board
PCF	Fluoropolycarbonate
PCFE	Poly(chloro-trifluoroethylene)
PCL	Poly (ε-caprolactone)
PCR	Postconsumer resin
PCT	Polycyclohexane terphthalate
PCU	Poly(carbonate–urethane)
PDMA	Polydimethacrylate
PDMDPhSi	Poly(dimethyl-*co*-diphenyl)silylene
PDMeDMPCl	Poly(1,1-dimethyl-3,5-dimethylenepiperidinium chloride)
PDMPhMS	Phenylmethylsiloxane
PDMPO	Poly(2,6-dimethylphenylene oxide)
PDMS	Polydimethyl siloxane(s)

PDMS-AO	Polydimethylsiloxane-*g*-alkylene oxide
PDMS-CO-PMPhS	PDMS-*co*-methylphenylsiloxane
PDMSiM	Polydimethylsilylenemethylene
PDMSiPh	Polydimethylsilphenylene
PDMS-NH$_2$	Aminoalkyl-terminated PDMS
PDMS-PDPhS	Poly(dimethyl-*co*-diphenyl)siloxane
PDMS-PMPhS	Poly(dimethylsiloxane-*co*-methylphenylsiloxane)
PDMS-PU	Poly(dimethylsiloxane-urethane)
PDPhS	Polydiphenylsiloxane
PE	Polyethylene
PEA	Polyester acrylate
PEAA	Polyethylene-*co*-acrylic acid
PE-AA	Poly(ethylene-ran-acrylic acid)
PEAc	Polyethyl acrylate
PEBA	Poly(ethylene butylene adipate)
PEC	Polyestercarbonates
PECH	Polyepichlorohydrin
PEEA	Poly(ethylene-*co*-ethyl acrylate)
PEEK	Poly ether ether ketone
PEG	Poly(ethylene glycol)
PEGA	Poly(ethylene glycol adipate)
PEI	Polyether imide(s)
PEK	Polyether ketones
PEM	Poly(ethyl metacrylate)
PEMA	Poly(ethyl methacrylate)
PEN	Polyethylene naphthalate
PENT	Poly(ethylene-2,6-naphthalene dicarboxylate)-*co*-poly(ethylene terephthalate
PEO	Polyethylene oxide
PEOX	Poly(ethyloxazoline)
PEPI	PA 6/polyepichlorohydrin

PES	Polysiloxane
PEST	Polyester (s)
PESU	Polyethersulfone
PET	Poly(ethylene terephthalate)
PET/I	PET-isophthalate
PET12	Polyethylene terephthalate-12
PET-OB	Polyethylene terephthalate-oxybenzoate
PEU	Poly(ether-urethane)
PF	Phenol formaldehyde
PGI	Polyglutarimide
PGPTA	Polyglycerylpropoxy triacrylate
PH	Phenolic resin
Ph	Phenyl
PhA	Phthalic anhydride
PHA	Poly (hydroxyalkanoates)
PHB	Poly-hydroxybutyrate
PHBV	Poly(hydroxybutyrate-*co*-hydroxy valerate)
PHEA	Polyhydroxyethyl acrylate
PHEE	Polyhydroxy ester ether
PHEMA	Poly(2-hydroxyethyl methacrylate)
PHMS	Polyhydromethylsiloxane
PHQ	Phenyl hydroquinone
phr	Parts per hundred rubber
PHS	Poly(styrene-*co-p*-hydroxystyrene)
PHT	Polyhexylene terephthalate
PHV	Polyhydroxyvalerate
PHXMS	Polyhexylmethylsiloxane
PI	Polyimide (s)
PIB	Polyisobutylene
PIBMA	Poly(isobutyl methacrylate)
PICA	Poly(itaconic acid)

PIR	Polyisoprene
PLA	Poly(lactic acid)
PLLA	Polylactide
PM6BA-OMe	Poly 4{[6-(methacryloyloxy)hexyl]oxy}-*N*-(4′- Methyl-oxybenzilidene) aniline
PM6Cz	Poly 4{[6-(methacryloyloxy)hexyl]oxy}-*N*-(carbazolylmethylene) aniline
PM6Q2	Poly 4{[6-(methacryloyloxy)hexyl]oxy}-*N*-(2-quinolinyl-methylene) aniline
PM6R	Poly 4{[6-(methacryloyloxy)hexyl]oxy}-*N*-(4′-R-benzilidene) aniline
PM6SBNO$_2$	Polymethacryloyloxyhexyloxynitrostilbene
PMA	Polymethacrylates
PMAA	Polymethacrylic acid
PMDA	Pyromellitic dianhydride
PMMA	Poly(methyl methacrylate)
PMMA-*g*-PS	Methyl methacrylate-*g*-poly(styrene) graft-copolymer
PMMA-OH	α-Hydroxy PMMA
PMO	Poly(2-methyl-2-oxazoline)
PMOC	Poly(2-methyl-2-oxazoline) copolymer
PMPI	Poly(methylene(phenylene isocyanate))
PnBMA	Poly(*n*-butyl methacrylate)
PNI	Polynaphthimidazole
PNVP	Poly(*N*-vinylpyrrolidine)
PO	Polyolefin(s)
POD	Poly(*p*-phenylene-1,3,4-oxadiazole)
POE	Polyolefin elastomers
POEOD	Poly(oxyethylene oxide diane)
POHMS	Polyhydroxymethyl-siloxane
POM	Polyoxymethylene
PP	Polypropylene(s)
PPB	Polyphenyl boronic compounds

PPE	Polyphenylene ether
PPG	Poly(oxypropylene)glycol
PPhMS	Polyphenylmethylsiloxane
PP-MA	Maleated polypropylene
PpMS	Poly-*p*-methylstyrene
PPO	Poly phenylene oxide
PPO-PC	Poly(propylene oxide)-polycarbonate
PPO-PhZ	Polybispropoxy-phosphazene
PPQ	Polypquinoxaline
PPr	Polypropylene oxide
PPS	Polyphenylene sulfide
PPT	Polypropylene terephthalate
PPV	Poly(*p*-phenylene vinylene)
PPy	Polypyrrole
PQ	Polyesters with quaternary ammonium groups in the side chains
PR^{1}R^{2}Si	Differently substituted polysilylenes (R^{1}R^{2}n-C$_5$H$_{11}$ R^{1}R^{2}n-hexyl R^{1}CH$_3$ and R^{2}C$_3$H$_7$ R^{1}CH$_3$ and R^{2}C$_{18}$H$_{37}$ R^{1}R^{2}n-butyl)
PS	Polystyrene
PS(OH)	Poly{styrene-*co*-[*p*-2,2,2-trifluoro-1-trifluoromethyl)ethyl-α-methylstyrene]}
PSA	Poly (sodium α,β-D,L-aspartate)
PSAAm	Poly(salicylidene allyl amine)
PSAc	Poly(Sodium acrylate)
PS-b-HPB	Poly(styrene-b-hydrogenated butadiene)
PS-b-PEO	Polystyrene-b-poly(ethylene-oxide)
PS-b-PMMA	Polystyrene-b-poly(methyl methacrylate)
PSC	Polysaccharide-chitosan
PSD	Polystyrene-d$_8$
PSD-PB	Polystyrene-d$_8$-polybutadiene diblock copolymer
PS-GMA	Poly(styrene-ran-glycidyl methacrylate
PS-g-PEO	Polystyrene-*g*-poly(ethylene-oxide)
PS-g-PMMA	Poly(styrene-*g*-methyl methacrylate) graft-copolymer

PSiαMS	Polysila-α-methylstyrene
PSiS	Polysilastyrene
PSLG	Poly(sodium L-glutamate
PSMA	Polystyrene-*co*-maleic anhydride
PS-P(MMA-CO-MAA)	Polystyrene-b-poly(methyl methacrylate-*co*-methacrylic acid)
PS-PEO	Polystyrene-poly(ethylene-oxide)
PS-PU	Poly(siloxane-urethane)
PSSA	Poly(styrene sulfonic acid)
PSSNa	Poly (sodium styrene sulfonate)
PSU	Polysulfone
P-t-BA	Poly-*t*-butyl acrylate
PTFE	Polytetrafluorethylene
PTHF	Polytetrahydrofuran
PTMA	Polytetramethylene adipate
PTMG	Poly (tetramethylene glycol)
PTMO	Poly(tetramethylene oxide)
PTMPS	Polytetramethylsil-phenylenesiloxane
PU	Polyurethane
PVA	Polyvinyl alcohol
PVA(B)	Oxidised PVA
PVA(D)	Acetalised PVA
PVA-F	PVA crosslinked with *p*-formaldehyde
PVC	Polyvinylchloride
PVC-PP	Polyvinyl chloride/polypropylene
PVDC	Polyvinylidene chloride
PVDF	Poly(vinylidene fluoride)
PVF	Polyvinyl fluoride
PVME	Poly(vinyl methyl ether)
PVMK	Poly(vinyl methylketone)
PVP	Poly(vinyl pyrrolidone)
PVPh	Poly(*p*-vinylphenol)

PVyAc	Polyvinyl acetate
ran	Random
RCP	Random copolymer resins
RGT	Reactor granule technology
RH	Relative humidity
RIM	Reaction injection moulding
rpm	Revolutions per minute
RRP	Rigid-rod polyesters
RTP	Room-temperature phosphorimetry
RTV	Room temperature vulcanisable
S	Styrene
SA	Sebacic acid
SAA	Poly(styrene-*co*-acrylic acid)
SAlg	Sodium alginate
SAM	Styrene-maleic anhydride copolymer
SAN	Styrene acrylonitrile
SANS	Small-angle neutron scattering
SAXS	Small-angle X-ray scattering
SB	Styrene-butadiene
S-b-B	Poly(styrene-b-butadiene) copolymer
S-b-B-b-S	Poly(styrene-b-butadiene-b-styrene) diblock copolymer
S-b-E	Poly(styrene-b-ethylene)
S-b-EB	Poly(styrene-b-(ethylene-*co*-butytlene)) copolymer
S-b-EB-b-S	Poly(styrene-b-(ethylene-*co*-butylene)-b-styrene)
S-b-EP	Poly(styrene-b-(ethylene-*co*-propylene)) copolymer
S-b-EP-b-S	Poly(styrene-b-(ethylene-*co*-propylene)-b-styrene)
SBH	SA (25%), DHBP (25%), HBA (50%)
SBHN	SA (10%), DHBP (10%), HBA (30%), HNA (50%)
S-b-I	Poly(styrene-b-isoprene)
SBR	Styrene butadiene rubber
SBS	Styrene-butadiene-styrene block copolymer

SC	Soluble collagen
SCB	Short chain branch
SCLCP	Side-chain liquid crystalline polymers (s)
SCORIM	Shear controlled orientation in injection moulding technique
S-co-TMI	Styrene-*m*-isopropenyl-α,α-dimethylbenzyl isocyanate random copolymer
SD	Spinodal decomposition
SE	Styrene-ethylene
SEBS	Styrene-ethylene-butylene-styrene block copolymer
SEBS-g-MA	Styrene-poly(ethylene-*co*-butylene) block, graft maleic acid
SEM	Scanning electron microscopy
SEN3PB	Three-point bend type tests
SEP	Styrene-ethylene/propylene copolymer
SEVOH	Starch/EVOH
SGMA	Styrene-glycidyl methacrylate copolymer
SIN	Simultaneous interpenetrating network(s)
S-IPN	Semi-interpenetrating network
SIS	Styrene-isoprene-styrene copolymer
SMA	Poly(styrene-*co*-maleic anhydride)
SMAA	Poly(styrene-*co*-methacrylic acid)
SMC	Sheet moulding compound(s)
SMMA	Polystyrene-*co*-methyl methacrylate
SNaMA	Poly(styrene-*co*-sodium methacrylate)
SPAN	Sulfonic acid ring-substituted polyaniline
sPMMA	Syndiotactic PMMA
sPP	Syndiotactic polypropylene
SPS	Sulfonated PS
sPS	Syndiotactic PS
sPVA	Syndiotactic PVA
SRF	Salicylic acid–resorcinol-formaldehyde polymeric resin
SS	Soft segment

SSC	Soft segment concentration
SSL	Sulfite lignin sulfonate
S-tBA	Poly(styrene-ran-*t*-butyl acrylate)
SVP	Poly(styrene-vinyl phenol)
TA	Terephthalic acid
TBES	Poly [terephtaloyl *N, N*-bis (2-hydroxyethyl)-2-amino-ethansulfonic acid
TBES-K	Potassium salt of TBES
t_c	Curing time
Tc	Time of crosslinking
Tdec	Decomposition temperature
TDI	2,4-toluene diisocyanate
TDO	Thiodiethanol
TDPA	Total degradable polymer additives
TEA	Triethylamine
TEGDM	Tetraethylene glycol dimethacrylate
TEM	Transmission electron microscope
TEMPO	4-Amino-2,2,6,6-tetramethylpiperidine-1-oxyl
TerPP	Poly(propylene-*co*-ethylene-*co*-1-butene);
TFA	Trifluoroacetic acid
T_g	Glass transition temperature (s)
TG	Thermogravimetric analysis
TGDDM	Tetraglycidyl ether of diphenyl aminomethane
TGMDA	Tetra glycidyl methylene dianiline
THF	Tetrahydrofuran
T_m	Melt transition temperature
T_m	Melting temperature
TMBA	Tetramethyl bisphenol A
TMHFPSF	Tetramethyl hexafluoro polysulfone
TMP	Trimethylpropane triol
TMPC	Tetramethyl polycarbonate

TMPC	Tetramethylbisphenol A polycarbonate
TMPSF	Tetramethyl polysulfone
TMQ	2,2,4-Trimethyl-1,2-dihydroquinoline
TOC	Theoretical Oxygen Demand
TPC	Terephthaloyl chloride
TPE	Thermoplastic elastomer
TPO	Thermoplastic polyolefin(s)
TPS	Thermoplastic Starch
TPU	Thermoplastic polyurethanes
TRAMS	Time-resolved anisotropy measurements
TSDC	Thermally stimulated depolarisation current
TSE	Twin-screw extrusion
tTS	Time-temperature superposition
U_b	Energy to break
UCST	Upper critical solution temperature
UHMWPE	Ultrahigh molecular weight polyethylene
UPE	Unsaturated polyester
UPS	Ultrafine PS particles
USDA	US Department of Agriculture
UV	Ultra violet
UV-VIS	Ultra violet visible spectrum
Va	4-Hydroxybenzoic acid (73%)/HNA (27%)
VA	Vinyl acetate
VAMAC	Ethylene acrylic elastomer
VB	2-Hydroxy-6-naphthoic acid 60%, terephthalic acid 20%, aminophenol 20%
VER	Vinyl ester resin
VLDPE	Very low density PE
VTBN	Vinyl terminated acrylonitrile butadiene copolymer
WAXD	Wide angle x-ray diffraction
WAXS	Wide-angle x-ray scattering

WM	Waxy maize
X_c	The weight fraction crystallinity for blends
XPS	X-ray photoelectron spectroscopy
XRD	X-ray diffraction
ZN	Ziegler-Natta catalyst (s)
ZNLLD	LLDPE prepared using a Ziegler-Natta catalyst
ZnPET	Zinc-neutralised PET
ZnSPS	Zinc sulfonated polystyrene ionomers

Index

Ekkcel 1-2000 659
Elastomeric polyurethane blends
    morphology of 495
Engineering plastics 47
EnPac 623
Enthalpic relaxation 451
Environmentally degradable polymers 615
Environmentally-friendly polymers 615
    legislation 626
Ethylene oxide/PVA blends
    viscosity 318
EPDM blends 123, 492
Epoxidated natural rubber 91
Epoxy resins
    composites
      toughening of 432
    engineering thermoplastics
      toughening of 425
    high performance
      toughening of 422
    mechanical properties of 420
    modification by rubbers 413
    PBT modification 426
    toughening 411-412, 418
    toughening additive 413
Epoxy rubber particles 425
Epoxy-CTBN
    properties of 415
Epoxy-modified lignin-based blends 569-570
Equation of state theories 530
Eucalin 571, 592-593
EVOH
    copolymer starch blends 340
    glass transition temperature 342
EVOH/low viscosity PA6 blends
    tensile properties 353
EVOH/PA blends 349, 355
    hybrid blends 354
    oxygen and toluene permeability 354
    T_g 350

EVOH/PEOX blends 358
    phase behaviour 357
EVOH/PET blends 355
    oxygen permeability 355
EVOH/PO blends 346
    blends morphology 347
    film
      permeability 348
    oxygen permeability 347
EVOH/poly(ethyloxazoline) blends 357
EVOH/poly(styrene-*co*-maleic anhydride) blends 344
EVOH/SMA blends
    melt flow index 345
    storage modulus 345
    viscosity 344
EVOH/starch blends 342
EVOH/starch/hydroxylapatite blends 342
    tensile test results 343-344

F

Filled blends
    properties 62
Filled IPN 511
Fillers 210
Film
    extrusion costs 644
Fishing line
    biodegradable 641
Flame retardant 67
Food-contact applications 637
Fracture toughness 412
Frontal polymerisation
    self propagating 371-372

G

Gas separation membranes 461
Gas transport membranes 458

745

General purpose polyester resin
    flexural properties 429
Glass reinforced, filament wound
composites
    properties of 431
Glass-epoxy composites
    effect of silicone on 433
Graft products 380
Grafting 379
Granlar 659
Granmont Inc. 659
Guggenheim quasi-chemical method 86

H

HDPE
    flame retardant 67
Heterocyclic polymers
    high temperature 441
Heveaplus MG 418
High impact polystyrene 123
Hivalloy 47-48
  resins 43
HNBR blends 492
Homopolyamides
    aminocarboxylic acid type 201
    diaminodicarboxylic acid types 201
HPC/EC mixture
    viscosity 688
HPC/lignin blends 601
HPL/PMMA blends 599
Hydrogel networks 374
Hydrolytically degradable plastic 628
Hydrophilic protein films 633

I

ICI Company 659
Immiscible blends
    rheological behaviour 685
Impact test 54

Indulin AT 592-593
Injection moulding 644
Internal shrinkage stress 510
Interpenetrating polymer networks 373,
526, 545-546, 549-551
    amidoximated PVP 374
    formation kinetics 509-510
    morphology and properties of 547
Intravascular stents 640
Ion-containing polymer blends 401
Ionomer polyblends 185
    coordination complex formation 186
    intermolecular attractions 186, 189
    ion-coordination interactions 186
    ion-dipole interactions 188
    ion-ion interactions 187
    polymer backbones 189
    similar ion pairs 188
Ionomer
    thermoplastic 431
Izod impact test 53

K

Kie Maru mulch 626
Kitchen and food waste 633
Kraft process 571

L

L-lactic acid 633
L/PVC
    weathering stability 597
Landfill
    harmonisation 627
Latex IPN 503, 506
Lignin caprolactone/PVC blends
    thermal analysis data 598
LCP/PA blends
    elastic modulus 695
    ultimate elongation 695

microphase separations 488
Scanning electron microscopy 387
Semi-IPN 503, 506
   tensile strength and elongation at
   break 325
SF/PVA blends
   cast films 336
Shear flow behaviour 683
Showa Highpolymer 643
Silicon-based oligomers
   low molecular weight 539
Silicon-based polymers 525-526
   biomedical applications 544
   blends systems 529
   components 527
   dynamic scanning calorimetry curves
   540
   electrical properties 544
   influence of additives 537
   interaction parameter 529
   mechanical properties 543
   miscibility 539
   properties and applications 541
   rheology 541
   surface/interfacial tension 533
   thermal behaviour 542
   thermodynamic aspects 529
   thermodynamically characterised
   blends systems 531
Silicones 419-420
Siloxane blends
   interfacial tension 534
Simultaneous interpenetrating networks
374, 503, 507
Slow puncture resistance 60
Smectic behaviour 666
Smectic layer 668
Smectic phase 667-668
Sodium lignosulfonate 581
Solid carboxyl terminated acrylonitrile
butadiene copolymer 418

Solvay 632
sPP
   markets for 64
sPP/aPP blends 69
sPP/iPP blends 68
sPS/EPR blends 126
Starch ester technology 630-631
Starch/EVOH films
   tensile strength 341
Starch/PE blends 622
Starch/PHEE blends 638
Starch/plastic blends 620-621
Starch/polycaprolactone blends 634
Starch/polymer blends
   biodegradability 630
Stress-whitening 61
   reduction 48
Styrene block copolymers 209
Styrene block copolymer/POE blends 68
Styrene copolymer blends 80, 121, 210,
252
   properties of 122
Styrene monomer/PB mixture
   fracture surface 128
Styrene polymers
   industrial applications for 171-173
   miscellaneous blends 154-168
Sulfonated PEEK 453
Sumitomo 659

T

Tapioca starch/LDPE blends 623
TBES
   structure 303
TBES/PVA blends 303
Tennessee Eastman Chemical Company
659
Tensile properties 356
Terephthalic acids
   2,5-disubstituted 705